Identification of Gemstones

Michael O'Donoghue

Louise Joyner

OXFORD AMSTERDAM BOSTON LONDON NEW YORK PARIS
SAN DIEGO SAN FRANCISCO SINGAPORE SYDNEY TOKYO

Butterworth-Heinemann
An imprint of Elsevier Science
Linacre House, Jordan Hill, Oxford OX2 8DP
200 Wheeler Road, Burlington, MA 01803

First published 2003

British Library Cataloguing in Publication Data
A catalogue record for this book is available from the British Library

ISBN 0 7506 55127

For information on all Butterworth-Heinemann publications visit our website at www.bh.com

Typeset by Avocet Typeset, Chilton, Aylesbury, Bucks
Printed and bound in Great Britain

Contents

Preface

This book is not intended to be a textbook on what has become the science of gemmology – details of the instruments in common use and notes on how to use them in gemstone identification can be found in a number of current texts, notably Peter Read's *Gemmology* (Butterworth-Heinemann) and in the journals of the different gemmological teaching bodies throughout the world.

From the middle of the twentieth century a number of gemmological textbooks were published, some giving an overview of gemstones as special types of mineral (the 14 editions of G. F. Herbert Smith's *Gemstones*), others giving a view of gemstones in the retail jewellery context. From 1942 until the early 1990s B. W. Anderson and Robert Webster produced editions of *Gem Testing for Jewellers* (first edition 1942, later published as *Gem Testing*) and *Gems*, respectively.

However fast the textbooks appeared developments in the world of gemstones constantly outstripped their contents: by the 1980s a watershed was reached: readers would increasingly find that many of the tests needed to distinguish not only the latest synthetic products but also the nature of treatments designed to enhance colour and clarify interiors were well out of the reach of the jeweller. In addition, the average gemmologist had become limited by increasingly stringent health and safety requirements: some of the liquids long used for determining refractive index and specific gravity have been abandoned on these grounds. The reader might ask whether or not there is still room for the traditional tests, but there is: even today a great deal of information can be found on first examination and we should remember that not all gemstones are unusual, synthetic (or natural) species and that not all have undergone treatment.

The organization of a text dealing with the determination of unknowns presents problems. Repetition is inevitable and even a certain amount of explanation of results is unavoidable. In the preface to *Synthetic, Imitation and Treated Gemstones* (Butterworth-Heinemann, 1997) we said that readers will find that information is sometimes repeated: we have allowed repetition for the sake of 'if you miss something in the diamond chapter you may find it in the glass chapter'. Different gem species are considered together, rather than different colours. No one method is the best but a colour-based approach would mean that red diamond would have to go under red stones (where a diamond would not be expected) or be repeated under diamond – there are other obvious drawbacks. It became apparent that most readers will know what the major stones are so that a text arrangement based on these species first would fit in with the expectations of the majority.

While it would be good to include photographs and diagrams of all the phenomena that a gem tester is likely to encounter, such an aim could never succeed. Instead we have selected examples of some of the most dangerous pitfalls that may be encountered by those charged with establishing the nature of a specimen and perhaps certifying it too. In the same way it would have been impossible to include full citations of papers, even to the limited extent to which this was done in *Synthetic, Imitation and Treated Gemstones.*

Like all determinative sciences, gem testing involves the handling and interpretation of a large number of numerical data. While the major figures appear in the text, additions and revisions are constantly occurring: these will be updated on the companion web page as they come to our notice.

While the details of gemmological testing are always stated first when a particular species is under discussion and are usually given in the same form, readers will find many references to testing procedures involving instruments and techniques to which they will not normally have access. Since many vital investigative points can be solved only by the use of advanced techniques they have been mentioned on all occasions where they have provided the diagnosis needed – on the compositions of filling materials, for example.

Looking more closely at the figures presented in the text, readers will find that refractive index (RI) is sometimes given as a single value and sometimes with figures bracketing a range: details of the position of different indexes within the range are shown on the website. These figures are taken, like all the other data quoted in the book, from the most up-to-date papers. We have not usually included notes for experienced gemmologists on whether or not a particular ray is being cited when an anisotropic material is being discussed, nor have we attributed a particular colour of a pleochroic effect to a particular ray, except where the case is unusual (Maxixe blue beryl is an example). These details can be found on the website.

Details of absorption spectra are always given with the higher wavelength values first, thus 700–400 nm and the terms 'bands' and 'lines' have been used interchangeably. The best depictions of absorption spectra (in black and white) can be found in Anderson and Payne, *The Spectroscope and Gemmology* (1998: ISBN 0 7198 0261 X).

Other values given are specific gravity and hardness: cleavage is noted where its presence is significant. Luminescence (fluorescence and phosphorescence) under long-wave and short-wave ultra-violet radiations (LWUV, SWUV) is noted when it helps the investigator: the fluorescent response of a specimen to the higher energies of X-rays is noted only where it may solve an otherwise intractable problem.

While the website contains updates of material in the present text, we have added a number of notes from the most recent current literature available at the time of writing in early 2002. The points made are placed at the end of the appropriate chapters.

Readers who regularly attend gem and mineral shows and who see there fashioned examples of minerals not usually used ornamentally will find details of most of the species that have been so used in the chapter dealing with the rarer gemstones. Those show visitors (the majority) who collect crystals will find descriptions of the more obvious crystal shapes shown by the major species and some less common ones.

In an ideal world it would be pleasing to include details of crystal form and symmetry and though this has traditionally been taught to students it should long have been realized that such teaching penetrates insufficiently far into the subject to make what is learnt of much practical use. General descriptions of shape are usually adequate enough.

In this context we should not forget that synthetic crystals have their own considerable scientific interest as well as beauty and we have included details of substances which are very rarely seen: the argument that they may one day materialize into fashioned gemstones holds little water – they are interesting for their own sake.

Certification of expensive diamonds demands that seller and customer understand the system of *diamond grading* which lies behind the colour and clarity grades upon which values are determined. This is a subject needing a monograph of its own and *for that reason it has not been treated in the present text*: the most recent textbook is by Verena Pagel-Theisen, *Diamond Grading ABC*, ninth edition, 2001, published by Rubin & Son, n.v., Pelikaanstraat 96, B-2018 Antwerp, Belgium: the ISBN is 3 9800434 6 0.

Michael O'Donoghue
Louise Joyner

Note: The gem trade does not always use new or fairly new political names for some major gem-producing countries though some auction houses have changed. Since this book is based on enduring gem-trade usage we have kept Burma rather than Myanmar and Madagascar rather than the Malagasy Republic.

Acknowledgements

The authors thank the many individuals and institutions who have helped us in the compilation of the book. We are grateful to GemA for the pictures and in particular to Steve Kennedy, Kenneth Scarratt and Alan Jobbins, on whose skills and experience the GemA photograph collections have relied for many years.

We are grateful to Peter Read for much advice given. as so often in the past, promptly and cheerfully.

Chapter 1

Gem testing: some disappointments, instruments and techniques

Throughout the book, in which a great number of facts and figures are presented, a number of technical terms will recur, some of which will be explained (and often repeated) on the spot. Some others will be found in this chapter. It is not possible to explain precisely how all the gem tests are carried out since developments in gem species synthesis, especially of gem-quality diamond, have taken gem identification and certification away from the simpler instruments, though these can usually give a very good start to your investigations and these techniques will be briefly explained.

Further details will be found in the second edition of Peter Read's *Gemmology* (Butterworth-Heinemann). *Gemmology* is designed to take students through the examinations of the Gemmological Association and Gem Testing Laboratory of Great Britain (GAGTL: Gem-A) and includes specimen questions to this end. Just as *Gemmology* has no room for details of the species you are likely to want to test, so this present book has no room for detailed explanations of the theories behind the operation of the testing instruments. Both texts, however, inevitably overlap to some extent.

Magnification

Magnification overall is the best testing medium for gemstones though some specimens don't appear to tell you very much at first. Either the 10× lens (or loupe) or the microscope will be the first place to start. Magnification of 10× is internationally agreed by diamond merchants and graders so that when a particular stone is being discussed at a distance both parties understand that its properties (inclusions, state of facets) are being reviewed at 10×. Today all lenses on the market are achromatic (no colour fringes round the image) and aplanatic (no edge distortion): but you may be given a lens from times long ago, so watch it! Handling the lens can be awkward: not the lens itself but the specimen, which needs to be held by tongs if fingers are too large for it.

Tongs are not at all easy to use and is one reason why the microscope is to be preferred. With even a 10× lens the distance between stone and eye (the focal length) is quite critical and the smallest movement can send the stone out of focus (this is another reason why lenses with higher magnifications do not help as much as might be thought). Holding a stone rather gingerly in tongs tenses the upper body and makes it more likely that the stone will be gripped too hard and escape. The inside of the tips is usually grooved at right angles to their length and some workers like to add a single groove at right angles to the rest, so that the table facet might more easily be accommodated. At one time a manufacturer of tongs embedded the inside of the tips with diamond powder – this is not to be recommended. Don't stand between the specimen and the light by which you are viewing it!

What you may see with the lens is described in the following chapters but the microscope needs further explanation: choice of light source and its positioning make a great deal of difference to observations. At certain times transmitted light will be preferred and at others indirect lighting from the side or above. The technique of dark-field illumination where the specimen, lit from the side, shows up against a dark background is particularly well suited for examining inclusions which appear light themselves. Details of the workings of the microscope can be found in *Gemmology* and elsewhere.

The refractometer

The refractometer measures the refractive index (RI) of the specimen and also its birefringence or double refraction (DR) when this property is present. While many stones have too high an RI (this is called a negative reading) for the glass of the refractometer or for the liquid used to make optical contact between instrument and specimen, most of the better-known gem species will give readings though some practice is needed. For observations to be possible a flat surface is required: a faceted stone whose table appears to be suitable may not in fact be flat enough and no reading is possible though the observer might expect one – this is often the case with glass which may be moulded rather than polished. Cabochons domed top and bottom may be hard to test though there is a method known as distant vision or the spot method by which readings can sometimes be obtained. The refractometer is in fact able to tell the skilled observer how some fashioned stones relate to the original crystal.

Throughout the main text refractive index figures may be given in more than one form: a single figure (for example, 1.718 for natural spinel), a range (always given with the lower value first – this indicates the full range of RIs possible with an anisotropic material) and two, sometimes three separate figures which in both instances indicate not only the two end members of the range but also the three possible refractive indices seen with some types of anisotropic material. I have chosen to present RI values in the form of the original papers in which they were first published.

Crystals

Crystals, while often offered very attractively in jewellery, are not covered, except in a few special cases, in this book. We should know that, apart from glass and opal, all known *inorganic* solid substances are crystalline and that the three-dimensional solids, many with recognizable and familiar shapes, can be divided into systems and classes on the basis of their internal symmetry.

While crystals are often collected by those interested in gemstones, they do not come closely within the scope of this book (though for the benefit of visitors to gem shows and gem crystal collectors they are described later): nevertheless we should remember that crystalline symmetry and properties affect the appearance of the fashioned stones we have to examine. Crystals with flat faces could be tested on the refractometer when no one is looking – don't try it at gem shows – but remember that many minerals are porous and may be permanently stained by the contact fluid. This, by the way, applies to opal, turquoise and some other gem materials.

Where stones come from – locality information

We shall see that locality (place of origin) information is very important for the major species, especially for diamond (though the information is exceptionally hard to find if it is obtainable at all), ruby, blue sapphire and emerald. Should gem-quality crystals of these species reach the tester before anyone can fashion them they can give a great deal of information about the places in which they were found but they hardly ever reach the gem markets in their unfashioned state.

Polarized light and what it can tell us

Gemstones when sufficiently transparent may be placed between two pieces of Polaroid so arranged that light attempting to pass through them both will not reach the observer – this is called the extinction or crossed polars position. Some gem species can be distinguished from one another by use of the simple polariscope which arranges two Polaroids above one another in the crossed position while leaving space for the specimen to be placed between them with a lamp below to illuminate the proceedings. Crystals with birefringence (anisotropic crystals) can be distinguished from those with none (isotropic) and from non-crystalline materials.

Specific gravity

Specific gravity is another important property of gemstones: the figures are the same as for relative density (weight per unit mass). Specific gravity (SG) is the weight of a substance compared with the weight of an equal volume of pure (distilled) water. Two methods of obtaining the figure have been used by gemmologists. Hydrostatic weighing is complicated and liable to system error, while the heavy liquid method employs chemical compounds which are now unfavourably regarded on health and safety grounds. Under laboratory conditions, though, they may perhaps still be used.

Colour

Colour is the most important property of a gem or ornamental material and although colour measurement is possible by a number of methods, reference to colour figures or colour wheels is not necessary to find out what an unknown specimen may be. The cause of colour in gemstones is a study in itself: an excellent reference book is Nassau's *The Physics and Chemistry of Colour*, see The literature of gemstone identification, p. 301.

The dichroscope

Two instruments, the dichroscope and the spectroscope, help the gemmologist towards identification of an unknown specimen by tests in which colour plays the leading part. Coloured crystalline anisotropic materials may show either different colours or more commonly different shades of the same colour when viewed in different directions; however, this will not be seen in coloured isotropic stones. Species like tourmaline may show the effect very strongly and the work of the lapidary must take this property of *pleochroism* into account when deciding where on the crystal he will place the main facets of the stone to be cut. Notable examples of pleochroism and how it affects testing will occur throughout the main text of the book.

While the observer can look along different directions in a gemstone and see the different colours, it is easier, certainly where small specimens are concerned, to use the small metal tube known as the dichroscope. Just to make things difficult, all anisotropic crystals have one, sometimes two but never more than two, directions in which they behave like isotropic ones. If the stone is coloured the dichroscope will usually show the pleochroic colours side by side but this will not be seen in those critical directions known as optic axes. Isotropic stones don't have any optic axes but neither do they show pleochroism. These and many other possible snags can be avoided simply by examining the specimen from different directions or by turning it round so that all possible directions are tested.

The terms dichroism and pleochroism tend to be used interchangeably today.

The spectroscope

The same interchangeability affects the use of the terms *lines* and *bands* when referring to absorption phenomena. These can be seen as dark lines of varying thickness, or as bands covering a greater area of the ribbon of spectrum colours displayed when a hand spectroscope is directed at a source of white light. Pointing the spectroscope at daylight will show only fine lines but many coloured and one or two colourless gemstones will also show absorption spectra indicating the presence of different chemical elements. Many of the absorption spectra of important species come close to being diagnostic on their own but for positive identification it is usually better to make some other simple test as well.

Throughout the main text of the book details of notable absorption patterns are given and in some cases measurements of wavelengths in nanometres as well ($1\,\text{nm} = 10^{-9}$ m). The absorption spectra of greatest use are easy to memorize and the technique of spectroscopy in the visible is not too hard to master once you have seen one or two really distinct examples. Choice of illumination is critical, some spectra being seen best by transmitted and some best by reflected light. Further details can be found in *Gemmology*.

Fluorescence and phosphorescence

The phenomenon of fluorescence means that a substance will emit radiations of a higher wavelength (=lower energy) when stimulated by energies of a lower wavelength (=higher energy). The emitted radiations are sometimes in the visible region so you can see them and when they continue to be visible after the stimulating rays have been turned off the substance is said to be phosphorescing. The name fluorescence is used in general to describe any visible effect.

Both long-wave ultra-violet (LWUV) and short-wave ultra-violet (SWUV) radiations can be used and convenient handsets have long been available among mineral collectors. While the effects seen are often beautiful they do not always diagnose the specimen, but there are exceptions to this: many Cape series diamonds will fluoresce blue and phosphoresce yellow and this response is a good indication of their nature. Some colour-enhanced diamonds also show fluorescent effects and these will be reviewed in the diamond chapter. The dangers of misinterpretation of fluorescent effects, though, can be seen in colourless synthetic spinel, often used as a diamond imitation and fluorescing sky blue under SWUV, and in some of the synthetic gem diamonds which behave in the same way.

Laboratory methods of testing

Throughout the main text we shall find many references to testing methods which are found only in laboratories – some of them research laboratories. Most of these methods are known by their initials but explanations of how they work are outside the remit of this book. Nevertheless we should know that when gemmological tests cannot provide answers to our questions, solutions should be found elsewhere. Methods of testing in this category include Raman spectroscopy, X-ray fluorescence (XRF) and energy-dispersive X-ray fluorescence (EDXRF). The existence of these testing methods and several others shows that no specimen can remain unidentified indefinitely. Work continues on locality determination for diamond.

Some disappointments and why they may occur

Those who buy a cut diamond, perhaps in a ring, may feel, after leaving the jeweller and when some of the excitement has worn off, that the stone is not quite so arresting in appearance as they thought at first. It may not seem to give off quite the same flashes of different colours on a sunny day nor does it look quite so fine under the office striplighting.

On the other hand visits to a cathedral and to the British Museum were enlivened by a sudden perception that the diamond had somehow regained its original appearance, as the flashes of spectrum colour seemed to have returned.

Similarly a ruby may look fine in the shop but less so under the office striplighting – even when the stone is examined in the street after purchase something seems to be wrong but it was hard to say what it was.

The buyer of a blue sapphire in Sri Lanka found his purchase entirely satisfactory until he got to England and looked at it again. Most of the colour seemed to have turned to grey.

Were all these experiences evidence of chicanery on the part of the seller? Or was something else happening?

The responsible agent in each of these cases was the light by which they were viewing their purchases. A diamond needs single spots of light rather than diffused light to show off its fire, so it is less arresting when the lights are placed to give an even illumination to the viewing area. The museum and the cathedral each had distant single spotlights which showed the diamond's fire to perfection.

Daylight is too diffused so the journey from skilfully lit jeweller's shop to street was enough to reduce the stone's magic. Purchase followed by a candle-light supper is an excellent plan with a diamond.

The ruby needs bright light to bring out the red colour – sunlight which is rich in the same wavelengths that give the red to ruby (and some blue to give an impression of purple) is the ideal medium for viewing – try looking at a ruby in Burma or Sri Lanka. Striplighting is richer in the blue wavelengths so a blue sapphire looks better here. The dark blue sapphire did not look too bad in the grey light of England but it needed greater strength of illumination for the best effect to be shown.

The light by which a stone is viewed is called incident light because it 'falls' on a specimen. Those selling stones need to make careful use of their shop or desk lights so that the stone being offered for sale looks at its best.

Troubles in testing

Students and experienced gemmologists have 'off' days when they 'cannot get their eye in'. This happens in fact to everyone sooner or later and I have often been told 'it's my glasses'. Glasses do have to be taken into account when using any optical instrument but it is untrue that wearing them rules out satisfactory observations. You can always take them off, switch to a different pair, get your eyes tested or try moving the specimen or source of light.

There seems to be an in-built unwillingness among English gemmologists not to move anything. One naturally hopes that this does not extend to every area of life: in a previous book I recalled an old and distinguished friend saying 'an Englishman needs time' – but he needs movement as well and moving the head when examining a specimen is essential if you are not to miss particular optical effects.

When using the spectroscope the eyes need to swivel and this can be practised at any time when nothing else is going on, though I do not recommend practising on the Underground, for example, if you do not wish for sudden enthusiastic but unwelcome company.

The light source can be moved closer or further away from the specimen: its nature can be changed (tungsten or fluorescent tube): the background on which the specimen rests may be black or dark, colourless or light and the illumination can be directly transmitted or reflected. Nothing should be ruled out, though considerations of heat should not be forgotten. Many stones (opal is one) may become dehydrated when overheated and this could affect their appearance. Cracking is possible in some cases.

Sometimes a stone needs a special type of lighting: the grading of diamonds for colour (not covered by this book) uses a near-daylight type of lamp which is pretty well standard throughout the diamond grading world. Combined with the standard 10× magnification for clarity and cut, any diamond should be 'placed' in its category with sufficient accuracy for a distant dealer to take for granted what the unknown stone looks like.

If any reader does not realize how great the difference is between well-lit and poorly lit goods, look in any up-market clothes shop (not in Savile Row, perhaps, where colours might be fashionably sombre). See how red–orange clothes are lit by tungsten bulbs to bring out their 'warm' colours and blue–green ones by 'cooler' striplighting.

It is quite amazing how students make things hard for themselves by getting between the incident light and the stone, by using an inappropriate type of room lighting without boldly marching to the switch and doing something about it, or by keeping their heads as stiff as their lips or at least swivelling their eyes!

Failure

This is an unfashionable concept today but there will always be times when a stone will just not tell you what it is. I am not referring to the very uncertain possibly treated diamond or ruby but to the ordinary stone which really ought to give you an answer without recourse to X-ray diffraction or some other expensive and probably unnecessary test. These obstinate stones are very often glass or some kind of composite, less often a species hitherto unknown to science. Cutting (or moulding) the facets may have been carelessly executed and the facet selected for testing may be nowhere near optically flat (this is not a prerequisite of a specimen chosen for test nor is the investigator likely to come across an example).

You may find that the usual tests give no more than a hint of the unknown's identity but that two or more tests will get you a little further, to a point at which you can say that some species at least are ruled out. It is quite true that experience will often give you an idea even when no easy test is available.

Chapter 2

Diamond

Diamond, the element carbon, can be mined and polished but it can also be imitated, enhanced or synthesized. The same can be said for many other important gem species but diamond is unique in its physical properties, in its appearance and in the special place it occupies in public esteem and tradition. Long-held views of diamond are deep-rooted and suggestions that a particular stone might not be what it seems (synthetic, colour-enhanced) are far more powerfully resisted than they might be for other gemstones. Gemmologists have to take this into account when they publish information on diamond: those handling diamond jewellery need at all times to know exactly what level diamond testing has reached because they will inevitably be asked about it sooner rather than later.

While the hardness (resistance to abrasion) of diamond places it above all known natural and man-made substances the gemmologist cannot make much use of it in testing, unless he also handles rough material. It has often been said that jewellers of the past reached for a file when faced with a possible diamond but we know that the pronounced cleavage of diamond can easily be initiated by such treatment. For the same reason a faceted diamond should not be used to write your name on glass. If a cleavage, for whatever reason, has started in your diamond, rainbow-like markings will be seen inside the stone. Glass or some other possible crystalline imitations may not show cleavage to the same extent but this may not be much consolation.

Some of the other standard gemmological tests are not applicable to diamond and those tests which can be used give results which need to be evaluated with very great care. The days in which jewellers could be certain of their judgement of diamond have gone, at least for a while.

The physical and optical properties of diamond are simply stated and there are few deviations from the standard: RI 2.417, SG 3.52, H 10. Its dispersion is 0.044. As we shall see below some diamonds fluoresce and some do not, some diamonds are coloured and others colourless, some conduct an electric current while others will not do so. Furthermore diamonds can be classified into types and those dealing with the stones ought to know what these types are even though they will not be able to make their own classifications.

The gemmologist will sometimes be able to distinguish diamond by the nature of its mineral inclusions, where they can be seen (there are no liquid inclusions to worry about), but in answer to the question (which we are bound to be asked in the course of our dealing with gemstones) 'can you determine the origin of a particular polished diamond?' we have to reply that we cannot at the present state of knowledge. The question of 'conflict diamonds' is yet to be resolved both politically and scientifically, though (very) preliminary efforts are being made to lay down some guidelines.

Diamonds have their own vocabulary which we should learn early on: 'Cape' diamonds are colourless with a hint of yellow, or they could be said to be off-white. On the other hand stones with a definite colour, yellow, pink, green, blue and so on, are known as 'fancies' (this usage has spread to Sri Lanka corundum where many beautiful colours apart from ruby and blue sapphire may be found).

The diamond types

Knowledge of the types of diamond is now becoming more common than before even though more types have been identified. It is quite sensible for the gemmologist or jeweller to have knowledge of them. The part played by nitrogen (N) will be understood as the text continues.

Type I diamonds contain nitrogen which is absent (or present only in very low concentrations) from Type II diamonds.

Type I diamonds can be divided into Types Ia and Ib and these can be subdivided into Types IaA and IaB.

Type Ia diamonds contain nitrogen either as pairs of atoms when they are Type IaA or as larger clusters containing an even number of N atoms (Type IaB). Over 98 per cent of large clear natural diamonds are Type Ia and they are near-colourless to yellow with some grey or brown examples. Type Ib diamonds are very rare (only about 0.1 per cent at most of all known specimens) and contain isolated N atoms. They are usually a deep yellow.

All Type I diamonds *absorb* in both the infra-red (IR) and the ultra-violet (UV) regions. They usually show a blue fluorescence and the Cape stones are of this type. In their absorption spectrum a characteristic band at 415.5 nm can be seen (you can see it in stones which began life as Capes and which have subsequently been treated to show a better colour). The band is called the N3 band and you may find unexplained references to this in the diamond literature.

With the hand spectroscope gemmologists should be able to see a weak absorption band at 504 nm in brown rather than yellow Type I diamonds and some examples may fluoresce a greenish colour: no Type I stones are electro-conductive.

Type II diamonds contain no nitrogen, do not absorb in the same region of the IR as Type I stones and *transmit* in the UV below about 225 nm. They are effective heat conductors. Type IIa diamonds are colourless and virtually impurity-free, transmitting in the UV down to about 225 nm. Type IIb diamonds contain boron and are most commonly blue: they are rare, comprising only about 1 per cent of all diamonds. After exposure to short-wave ultra-violet (SWUV) radiation they may give a blue phosphorescence and some have a grey or blue body colour. Type IIb diamonds are electro-conductive and transmit down to about 225 nm. The blue stones do not show any absorption in the visible region so the hand spectroscope cannot help. Some may absorb faintly in the red but this is hard to see.

Identification of diamond

Before you do anything and whatever you think the unknown specimen to be – clean it first! Diamond in particular is attracted to grease (there is plenty on the skin) and its optical properties are impaired by even a thin coating. Grease is easy to remove – a soft toothbrush will shift it with the assistance of warm (not hot) water and a mild detergent. The change in appearance is amazing – dispersion, which may seem to have vanished, will once more be seen. Cleaning or neglecting it is one of the reasons why a gemstone (not just diamond) will over time appear 'not so good as I remember it in the shop'. Very hot water is dangerous for the observer and thermal shock may extend an unsuspected surface-reaching cleavage. Don't use your own toothbrush – borrow a partner's!

When faced with a polished (faceted) diamond remember why a colourless (or near-colourless) stone should be so valued. You cannot see its hardness but only the unique (adamantine) lustre which that hardness allows the polisher to exploit, giving one of the two chief beauties of diamond, its brilliance. The other is known as 'fire' (dispersion) and shows itself by the flashes of spectrum colour from the small facets surrounding and at an angle to the table facet. A gemstone in which these two properties are uniquely combined is bound to be imitated by whatever material that can be imagined to approach diamond's appearance. The desire to achieve this has allowed glass – cheap to manufacture, easily available and doesn't look too bad – to take first place in the table of diamond's imitators.

Glass

To imagine glass as diamond needs some imagination, good lighting and some distance. Touching glass after diamond shows that glass is a far poorer conductor of heat and the eye may even be able to see rounded facet edges (if the specimen has been made in a 'faceted' mould) and shell-like fractures on those edges. Large faceted pieces of glass will not usually show the extra facets placed on a similarly large diamond as to polish them would add greatly to the final price. The girdle of a diamond, if left unpolished (the custom today is to polish the girdle), may show minute triangular markings, trigons, which are usually taken as proof of diamond: though there have been reports of trigons being placed on the roughened girdle of cubic zirconia and other imitations the practice cannot be common. Touching the unknown with the tongue to determine whether or not it is crystalline is not a good idea even though crystalline substances feel distinctly colder than non-crystalline (amorphous) ones – the best example is glass. The stone may previously have been tested with chemicals and inadequately cleaned and in any case most gemstones, natural and artificial, are crystalline. A drop of water placed on the surface of a diamond will stay coherent and not break up into beads: a pen has been devised to draw a line on the surface with a special ink: if the unknown is diamond the line will remain homogeneous. We also need to remember that glass is nearly always one of the components in a composite specimen so that even when observation through the table may reveal apparently natural features that glass would not show, the base needs examining as well.

Use of the 10× lens will almost certainly detect notably near-spherical bubbles, randomly distributed, and a swirliness of colour in coloured glass. When learning to use the lens it is easy to mistake tiny crystals (crystallites) in natural minerals for 'bubbles' which will only be found in some synthetics (though usually not those which could easily be confused with diamond) or in glass.

Habitual wearers of diamond necklaces will notice a glass one at once – it will not hang properly as most glass has a lower specific gravity than diamond.

Nonetheless, to confuse another substance with diamond could lead to considerable financial disturbance so that few will rely only on touch, unaided sight and the 10× lens. The standard refractometer will give no reading for diamond since its RI of 2.417 is higher than that of the refractometer glass (near 1.962) – this absence is called a negative reading. While the SG could be obtained by hydrostatic weighing, system error is almost unavoidable – the method cannot easily be used for stones of less than 1 ct and the time taken would be far too long. The use of heavy liquids is now less useful on account of the danger of most of those once used.

While we cannot easily test 'diamond' by simple methods to see if it really is a diamond, glass, at least among its simulants, may be easier. Providing the facet tested is polished well enough to approach optical flatness most glass will give an RI between 1.50 and 1.70, though as we shall see there are exceptions. The RI does not vary with colour, there is no birefringence and between crossed polars it gives an uneven or stripy appearance of light and dark (some natural stones give this too and even diamond may do so). A parcel of stones with closely matching colours may well be glass or synthetics because colour can be controlled better by man than by nature.

Inclusions

Most important of all, artificial glass contains no natural solid inclusions: diamond hosts a wide variety of minerals. Gemmologists learn about them but are not usually equipped to identify them, though this does not matter much – neither glass nor the many other man-made materials which may be used to imitate diamond contain natural solid inclusions so that a diamond-like specimen containing obvious crystals (especially if they are coloured) has a good chance of being diamond. It is worth remembering that glass can be given any colour by the manufacturer whereas the colours found in natural diamond are quite restricted: when diamonds are treated to alter or improve their colour the final colour is usually one associated with natural diamonds and not a new, easily recognizable one. Gemmologists have sometimes been deceived by a process called devitrification which frequently occurs in glass: rather natural-looking though small and usually shapeless solids congregate in patches but familiarity with natural inclusions should soon enable the investigator to recognize signs of devitrification as other glass properties will be found in the specimen.

The solid inclusions in diamond may not be identifiable by simple magnification but from time to time a stone hosts coloured or otherwise notable crystals. The commonest mineral inclusion in diamond is diamond itself and included crystals can often be identified by their characteristic adamantine lustre. Coloured mineral inclusions may be pale green (probably olivine), a darker green (chrome diopside or chrome enstatite), or red (garnet): others are dark – ilmenite, pyrrhotite or bright metallic (pyrite). One example of a ruby inclusion in diamond is known. Rutile crystals may be brown or yellow, calcite whitish, quartz and hematite light and dark respectively: black 'carbon' inclusions will be graphite. Some chrome spinel (reddish-brown) has turned up in

Siberian diamonds. Garnet seems to be fairly plentiful in diamonds from South Africa and a structural defect in some Indian diamonds resembles the Maltese cross (this type of cross has splayed ends).

Whether or not you can identify what the mineral inclusions in diamond actually are is immaterial: no other diamond-like species shows so great a range of coloured solids and when coloured diamonds are the subject of testing the body colour of the host, while it may mask any colour shown by the inclusions, will not affect perception of the crystal forms on view. No man-made imitation of diamond will contain solid mineral inclusions, nor will natural minerals fashioned to look like diamonds include so great a range of species.

Having read about solid inclusions in diamond the reader should not begin to develop the idea that it is possible from the inclusions alone to determine the origin of a particular specimen. At the time of writing the vexed question of 'conflict diamonds' is occupying the media. We should be honest and tell clients that while it may be possible to assign some diamonds to their place of origin, this cannot be carried out by standard gemmological tests: in the future it may become possible for specimens to be placed in their geographical context by the use of what are now highly sophisticated tests, but geology is unaware of political boundaries. Diamond workers dealing with diamond crystals will have clearer ideas since they often handle them in groups emanating from particular countries.

Imitations of diamond

Rutile and strontium titanate

In the years immediately after the Second World War the flame-fusion method was used to grow two diamond-like substances, rutile and strontium titanate. Both were attractive and interesting in themselves and while rutile was easy to distinguish from diamond on account of its very high birefringence (doubling of back facet edges when viewed through the table facet was very easy to see), its colour, when undoped, was very much the same off-white to yellow as the Cape diamond. High birefringence can go unnoticed if you are unaware of such things. Though the flame-fusion method produces corundum crystals showing clear signs of growth, the near-colourless rutile and the completely colourless strontium titanate showed no curved growth areas. In both cases the RI is too high to be measured on the refractometer and strontium titanate shows no birefringence (like diamond). The birefringence is indicative of rutile which in any case shows so high a dispersion that most people viewing it would think it too high for diamond, even though orange–yellow to blue fancy colours obtained by doping are a little less common and regarded with less certainty.

There should be little trouble identifying rutile providing you are aware of its existence. The commonest trade name for rutile is Titania which of course was quite appropriate for its composition. The name is found only in textbooks today. Fabulite (the name given to strontium titanate) was (and is – these things don't disappear just because fresh imitations come onto the market and threaten to deceive for a while) a different matter. It shows no birefringence and no 'Cape' off-white to yellow colour but is a clear unincluded stone. By far the majority of stones are colourless but coloured varieties are around. The dispersion is high though not so high as rutile and stones look very attractive. There are two areas in which strontium titanate can be dangerously deceptive:

one is when a number of small polished matching stones (melée) are together in a parcel – somehow the dispersion seems less obvious or you can show them to a client by diffused light (out of doors will do very well) when again the very high dispersion seems acceptable (that is, as diamond).

The second instance is when strontium titanate forms the base or pavilion of a fashioned stone, the top being manufactured from synthetic colourless corundum or spinel which costs relatively little. The lower dispersion of the top (crown) lessens the high dispersion of the base and again makes 'diamond' appear a more possible identification. To make matters even harder, the astute gemmologist may be deceived by not finding the join of the two parts exactly at the girdle (this is where honest doublets are joined). As always when faced with an uncertain stone, think composite before you have investigated too far along some more recondite line of enquiry.

At this point the constants of synthetic rutile and strontium titanate can be given for the record although they cannot be easily tested by standard gemmological methods:

rutile: RI 2.62–2.90 DR 0.287, dispersion close to 0.3 (diamond is 0.044), SG 4.25, H 6.5. Coloured (doped) specimens show no useful absorption spectrum and off-white to yellow ones show no 'Cape' diamond spectrum which has an unmistakable absorption band at 415.5 nm.

strontium titanate: RI 2.41, no DR, dispersion 0.19, SG 5.13, H 5.5 to 6 and specimens are brittle: they can quite easily be marked by a needle-point and may for this reason show ladder-like markings. Coloured (doped) specimens show no useful absorption spectrum.

The synthetic garnets

While rutile and strontium titanate make interesting if not entirely convincing diamond substitutes, the synthetic garnets which first appeared around the 1970s caused more trouble in the diamond trade (at first). While die-hard gemmologists might object to a non-silicate material being called garnet, crystal growers use the name because the materials, which are oxides, display the garnet-type crystal structure. The names YAG (many trade names, including Diamonair) and to a lesser extent GGG (three Gs, no trade name) became well known and would probably be holding prime position among diamond substitutes today were it not for the coming of cubic zirconia (CZ) which does not show the garnet structure.

While the synthetic garnets are most often colourless, doping (the addition of another element or elements to the crystal at impurity rather than compositional level) can produce a number of quite spectacular colours, none of which resemble any colour shown by diamond. Some of the doped stones show a sharp multi-band absorption spectrum which immediately confirms artificial origin but the other standard gemmological tests cannot be used.

The synthetic garnets have the general formula $A_3B_5O_{12}$ where A could be yttrium (as in YAG), B aluminium (again as in YAG) and O = oxygen. GGG stands for gadolinium gallium garnet: G is used for garnet rather than O for oxygen merely for reasons of pronunciation. Looking at 3Gs first as it is much the rarer of the two (the starting materials are more expensive) its high specific gravity at just over 7 puts it well ahead of diamond's 3.52. Unless the specimen has a colour giving an absorption spectrum in the visible region the tester needs to use a reflectivity meter or thermal conductivity tester. The same can be said of the various colours of YAG which is less dense than GGG but still denser, at 4.55–4.6, than diamond.

The synthetic garnets are usually grown on seeds in an open, Replenishable, noble metal crucible with a flux (compound) which is melted first as it has a lower melting point than the substance which it is desired to grow. The molten flux then helps dissolve the starting material of the desired substance, the lower temperature of growth reducing wear on the very expensive crucibles. It is worth knowing something about growth methods, which will be referred to often in the course of the book as the nature of the inclusions in crystals grown in this way is highly characteristic: furthermore the same method is used to grow high-quality rubies and emeralds.

While this does not affect diamond identification, a number of analogues of YAG and GGG exist, having been grown for experimental use. Some of these, even though they are very expensive, may escape from research bodies (from which they cost very large sums when sold legitimately) into the crystal/mineral markets and defy identification. These are rare items and are found in very few collections.

Cubic zirconia

Diamond's most effective simulant at the time of writing is cubic zirconia (CZ). The adjective 'cubic' is there to show that there is a (rare) natural form which does not belong to the cubic crystal system: zirconia indicates the oxide of zirconium, as silica/silicon, beryllia/beryllium. The melting point of zirconia is 2750°C, far too high for any crucible material to withstand, so another method of growth had to be sought. Crystals were eventually grown within a block of their own powder, the outside being kept cool by circulating water. This method of growth became known as skull-melting from the shape of the apparatus.

CZ is colourless as grown but crystals are routinely doped to give a variety of colours. Even today, despite claims, no really convincing ruby or emerald imitation has been achieved. Identification cannot be successfully achieved by standard gemmological tests but the reflectivity meter and thermal conductivity probe can be used with good results.

Zirconia as found in nature is not cubic, so, to attain this form for the artificial material and thus avoid birefringence, two different stabilizers may be used: when yttrium is added the RI is 2.171, SG 5.95, dispersion 0.059, hardness 8.25. There may be some greenish-yellow or reddish fluorescence and parallel rows of small semi-transparent cavities extending into hazy stripes of minute particles may be seen inside the stones.

The calcium-stabilized zirconia has RI 2.177, SG 5.65, dispersion 0.065, hardness 8.5. There may be some distinct yellow fluorescence and stones are virtually inclusion-free.

Testing instruments

The reflectivity meter and thermal conductivity probe

This is a convenient place to discuss these two instruments as they are almost invariably used to test diamond rather than coloured stones (anything that is not diamond). Though multi-purpose tools have been devised to test all stones working on their reflective powers, over the years the demand for 'diamond/not diamond' instruments which are simple to operate has been too great for manufacturers to spend time developing more sophisticated apparatus. Diamond's hardness gives it unique powers of reflectivity when a specimen is polished and no likely imitation comes anywhere near. Operation is simple: whatever the housing looks like (black box or pen/probe) a beam of infra-red radiation is emitted and is reflected from the test surface. The reflected beam is picked up by a sensor which operates a lamp or sounds a bleep: this is usually a positive reponse in that the lamp or bleep indicates that the specimen is diamond. As always the operator needs to test the presented surface more than once as the angle between instrument and test surface should be as close to perpendicular as possible. Even then the specimen or instrument may behave eccentrically once in a while but as the test takes very little time it doesn't matter. The test does not distinguish diamonds which have been treated to change their colour from untreated ones, nor would it help much with the very rarely encountered diamond doublet (a 2-component composite in which both parts are diamond).

The thermal conductivity probe is equally simple and quick in operation. Making use of diamond's unique speed in conducting heat, a lamp/bleep signal is produced when a pulse of heat is supplied to the test surface and the speed of its dissipation registered by a sensor. The pen/probe employed is specially useful in testing large pieces of apparently diamond-set jewellery in which individual stones may be hard to reach by any other method.

Use of the 10× lens

What can the 10× lens do? Look at your specimen first and if it is small (say 0.25 ct and below) go straight to the microscope. Too small a stone makes observation difficult and the results hard to interpret – don't make the test hard for yourself. In any case small stones love to spring away from the tongs.

If the supposed diamond is polished, look inside: there may be characteristic solid inclusions as described above (no liquid ones in diamond and no gas bubbles). Look through the top (table) facet at the opposite facet edges: if they appear to be doubled (like tramlines) look in two or three other directions. If you see a similar effect in a direction at an angle to the one you first examined then the specimen is showing you the property of birefringence or double refraction. This property can be seen only in anisotropic crystals, never in glass, but some gem crystals (isotropic) do not show it – diamond is one of them. They are called isotropic or singly refractive; birefringent (doubly refractive) crystals are anisotropic.

Having established for the sake of argument that your unknown specimen is isotropic so that diamond is still a possibility, look to see how sharp the facet edges are. If they are rounded, diamond is unlikely since rounding suggests that the stone has been made 'faceted' from the molten state in a mould and that it is almost certainly glass. Sharp facets, though certainly a diamond property, may still be seen on the harder diamond imitations YAG and CZ so your test is not over. If the stone is large and there are more facets than the 58 of the standard brilliant, remember that lapidaries (they polish any stone but diamond) will not find it worth the cost of placing extra facets on a relatively cheap synthetic.

All the previous paragraph is interesting but the points made add up only to a general, if strong, impression. It is true that a diamond merchant or polisher will find 10× magnification quite enough to spot a diamond, but the jeweller or gemmologist will need further tests and this is where the thermal conductivity probe and reflectivity meters come in. By the end of this paragraph YAG and CZ are still in the field.

Though we shall meet them later, stones with marked birefringence include zircon, whose colourless variety was often mistaken for diamond, and the rather uncommon mineral scheelite, which has been synthesized in colourless form. In synthetic corundum the birefringence is very small but it is so common that you could be caught out.

The lens will help to identify some treated diamonds with altered colour: umbrella-like markings seen encircling the culet (bottom flat facet of a brilliant-cut stone) indicate cyclotron treatment – the stone will usually be green or blue.

Tricky colourless imitations of diamond, apart from the synthetic garnets, CZ and anisotropic specimens, include synthetic colourless spinel. This material fluoresces a bright sky blue under short-wave ultra-violet (SWUV) radiation and until recently no natural diamond behaved in this way. Now some synthetic diamonds will do this so the test has to be carefully applied in conjunction with other tests. Fortunately, it is possible to measure the refractive index of synthetic spinel which is pretty constant at 1.728.

Fluorescence is still useful, however, when you encounter a complicated piece of jewellery studded, apparently, with many small diamonds. If, when examined under SWUV all the stones fluoresce bluish-white there may be something wrong as statistically it would be unlikely that each specimen would be diamond. If they were members of the Cape series they would show the characteristic yellowish cast. If all the stones fluoresced a sky blue under SWUV it is most likely that they would all be synthetic spinel. If the stones were all diamonds the most likely case would be for some of the stones to fluoresce under LWUV and some to remain inert. The chances are that all will be diamonds – but test them with a thermal conductivity probe to be sure!

Review so far

While the commonest imitation of diamond is glass, other natural and synthetic materials have long been used in hopefully vain attempts to deceive: once again here are the main points about them.

Glass is a poor heat conductor, its facets are often moulded rather than cut, it has a single RI varying most commonly between 1.50 and 1.70 and gives a striped appearance between crossed polars. A set of closely matching coloured stones could easily turn out to be glass, which contains swirls of colour together with randomly placed well-rounded gas bubbles. It is brittle and its facet edges are easily chipped leaving shell-like (conchoidal) fractures.

Colourless synthetic spinel has a single RI of 1.728 and fluoresces quite a bright bluish-white under SWUV. There is a striped effect between crossed polars: these stones are dangerous in small sizes and especially as replacements of single stones in eternity rings and pieces set with many small colourless stones. Synthetic gem-quality diamonds often fluoresce a bright sky blue under SWUV, though some recent reports suggest that there may be a response to LWUV from some diamonds, synthetic and/or treated.

The synthetic garnets, like diamond itself, give no RI reading on the standard refractometer (negative reading) since their RI exceeds that of the refractometer glass and of the contact liquid used. When coloured, some of the synthetic garnets give a very marked multi-line absorption spectrum quite unlike anything seen in coloured natural diamonds – which in any case do not show these kinds of colours. This is not to say that a client cannot be deceived into thinking that diamonds might occur in these colours. There seem to be no 'Cape', off-white to yellow synthetic garnets. The garnets show no natural mineral inclusions and no triangular markings (trigons) on the girdle surface: though girdles have been roughened and trigons filed on them this is not a common practice.

The remarks on synthetic garnets apply equally well to cubic zirconia which is best distinguished from diamond by use of the reflectivity or thermal conductivity testers. Groups of matching colours are equally suspect.

Composite stones with a base of colourless highly dispersive material and a hard top still show more fire (dispersion) than most natural diamonds and natural solid inclusions will not be present. The only exception is the rare diamond doublet in which the junction layer shows the reflection of a pencil point, for example, placed on the 'real' table facet.

None of the major natural gem species really look like diamond, either colourless or coloured. Of those whose dispersion is notable, most if not all show double refraction.

Synthetic rutile and strontium titanate make dangerously convincing diamond simulants: the strong double refraction of rutile and the spectacular dispersion of both could make 'diamond' a possibility for those who have not examined many diamonds, though as we noted above 'Cape' diamonds are not imitated by the synthetic garnets or by cubic zirconia. A doped strontium titanate has been made with a straw-like colour which could be mistaken for this variety of diamond, though it will not show the Cape absorption spectrum at 415.5 nm. This is quite a dangerous stone! The yellow synthetic rutile could, I suppose, be mistaken for diamond but this, too, lacks the Cape absorption band.

Imitations of coloured diamonds may show a sharp multi-line absorption spectrum never seen in diamond. A stone resembling Cape diamond with a pale yellow or off-white colour, fluorescing blue and phosphorescing yellow, is very likely to be diamond – if these effects are seen first try the spectroscope to see if the 415.5 nm absorption band is there. Some diamonds may show a greenish-yellow fluorescence with yellow phosphorescence; they are not common and the effects are hard to detect.

If it turns out that a specific gravity test is needed (the specimen has to be a fair size for testing by the hydrostatic weighing method) diamond's SG of 3.52 separates it from most of the more dangerous imitations (CZ 5.54–6.06, YAG 4.57–4.60, GGG up to 7.05). It is worth remembering that CZ is 1.7 times denser than diamond so that when someone substitutes a CZ for a 1 ct diamond the CZ will need to weigh 1.70 ct to fulfil the same size requirements and in similar circumstances a YAG would need to weigh 1.30 ct and a strontium titanate 1.45 ct. Of course no substitution of this kind goes on!

Other specific gravity figures that the investigator may find useful include synthetic spinel 3.61–3.65, synthetic rutile 4.25, strontium titanate 5.13.

Immersion techniques

If you are in a laboratory and the appropriate RI liquids are available (they are not likely to be unless you work in a university or research laboratory) a close approximation to the refractive index of an unknown can be found by immersing it (completely) in a liquid which is clear, as colourless as possible and whose refractive index you know. When the specimen and liquid have RIs close to one another both the edge of the stone and the facet edges become faint and vague like ice in water. The liquid di-iodomethane with an RI of 1.74 (this is the major component of the liquid used in refractometer testing) serves very well as a testing medium for 'immersion contrast'. If the specimen has the higher refractive index, as in diamond (2.417), the outline will show up boldly while the facet edges will appear faint: with synthetic spinel (1.728) and with most glass (RI in the range 1.50–1.70) will show a similar but less marked effect. On the other hand CZ (2.09–2.18) and YAG (1.83) will appear quite like diamond. If the immersed stones are in a flat-bottomed glass dish the effect can be photographed from below but in this case the reverse picture will be seen on the negative and the 'right' one on the print.

When the RI of the unknown is higher than that of the liquid and the experiment is projected by the desk-lamp onto a white background the facet edges will appear white and the outline dark. The outline will show bright and the facet edges dark when the RI of the specimen is lower than that of the liquid. The appearance of a colour fringe indicates that the RIs of specimen and liquid are a close match.

Using a microscope, the specimen can be placed in a transparent glass cell for immersion purposes: the RI of the liquid has to be known. When the experiment is lit and the focus altered a bright line of light will be seen passing from stone into liquid or vice versa depending upon which of the two has the higher RI. The line of light will pass from specimen to liquid when the liquid has the higher index when the focus tube is raised and from higher to lower when the focus is lowered. You may work out why a horizontal microscope performs this test better than a tilted one!

Liquids with an RI higher than that of diamond are not readily available and these are mainly laboratory tests but if a liquid with an RI lower than that of diamond but higher than the average RI of glass could be used the two stones could be distinguished from one another: in fact di-iodomethane with an RI of 1.74 is about as good as you could get.

Gemmologists ask if it is possible to measure dispersion as this is so important a feature of diamond and of course of its simulants. It is possible to carry out this test but it is so liable to system error (mistakes): those interested should consult a mineralogical rather than a gemmological textbook. The hard imitations of diamond (the synthetic garnets, CZ) have dispersions equal to or higher than that of diamond.

Synthetic moissanite

The synthetic material silicon carbide, SiC, has a hardness of 9.25, RI 2.648 and 2.691 with a birefringence of 0.043 and dispersion 0.104 – more than twice as high as that of diamond. The SG is 3.22. This material is the mineral moissanite (some naturally occurring types exist) and in recent years it has been synthesized as a possible diamond substitute on account of its hardness and high dispersion. As the figures above show, the birefringence should be enough to distinguish moissanite from diamond and the characteristic inclusions of groups of sub-parallel needles and stringers perpendicular to the table facet in fashioned stones make life easier than it might have been. In addition some specimens have been found to show uni-directional polishing lines and rounded facet edges. All suspected moissanites (or any diamond imitations) should be examined at right angles to the table as this is the direction in which the birefringence shows.

There have been suggestions that the RI could be lowered to bring it into the 'diamond' area of reflectivity meters. If this is possible repolishing might restore the original RI. Specimens appear to have been heat-treated and an oxide layer formed on the surface. A sample of heat-treated near-colourless moissanite showed a brownish colour across the surface of all the facets. When the samples were cleaned and hand-polished using cerium oxide on leather the reflectivity was restored to 98 per cent of that of the non-treated material. Those using heat as part of the testing of a suspected moissanite should be aware of this possibility of surface oxidation and should use no more heat than is necessary. The colour of the surface may be affected.

The thermal conductivity tester will give a 'diamond' response for both moissanite and diamond but this at least will distinguish the two species from all others.

In di-iodomethane diamond will sink and moissanite float. Not only colourless but green, yellow and blue types of moissanite have been manufactured but the colours are not arresting. A brown version is reported to have been grown in Russia but specimens are said to be almost opaque: the crystals were apparently grown by chemical vapour deposition.

At the time of writing synthetic moissanite was being grown and distributed by the firm C3 Inc. of North Carolina, USA. They have produced a detector, Tester Model 590, which distinguishes moissanite from diamond by examining the blue and near-visible UV areas of their absorption spectra. Moissanite shows an intense region of absorption extending from approximately 425 nm down to the UV region while colourless diamond transmits well down into the UV. Light from a halogen source is reflected from the table and if it transmits wavelengths from the blue to the UV region the observer is advised optically and aurally that the stone is a diamond. If there is no response the specimen will have absorbed this range of wavelengths and will be a moissanite.

The Presidium Moissanite Tester is designed to detect the very small current passed by a semiconductor type material. It informs the operator whether the unknown is moissanite or diamond. Peter Read, however, makes the point that synthetic moissanite is not a semiconductor and that the current detected by this instrument and by other moissanite detectors may arise from impurities in the synthetic material so far produced.

Synthetic gem diamond

General Electric

The most significant development in the gemstone world in the past few years has been the successful synthesis of gem-quality diamond of jewellery size. In 1970 the General Electric Company (GE) announced the manufacture of gem-quality stones as a spin-off from research work on high pressure, high temperature (HPHT) materials. Some of the research crystals were polished and thus entered the market.

They did not enter the market as quickly as rumours of important developments in the diamond world – this is always the case and it is good for the trade to be constantly vigilant. The original GE diamonds (three polished stones and five unpolished crystals) were fully reported in the fall 1984 issue of *Gems & Gemology*, the paper stating that it would take about one week to grow a crystal from which a polished stone of about 0.5 ct could be fashioned. Of the original faceted stones, all brilliants, one was near-colourless, one bright yellow and one greyish-blue. All were inert to LWUV but showed different responses to SWUV.

The near-colourless stone gave a very strong yellow fluorescence and persistent phosphorescence of the same colour. The yellow stone and the yellow crystals were inert to SWUV but the greyish-blue stone gave a strong slightly greenish-yellow fluorescence with persistent phosphorescence of the same colour. In the blue and near-colourless stones a cross-shaped pattern could be seen under SWUV.

These stones were to a large extent experimental productions and while some were sold, presumably to collectors, there cannot be a great number circulating in the diamond trade. None of the original GE stones showed any absorption in the visible and as many Cape diamonds are of similar colour the absence of the absorption band at 415.5 nm could be considered suspicious. Even under cryogenic conditions, when the band usually becomes visible (where it existed at all – many yellow diamonds began life as Cape stones and were treated to give a better yellow), the GE stones showed no visible absorption. One small yellow GE crystal did show the band but this was attributed to the presence of a small natural diamond used as a seed upon which the host crystal grew.

Only the blue and near-colourless GE diamonds were electrically conductive, a property of Type IIb stones. Natural Type IIb diamonds are colourless unless boron is present. It is believed that the presence of aluminium allowed the colourless specimen to be electrically conductive. Up to 1984 no natural near-colourless diamond had been found to be electro-conductive.

While the SG of some of the GE diamonds, measured hydrostatically, averaged 3.51, some specimens sank rapidly in a solution of approximately 4.0. These were found to contain inclusions of metallic iron from the growth process. The blue faceted stone and the near-colourless stone were strongly attracted to a pocket magnet while the attraction was less for the yellow diamond. A superconducting magnetometer showed that the GE stones could be separated from natural ones on the basis of their magnetic properties. Between crossed polars there were none of the characteristic signs of strain – natural diamonds usually show a somewhat confused mixture of light and dark absorption patches and it would be unusual for a natural diamond not to show them.

While there are not many GE stones around, they were the first to appear and since at least some of their characteristics are echoed by later productions a summary may be found useful as a starting point for the consideration of synthetic gem-quality diamonds in general:

- marked magnetic properties
- absence of strain between crossed polars

- electro-conductive near-colourless examples
- strong reaction to SWUV with no reaction to LW
- a near-colourless diamond without a blue or grey tint, strongly fluorescing and phosphorescing in SWUV while remaining inert to LWUV will almost certainly be synthetic.

Fancy yellow diamonds with no visible absorption bands seen with the spectroscope usually fluoresce and phosphoresce under UV: those showing neither of these responses will usually display the 415.5 nm absorption band. A yellow diamond with neither of these features will probably be synthetic: a near-colourless diamond with no tint of blue, grey or brown and no Cape absorption band will also most probably be synthetic. A strong fluorescence under X-rays with strong and persistent yellow phosphorescence also suggests a synthetic diamond as most natural stones fluoresce blue under X-rays.

Sumitomo diamond

In 1985 the Sumitomo gem-quality synthetic diamond was reported. Yellow crystals up to 2 ct were manufactured in some quantity and though the firm claimed that the diamonds were grown for industrial use we all know what can happen to 'industrial' crystals! The promise seems to have been kept but some sawn and partly polished yellow crystals are in major gem collections and undoubtedly some have been fully polished. The strong yellow colour is attributed to nitrogen and since it is vital to know something about the part played by nitrogen in diamonds here is a summary:

A high proportion of natural diamonds are Type Ia with a fairly large nitrogen content (around 0.3 per cent), the nitrogen being disseminated within the crystal as aggregates of small numbers of atoms substituting for adjacent atoms of carbon. Stones are usually near-colourless to yellow with some brown or grey examples.

Type Ib stones are so rare in nature (up to 1 per cent of diamonds) that a yellow Ib is likely to be synthetic. Though the crystals contain nitrogen the content is less than in Type Ia specimens and is dispersed rather than aggregated, substituting for carbon atoms. Diamonds of this class are usually a deep yellow.

Type IIa diamonds are usually colourless and the exact nature of their nitrogen content is uncertain, making it hard to detect. Only a few natural specimens have been reported. Type IIb diamonds are also very rare in nature and appear to contain more boron than nitrogen. They are usually blue or grey, sometimes colourless and are electro-conductive.

Infra-red spectroscopy will distinguish the different diamond types but this technique is not available to most gemmologists. Should it be necessary to distinguish between a natural diamond, a GE synthetic or a Sumitomo synthetic gemmologists can at least look at the following.

Under LWUV Type Ia diamonds may be inert or fluoresce an intense orange, yellow, green or blue. Type Ib diamonds will show the same colours but with a variable intensity. The GE and Sumitomo stones are inert. The same types respond in a similar but more variable way to SWUV while the GE stones are inert and the Sumitomo diamonds show a moderate to intense yellow or greenish-yellow fluorescence. Neither the GE nor the Sumitomo stones show phosphorescence while Type Ia and Ib diamonds either do show the effect in different colours or do not phosphoresce – this applies to both LW and SWUV.

Types Ia and Ib phosphoresce with similar variability under X-rays or do not phosphoresce. GE diamonds do not phosphoresce while Sumitomo crystals show a weak to moderately intense bluish-white phosphorescence. With the hand spectroscope gemmologists may find that Type Ia diamonds may be seen to show some sharp absorption bands while Type Ib stones show none: neither of the synthetic diamonds shows absorption in the visible.

Colour zoning is not rare in natural diamonds but in the Sumitomo yellow diamond a deep yellow inner zone and a narrow, near-colourless outer zone can be seen: there may be colour variation in the yellow zone. Inside the Sumitomo diamonds GIA found randomly distributed whitish pinpoint inclusions and opaque black metallic inclusions from the flux used in crystal growth – these are never present in natural diamond. Crystals have been polished in such a way as to avoid the inclusions appearing in the finished stone. Veil-like colourless areas could be seen in some crystals but these also could be avoided by polishing as they extend only a short distance into the crystal. Graining is also prominent, one type consisting of sets of lines both inside and outside the crystal and the other type, seen only inside the stone, consists of sets of straight lines radiating outwards from the centre of the crystal in four wedge-shaped formations resembling a cross with splayed ends. When the crystals were polished this changed to an hourglass shape which could be seen through the pavilions of the stones examined. A phenomenon which has come to be known as the 'bow tie' effect can be seen between crossed polars on the polariscope: this is an interference pattern and shows up best when viewed through two parallel sides of a rectangular (sawn) crystal.

De Beers diamond

The GE and Sumitomo diamonds were followed by a production from the De Beers Research Laboratory in Johannesburg. Diamonds ranged from light greenish-yellow through yellow to dark brownish-yellow.

The brownish-yellow diamonds were inert to LWUV but fluoresced moderately strongly under SWUV in yellow or greenish-yellow colours with inert zones: the yellow stones were inert to both types of UV. The greenish-yellow diamonds were inert to LWUV and gave a weak, yellow-zoned response to SWUV. Only the greenish-yellow stones showed phosphorescence but the effect was persistent. No absorption in the visible spectrum was reported for any of the stones. All stones showed distinct colour zoning (the greenish-yellow stones less strongly than the others) and the hourglass pattern could be seen in the brownish-yellow and yellow specimens.

Dense clouds of tiny white pinpoints could be seen in the brownish-yellow stones and to a lesser extent in the yellow ones – the pinpoints were isolated in the greenish-yellow stones. All three types contained metallic inclusions and grainy structures could be seen on the surface of faceted stones. No absorption was present in the visible. Some of the greenish-yellow colour may arise from nickel present during crystal growth.

IR spectroscopy has shown that the De Beers stones can be classed as Type Ib as they contain nitrogen dispersed as single atoms. Most natural yellow gem diamonds are Type Ia and yellow Type Ib stones are rare in nature. None of the De Beers stones were found to be electro-conductive and, like the GE and Sumitomo diamonds, showed a high thermal conductivity. The stones were weakly magnetic but overall the properties do not markedly differ from those of other synthetic diamonds of that period.

Blue diamonds were produced by De Beers for research purposes in 1993, the stones being doped with boron. A sample studied by GIA included three polished diamonds coloured light bluish greenish grey, dark blue and near-colourless with clarity grades ranging from VS1 down to SI. Fluorescence tests showed: light bluish stone in LWUV-weak orange with uneven colour distribution with a weak orange phosphorescence of about 1–5 minutes' duration. The same sample under SWUV gave a slightly greenish-yellow fluorescence of strong intensity and uneven distribution: very strong greenish-yellow phosphorescence persisting for 1–5 minutes.

The dark blue stone under LWUV gave slightly yellowish-orange fluorescence of moderate intensity with strong persistent phosphorescence of the same colour. This stone fluoresced slightly greenish-yellow under SWUV with a very strong yellow persistent phosphorescence.

Near-colourless stone is inert to LWUV but gave a slightly greenish-yellow fluorescence under SWUV with a very strong yellow phosphorescence. None of these effects would be expected from a natural diamond. All three stones under magnification showed distinct internal growth sectors and pinpoint inclusions similar to those in other synthetic diamond, and metallic flux inclusions were found. Colour zoning was pronounced and strain birefringence (randomly mixed light and dark) areas could be seen between crossed polars. The stones were found to be a mixture of Types IIa, IIb and Ia, and such a combination had not been previously identified in single diamonds. General properties were in the diamond range and there was no visible absorption. The danger of these early synthetics is in specimens of very small sizes which tend to be overlooked in testing and valuation.

Up to 1993 synthetic gem diamond was not too great a threat to the trade in that most examples showed sufficient features to distinguish them from natural diamond.

Russian diamonds

The next synthesis of gem-quality diamond took place in Russia, the stones being grown specifically for use in jewellery. GIA tested five faceted stones coloured yellow to orange or brownish-yellow and three apparently treated yellow to greenish specimens. The latter probably represent early examples of the now prominent high pressure/high temperature treated stones. In this instance the non-treated polished stones showed a greenish-yellow fluorescence under LWUV with varying weak to strong intensity but with no phosphorescence. Results observed under SWUV gave a weak to strong intensity yellowish-green to green with no phosphorescence. The treated diamonds gave a very strong greenish-yellow response to LWUV with a moderate to strong yellow phosphorescence. Under SWUV the treated stones showed a strong yellowish-green fluorescence with a moderate to strong yellow phosphorescence.

With the hand spectroscope some of the Russian synthetic diamonds showed absorption features when they had been cooled with a spray refrigerant. The non-treated stones showed a sharp absorption band at 658 nm with a weaker sharp band at 637 nm in one stone and at 527 nm in another. Treated stones were cooled in the same way and showed several absorption bands between 600 and 470 nm, with less absorption below 450 nm.

Several of the samples showed an uneven weak to moderately intense luminescence when viewed under a strong source of green light. Such an effect has not been reported for natural diamonds of this colour. IR spectroscopy showed that the Russian stones combined features of both Types Ia and Ib diamonds: usually yellow synthetic gem-quality diamonds have been pure Type Ib.

The properties of the Russian diamonds can be summarized: they were the first to show a response to LWUV and the first in which heat-treated examples fluoresced more strongly under LWUV than under SWUV. All the heat-treated stones gave a yellow phosphorescence so that a diamond which responds in any way to either type of UV needs careful investigation. Absorption bands between approximately 658 and 637 nm and between 560 and 460 nm appear to be confined to Russian diamonds and some of the bands can be seen with the hand spectroscope. Variability of the different responses to testing are a feature of the Russian product. It is likely that during production a number of different methods were tried and that temperatures and pressures also varied.

In 1993 GIA made the point that it was only to be expected that other colours of synthetic gem diamond would join the yellow, blue and near-colourless productions and in the same year they reported two red diamonds which had reached the trade. Both stones were a dark brownish-red and weighed about 0.5 ct. They both showed very distinct colour zoning and through the crown facets of one of them could be seen the outline of both square-shaped and superimposed cross-shaped light yellow areas surrounded by much larger areas coloured red. The cross-shaped pattern was more or less directly under the table facet and viewed through the pavilion facets the colour zoning was seen to be in four positions round the girdle and showing as narrow light-yellow zones surrounded by larger red areas. The other stones showed similar effects.

Using reflected light only one graining line could be seen on the table facet of one of the stones, the other one showing a faint surface graining pattern on the table facet with some parallel polishing lines. Large metallic-like inclusions could be seen in both stones which were also attracted by a pocket magnet.

Under both types of UV the crown facets of one of the red diamonds showed unevenly distributed very intense green and moderately intense reddish-orange regions, the green fluorescence corresponding to the light yellow zones described above. Under SWUV these sections gave a phosphorescence lasting several seconds. The reddish-orange fluorescence was seen only at one small point near the girdle under LWUV but under SWUV this colour was seen on all the large dark-red areas. One of the stones showed uneven luminescence of a different pattern: looking through the crown facets under either type of UV a very small area of red fluorescence could be seen near the centre of the table, this area being surrounded by a narrow zone of green fluorescence. Narrow bands of an orange fluorescence pointed from the green-fluorescing area towards the four corners of the table facet where there were areas of stronger orange fluorescence. The rest of the stone fluoresced a weaker orange-red and there was no phosphorescence. When the other stone was illuminated by strong visible light a moderate green luminescence was seen in the yellow area. The other stone did not show this effect.

Several sharp absorption bands could be seen between 800 and 400 nm in the spectrum of the 0.55 ct stone. Some of the bands between 660 and 500 nm could be distinguished with the hand spectroscope, especially when the stone had been cooled. Fewer bands could be seen in the 0.43 ct stone but they occupied similar positions. Both stones absorbed increasingly towards the violet and a showed a broad absorption region extending from 640 to 550 nm. IR spectroscopy showed that the stones contained elements of both Types Ia and Ib. Compared to other synthetic diamonds the two stones show an apparently unique fluorescence with a particularly unusual response to LWUV. Nickel was shown to be present and some of the absorption bands in one of the stones showed that it had been irradiated and heated.

Compared to natural diamonds the red stones can be distinguished by several different features: some Type IIa diamonds of natural pink colour show an orange fluorescence under UV but GIA reported that they had never seen a known treated pink specimen. Some pink to red diamonds with known natural colour show a blue fluorescence and are classed as Type Ia. Some pink to red treated natural diamonds of Type Ib show absorption bands at 637, 595 and 575 nm that are characteristic of treatment by irradiation and heating. Neither these stones nor two other natural treated diamonds of mixed Types Ib and Ia with pink to yellow colour and a moderately strong orange fluorescence under LWUV and SWUV looked anything like the two red stones as the pattern of colour zoning was not present nor the absorption bands ascribed to nickel.

Isotopically pure carbon-12 diamond

In 1990 GE Research and Development Center, Schenectady, New York, reported that they had made near colourless isotopically pure carbon-12 synthetic diamonds. Isotopic purity is very useful in a variety of research applications and it is not unlikely that polished stones of this composition will appear on the market at some time. Crystals were reported to show signs of strain birefringence, not usually seen in synthetic diamonds. They also contained rod-like metallic-looking inclusions and clouds of very small triangular or lozenge-shaped tabular inclusions beneath the octahedral crystal faces, with scattered pinpoint inclusions with a white, bright metallic appearance in reflected light and a brown appearance in transmitted light. No graining could be seen in the crystals but strain birefringence was apparent with a weak pattern of grey or blue. Both crystals examined were inert to LWUV but fluoresced a weak yellowish-orange under SWUV: when the radiation was switched on the colour seemed to increase in intensity before remaining at a fixed level. Cathodoluminescence turned out to be a more useful test with specimens showing a zoned pattern of cathodoluminescence corresponding to different internal growth sectors. The colour was a slightly greenish-blue. Under X-rays the stones fluoresced yellow with a very persistent yellow phosphorescence (up to 10 minutes in one crystal). No specimens were electro-conductive and none showed absorption bands in the visible. The crystals were assigned to Type IIa compared to the earlier GE-grown crystals which were electro-conductive and assigned to Types IIa/IIb. The newer GE crystals fluoresced strongly under SWUV and contained metallic inclusions.

Summary of synthetic colourless or near-colourless diamonds

Check: response to SWUV; for metallic inclusions; magnetism, especially when many metallic inclusions are seen or suspected; presence or absence of anomalous birefringence between crossed polars (the effect is more commonly seen in natural diamonds); signs of the Cape absorption band at 415.5 nm which is often seen in treated diamonds. *Look first for natural solid inclusions.*

Composition and colour: stones will be Type IIa or mixed Types IIa + Ib + IIb and will appear colourless or light grey or very light blue, yellow or green with some of the colour in sectors usually in sizes of less than 1 ct.

Optical phenomena: black cross between crossed polars: reflected light shows up metallic inclusions which may appear black by transmitted light. They may appear long and rounded or occur in groups or on their own. Clouds of pinpoint inclusions are also common – you need to ensure that the stone is very clean before testing since dust on the surface may be misleading.

Fluorescence and phosphorescence: colourless to near-colourless synthetic diamonds do not usually respond to LWUV but are much more likely to show fluorescence and sometimes phosphorescence under SWUV: the colours may be yellow, greenish-yellow or orange–yellow, weak to strong. What fluorescence is seen may be uneven though it may assume a square, octagonal or cross-shaped pattern. Most show phosphorescence after irradiation by SWUV with the effect lasting up to 1 minute: colour is usually yellow or greenish-yellow. There is usually no fluorescent response to visible light and no sharp bands can be seen with the spectroscope.

Magnetism and electro-conductivity: some specimens may be attracted by a small magnet and some are electro-conductive.

Summary of synthetic blue diamonds

Composition and colour: these are Type IIb or mixed Type IIb + IIa. Colour is anywhere in the range light to dark blue with some stones showing a greenish- or greyish-blue when there are small yellow growth sectors.

Optical phenomena: internal growth sectors with some showing a central octagonal shape surrounded by octahedral and cubic faces. Growth sectors are seen best under UV. The distribution of colour is usually quite even. Graining planes intersecting in patterns are characteristic and hourglass shapes common. Weak, cross-shaped anomalous birefringence is evident. Opaque black metallic inclusions and clouds of very small pinpoint inclusions visible.

Fluorescence and phosphorescence: the blue diamonds are inert under LWUV but show a yellow or greenish-yellow response to SWUV – like most synthetic diamonds the response to SWUV is usually greater than to LWUV (if there is any response). Fluorescence is unevenly distributed and follows internal growth sectoring with octagonal or square patterns. There may be mixed responding and inert sectors. Usually there is a yellow moderate to strong persistent phosphorescence lasting in some cases up to 1 minute. Any cathodoluminescence shows a distributed effect. There is no luminescence under visible light nor any absorption visible with the hand spectroscope. Blue synthetic diamonds may be attracted by a strong magnet and are electrically conductive.

Summary of synthetic yellow diamonds

Composition and colour: these may be Type Ib or Ib + Ia. If irradiated they are Type Ib or IaA, or if heat treated at high pressure (HPHT) they will be Type IaA. Colour ranges from greenish-yellow to orange–yellow to a brownish-yellow. Type Ib or IaA treated stones show an orange to pink or even red colour after irradiation and heating to about 800°C while Type IaA treated stones heated from 1700 to 2100°C at high pressure show a yellow to greenish to yellow–green colour.

Optical phenomena: growth sectors can be seen and are often octagon-shaped with additional cubic sectors. Colours usually show light and dark yellow sectors: stones that have been irradiated and heated to give a pink or red colour show pink or red zoning with some yellow sectors. Graining planes show between internal growth sectors, the planes sometimes intersecting to form patterns. Transmitted light may show up hourglass or other effects, and between crossed polars a black interference-caused cross is sometimes seen. Metallic inclusions of flux, some reaching 1 mm in length, and small pinpoint inclusions can be seen.

Fluorescence and phosphorescence: type Ib stones show no visible response to LWUV while Type Ib + IaA diamonds show weak to strong yellow or yellow–green fluorescence. Type Ib or IaA treated yellow stones show a strong green with a weak orange fluorescence in different growth sectors while only a weak orange is seen in some specimens. Type IaA treated diamonds show a notably strong greenish-yellow or yellow fluorescence. Under SWUV Type Ib stones show either yellow to yellow–green fluorescence of weak to moderate intensity while Type Ib or IaA stones give a weak to strong yellowish-green response. Type Ib or IaA treated diamonds give a strong to very strong green plus a weak orange fluorescence in different growth sectors with some specimens giving an orange fluorescence only. Type IaA treated yellow diamonds give a strong greenish-yellow response. The fluorescence can vary in intensity: in Type Ib or Ib + IaA stones it is stronger under SW than under LWUV and with Type Ib or IaA treated stones the response is either of equal strength or stronger under SWUV. In Type IaA treated stones the intensity is stronger under LWUV.

Fluorescent colours are unevenly distributed and duplicate the growth sectors. In Type Ib or Ib + IaA diamonds phosphorescence is rare but sometimes the observer may see a weak yellow or greenish-yellow effect persisting for some seconds. In Type Ib or IaA treated stones there may be a weak orange phosphorescence for several seconds: in Type IaA treated diamonds there may be a strong persistent yellow phosphorescence lasting for up to 1 minute. Visible light luminescence may be seen only in Type Ib + IaA diamonds with a weak to moderate green colour. Similar effects may be seen in other types, with a weak orange colour. Type Ib diamonds do not respond to visible light.

In Type Ib diamonds no sharp absorption bands are usually seen but with cooling a specimen may show the band at 658 nm. In Type Ib + IaA diamonds, when cooled, bands may be seen at 691, 671, 658, 649, 647, 637, 627 and 617 nm. In Type Ib or IaA treated stones several sharp bands may be seen at 658, 637, 617, 595, 575, 553, 527 and 503 nm. Type IaA treated stones may show sharp absorption bands at 553, 547, 527, 518, 511, 503, 481, 478 and 473 nm. Stones may be attracted by a strong magnet.

Diamonds with enhanced colour

Added to the general anxiety about synthetic diamonds is the growing awareness that any particular stone may not be showing the buyer or wearer the colour with which it left the earth. It may be argued that as long as the stone looks just as attractive (or more attractive) the origin of the colour that you see (as long as it is permanent) does not really matter very much – or that only large, important stones are affected.

This has not proved to be the case with the gem-buying public. It is true that the larger and more significant the stone, the more likely it is that it will need to be accompanied by a certificate showing its country of origin. This is already the case with fine rubies, blue sapphires and emeralds. Ascertaining the origin of a diamond is well out of the competence of the gemmologist or even of the laboratory, apart from those which have long dealt with diamond crystals and which have constantly received supplies of rough diamond from known locations.

Nonetheless a start has been made: Ekati diamonds from Canada are being sold in London by an old-established and highly respected firm of manufacturing jewellers who are issuing certificates giving not only the place of mining but also the weight of the original rough. This excellent enterprise will surely be followed by other firms with diamonds from other places.

Such certificates will also state whether or not the colour of the diamond has been enhanced. Why should a diamond's colour be altered? Most diamonds appear colourless (or 'white') but if many stones are examined it will be obvious after a time that there are a thousand shades of white! Many are off-white or yellowish and we know that these diamonds have traditionally been known as Capes. It would not be too easy to raise an advertising campaign to make them popular – their colour is too indeterminate – but it is possible to change them, or some of them, to more desirable colours.

This has to be a good thing, surely? The certificates already mentioned carry statements of the diamond's colour, clarity and weight, and colours (we are still discussing shades of white, not the stronger 'fancy' colours) are graded so that a diamond whose certificate bears a high colour grade will carry a higher price. The difficulties arise when the stone has been artificially treated to raise the colour grade – should the seller tell the buyer ('disclosure')? The seller may not know in the first place. It is left to laboratories and qualified diamond graders in the trade to try to establish a code of practice within which acceptable standards are agreed. The various jewellers' associations are agreed that this should be done and by and large general agreement has been reached internationally.

It is quite easy to improve the colour of a Cape stone but less easy for it to stand the test of time. The polished stone can be offered for sale in a stone paper with a blue liner, the blue deceiving the eye into thinking that the yellow colour is not there. When copying ink or indelible pencils were used more than they are today – one practice was to darken the back facets with such a pencil, in the hope that the setting would conceal the deception, at least for the time that the sale negotiations were in progress. Suspicion is the best test and any careful dealer will look automatically for what are really fairly obvious treatments. While small stones passing as diamond are possible intruders into a parcel of genuine small diamonds, such stones are unlikely to have been interfered with: small stones in second-hand jewellery deserve much more attention.

The suspicious mind should always remember that any diamond selected for *treatment* in order to improve its colour will not already be of fine colour! It is almost always the Cape stones, which might otherwise take a long time to sell, that will be more useful when treated. We should also bear in mind that the majority of treated diamonds are yellow as this colour sells much better than the rather dark greens and blues which can also result from treatment.

While many experiments have been made to alter the colour of diamond one early process needs particular care – if it is suspected. Radium salts were used to bury diamonds for periods up to one year: the colour changed to a dark green (from a Cape colour, presumably) and the stones still show quite high radioactivity after something like 100 years, this resulting from the implantation of fast-moving nuclei recoiling from their disintegration and entering the surface of the diamond (Nassau, 1994). Some of these treated stones show spots on the surface of the pavilion where radiation has been localized. Polishing the stones does not affect the radioactivity nor is the green colour, caused by α-particles, affected by heating. Americium has also been used experimentally, with similar effects.

A number of different irradiation methods have been used to alter diamond's colour: neutrons from a nuclear reactor, α-particles, protons and deuterons. These sources of energy all produce an absorption band extending from the infra-red to the yellow–green area of the spectrum. This band causes the stone to appear as dark shades of green and sometimes black and is of such significance that it is known as the general radiation band, GR1.

The sources just listed are not powerful enough for the colour of the diamond to be changed throughout: instead the stones show a surface coloration extending only to a small distance beneath. If a diamond when examined from below shows umbrella-like markings above the culet, this is highly characteristic and proof of irradiation via the culet. Irradiation through the table facet produces dark patches in different positions and if the stone is placed table-down on a light surface a dark ring will be seen.

Treatment by high energy electrons develops heat in the specimen and this may have an effect on its colour. Cooling during treatment may be necessary: the colour produced is usually blue to blue–green but stones are harder to test than those coloured by the action of the heavier α-particles. The colour produced is a function of the energies used during treatment: some diamonds may turn blue with lower energy electrons and a more greenish colour with higher energies. In any case the GR1 absorption band will be present.

While none of the processes used to irradiate diamond lead to transient coloration it is clearly best to settle for a method of treatment which will completely colour the stone. Despite the work of Crookes with radium salts, diamonds are unlikely to become radioactive as a result of treatment: however, Nassau (1994) describes a very dark green, black-appearing diamond which contained surface-reaching fractures which harboured metallic particles from the polishing process: the particles became radioactive on irradiation. Boiling the stone removed the radiation which passed through to the acid. Though reports do not state this it may be that the boiling removed the particles.

Irradiation and in fact any form of treatment may cause the occasional exotic colour to appear in diamond. Nassau (1994) reports a diamond which before irradiation fluoresced a greenish-yellow (it was hoped that a chartreuse colour would develop after treatment): neutron irradiation gave an orange–red colour with a bright orange–red fluorescence. This specimen may have contained some Type Ib characteristics. Nassau reports another diamond which was also neutron-irradiated: previous to treatment it appeared cloudy but afterwards showed a rich sky blue colour.

Nassau (1994) makes an interesting point: can a natural blue diamond (Type IIb semiconducting) turn a darker blue on irradiation? If this is possible, testing would be difficult because a blue electro-conductive diamond might be thought natural if the first test used was on electro-conductivity.

The colour produced by heating subsequent to irradiation (the term annealing is often used) is often more acceptable than that produced by irradiation alone, which, whatever the method used, is usually dark and not particularly attractive: dark green or dark greenish-blue are the commonest irradiation colours. Increasing heat generally alters these colours to green to brown and then to yellow – increasing heat after this point returns the specimen to its pre-treatment colour. Heating is stopped when the diamond reaches a colour considered suitable for the market.

While yellow diamonds are the commonest products of irradiation and heating, other colours do turn up. Pink or red diamonds can be the result of the double treatment and in these cases the diamond will have been Type Ib. Crowningshield via Nassau (1994) reports these diamonds to show a strong orange fluorescence which can also be shown by natural pink diamonds.

Heating from about 400°C upwards can destroy the GR1 absorption band and in addition forms an absorption band at 595 nm (sometimes cited as 594 or 592 nm), 503 and 497 nm or strengthens them if they are already present. The treatment causes the diamond to show colour: the original irradiation-caused blue may develop into orange, yellow, green or brown, depending upon the intensity of the heating. If a treated diamond is heated to 1000°C the 595 nm band will disappear, though the colour of the stone will be unaffected. It is possible that a gemmologist may encounter a yellow diamond without the 595 nm band: it will not be possible to say that the stone has not been treated simply because the band does not appear to be present. In natural yellow diamonds the band at 504 nm is often easier to see than the band at 497 nm but after irradiation the situation may be reversed or the strengths of the two bands may be equal. There should still be the Cape absorption bands present in the stones, however, and cooling can often reveal them.

The celebrated yellow Deepdene diamond, which now weighs just under 104.52 ct, was originally a Cape stone that was turned a deep green by irradiation and then to a golden yellow by heating. Some Type Ia diamonds may change to a bright yellow when heated close to 2000°C for a few minutes only: some polishing processes have involved sufficient heat to cause a colour change in diamond and sometimes a fancy light yellow can develop in conjunction with the absorption band at 503 nm.

Diamonds with enhanced clarity

As with colour, the clarity of a polished diamond forms the basis of a grading system and all stones of any size, and certainly important ones, will have details of their position on a clarity scale recorded on their accompanying certificates. Clarity means relative or apparently complete freedom from inclusions; when there are inclusions their size and position affect the clarity grade, since a large dark inclusion directly below the table would not attract many customers. These have sometimes been removed by lasers and the traces are prominent

Inclusions are not always fragments of other minerals (or of diamond) incorporated into the polished stone: the term includes fractures which can be caused by diamond's relatively easy cleavage. These can reach the surface and look unsightly so that it is not surprising that efforts have been made to diminish their effect.

Some of the first attempts to fill fractures and improve clarity grade were made in the 1980s by Zvi Yehuda of Ramat Gan, Israel. It was reported that after cleaning the stones were filled with a molten glass, the process being carried out at high temperatures. Glass remnants on the surface are removed. Fracture filling has been carried out at least since the 1980s and has occurred with stones as small as 0.02 ct and as large as 50 ct, according to publicity material. Even after more than 20 years there are claims and counterclaims about the durability of the process as well as discussions on disclosure.

In the 1990s the Israel-based firm of Koss & Schechter tried two different processes of filling diamonds. In both methods glasses were used, one a halogen glass, the other a halogen oxide glass. At that time halogen glasses were in general use as fillings. Koss stones showed internal orange and yellow flashes when the diamonds were rotated under dark-field illumination. Under bright-field illumination the flashes were blue and violet. Not all filled diamonds showed these colours. All the filled areas contained gas bubbles and some diamonds showed flow structures in the fillings. Filling material was sometimes a distinct yellow but fine crackled lines seen in some other fillings by the same firm were not apparent in stones filled in the experiment under discussion. Lead and bromine were found in all the stones when examined by energy-dispersive X-ray fluorescence analysis.

Stones filled with halogen oxide glasses did not fill successfully. By 1996 a claim by the firm that they would incorporate a fluorescent additive in their filler had not yet been fulfilled and cathodoluminescence did not reveal the presence of any additives. A wide range of flash colours was observed in the diamonds that would not fill satisfactorily – the colours included red, orange, yellow, blue, purple and pink. These colours were seen under dark-field conditions while under bright-field illumination the colours seen were bluish-green, green and greenish-yellow.

These facts were made known to the trade by Koss & Schechter in an unusual disclosure of working practices. The point was made that the RI of the filler should approximate that of the diamond host so that the filled area would be difficult to identify.

Stones belonging to and tested by GIA and which had been filled by Dialase Inc. of New York using a Yehuda-based filling process showed no sign of weight gain (the filling would be in the form of a thin film) while their appearance seemed to be enhanced in some cases and unchanged in others. Four out of the six stones tested showed a drop of one full colour grade after treatment and two stones showing no improvement in clarity also dropped to a lower colour grade.

Under magnification the filled diamonds showed a slightly greasy appearance with a slight tinge of yellow. The lens showed a large number of surface fracture signs, quite enough to alert a diamond dealer. The Dialase-filled diamonds showed the flash effect with a characteristic yellow–orange colour seen under dark-field lighting. Under bright-field lighting the same area changed to an intensely vivid electric blue: tilting the stone backwards and forwards showed the flash colours changing from orange to blue to orange. The effect is best seen when the stone is examined at a steep angle and close to a direction parallel to the plane of the treated fracture. The lighter the colour of the host the easier it is to detect the flash colours, which are hard to see, for example, in a dark-brown diamond where only the blue colour will be apparent.

In very small fractures the flash effect may not be visible. Care should be taken not to mistake the orange flash for the iron-staining patches often seen in diamonds and other stones. Since the filled areas are in fact thin films they may show interference colours which will appear as multiple colours resembling rainbows. Flow structures and gas bubbles are common in glass-type fillings and if the filled area is carefully examined with the lens or microscope it may be seen to show a light-brown, light-yellow or orange–yellow colour. It has been suggested that the filling material used in the Dialase experiment is a lead–bismuth oxychloride with an RI close to that of diamond.

The Dialase filling was not affected by ultrasonic cleaning, steam treatment or boiling in a detergent solution. The filling was unharmed by thermal shock and by stress induced by the setting process. It was damaged by repolishing in some instances and when the flame of a small torch used in repair was brought close to the filling signs of sweating in the form of beads on the surface were apparent. The best illustrated summary of these experiments can be found in *Gems & Gemology* (Fall, 1994). The paper makes the point that flash colours are the best clue to fracture filling. Readers can also be referred to a most useful chart, also published by GIA, which indicates in colour the main features of filled diamonds, covering flash colours, trapped bubbles, flow structure, misleading features such as interference colours, feathery appearance in unfilled breaks, natural iron staining, brown radiation staining and burn marks on the diamond surface left by the polishing process (these can be mistaken for remnants of filler left on the surface). The chart accompanied *Gems & Gemology* for Summer 1995.

The use of diamond thin films as improvements to diamonds

While thin films can be applied to any gem species, in the case of diamond they could be used to alter the colour of a polished stone. The technique to some extent parallels the blooming of camera lenses, now carried out as a matter of course and the thin films used have a thickness of around 1 μm (0.001 mm). The films are applied to the surface by chemical vapour deposition (CVD). The thin layer helps to resist abrasion and to improve the stone's appearance.

The presence of thin films can usually be spotted by interference colours on the surface (this is due to the difference in RI between film and host) and is particularly notable when in rare cases a film of air occurs between surface and coating. The films cause an unusual haziness best seen in dark-field illumination – this is due to light scattering within the film. When a coated stone is examined in diffused light and held against a white background the film causes the surface to appear brown. The coating has a polycrystalline composition so that it does not show extinction between crossed polars and it shows no absorption in the visible.

Thermal conductivity tests have given similar readings for the coating and a silicon substrate so that coating a non-diamond with this type of film would not deceive the tester. For a coated non-diamond to pass as diamond the diamond thin film would have to be at least 5 μm thick.

The colour of a polished diamond might be altered by a thin film coating but the technique seems not to have proceeded beyond the experimental stage, in which a blue film of 20 μm was deposited on a natural near-colourless diamond octahedron, giving a blue electro-conductive crystal. Immersion should reveal this kind of treatment since natural blue diamond has a notably patchy colour and the film would show sharp edges. It is possible that a film could be applied to a Cape diamond substrate and the yellow and blue cancel one another out to give an 'improved' stone.

Coating by blooming as in camera lenses has been used with Cape diamonds. The coating can be seen when the stone is rotated in a strong light (the bluish bloom will then be seen) and there is a spotty or granular appearance in the girdle region, with pitting. Some coatings can be dissolved in sulphuric acid when the yellowish colour of the stone may be seen beneath.

Diamond crystals have been burnt to oxidize the surface and produce a whitish appearance, so that a yellowish crystal may resemble a whitish one with a frosted surface. Rumour has it that a diamond polisher was deceived by a parcel of crystals of this type which when cut returned to their yellow colour.

General survey of diamond, its simulants, syntheses and treatments

Glass is still the commonest and in some ways the best imitation of diamond. It can show adequate dispersion and when cut rather than moulded can look very good when set appropriately. Swirliness of colour and well-rounded thick-walled gas bubbles, with the absence of natural mineral inclusions, have been mentioned several times above as characteristic of glass. A piece of jewellery with well-matching coloured stones should be carefully examined. RI readings will very often be in the range 1.50–1.70 and there will be no birefringence. Glass shows no crystalline properties, which will be shown by all its natural imitators and by such artificial materials as the synthetic garnets and CZ.

CZ and the synthetic garnets may be doped to give colours which may be deceptive but which often show absorption spectra which would never be seen in diamond. They, like diamond, can be satisfactorily tested only by thermal conductivity testers or reflectivity meters. Fluorescence can be useful but may be deceptive, especially as synthetic diamonds fluoresce under SWUV. SG tests are slow and prone to system error (arithmetic shaky, quite often).

Synthetic gem-quality diamonds may be almost any colour once irradiation and heating have been in action.

Yellow, blue, colourless and near-colourless synthetic diamonds have been made by several manufacturers: stones are often magnetic due to their metallic inclusions, may fluoresce under SWUV, show characteristic graining patterns and combine more than one of the diamond types. Hourglass and similar patterns may be seen between crossed polars. Fluorescence or cathodoluminescence colours may echo growth sectors and synthetic diamonds may be irradiated and heated to give enhanced colours. As long as the possibility is borne in mind the gemmologist has enough clues to make a correct identification of a synthetic diamond.

High prices are asked for coloured diamonds so it is worth while trying to alter indifferently coloured Cape stones by irradiation and heating to give fancy colours. Yellow stones give the most trouble and gemmologists should remember that elements of the Cape absorption spectrum may still be present in the treated stone – they are better seen when the diamond is cooled.

Surface-reaching fractures are now routinely filled with glassy material in which gas bubbles can be seen. Unexpected flashes of orange or blue are a clue to this treatment and when accompanied by signs of a glassy flow structure can be satisfactorily identified. The question of disclosure can pose problems for jeweller and customer and is still not quite resolved.

It is easy to forget composites and some examples with a highly dispersive base and hard transparent top can be particularly deceptive especially in small sizes. Some diamonds have been coated with thin films and may then show interference colours on the surface.

High pressure/high temperature treated diamonds

In 1999 it was reported that natural high purity diamonds were being processed at high temperatures and high pressures (HP/HT) by General Electric Company to make them more nearly colourless. Agreement was reached at the time with the firm Pegasus Overseas Ltd (POL) that the girdle surfaces of all the HP/HT treated diamonds would be inscribed GE POL. GIA found that 99 per cent of the treated stones were Type IIa and were originally brown or brownish. Specimens show internal graining and partially healed cleavages: some stones have a hazy appearance. The lettering can be removed but traces can still be detected.

When tested in the laboratory the GE POL diamonds showed a low concentration of single nitrogen compared to untreated diamonds of similar type and appearance. Absorption spectroscopy showed that any yellow coloration in HP/HT treated diamonds was due to low concentrations of single nitrogen, not observed in untreated Type IIa diamonds. Laser-excited photoluminescence spectroscopy revealed the presence of nitrogen-vacancy centres in most but not all of the HP/HT treated diamonds. When such centres are present the ratio of the 637:575 nm luminescence intensities gives a possible method of distinguishing HP/HT stones from untreated Type IIa diamonds.

Indications of HP/HT treatment in yellow to green diamonds include a highly saturated body colour, well-defined brown to yellow octahedral graining, moderate to strong green 'transmission' luminescence and visible evidence of heating. Strong absorption bands at 515 and 505 nm can be seen with a hand spectroscope. Near-IR spectra show a peak at 985 nm, an indication of exposure to high temperatures.

Bearing in mind that most HP/HT treated diamonds are Type IIa the SSEF Swiss Gemmological Institute introduced the SSEF Type IIa Diamond Spotter to distinguish Type IIa stones from other diamonds of similar colour. A Raman spectrum is then obtained using the 514 nm laser line of a Raman spectrometer. The presence of peaks at 3760 cm^{-1} shows that a small number of N-V centres exist in the stones and this appears to be a characteristic of the GE process since up to the date of this study (2000) all GE POL diamonds have been found to show this emission. A near-colourless Type IIa diamond with an N-V centre emission is thus very likely to be a GE POL treated specimen.

Examination of seven Russian gem-quality synthetic yellow diamonds and three treated by post-growth annealing in HP/HT conditions showed distinctive properties of colour and luminescence zoning, metallic inclusions, graining and sharp absorption bands in the violet. UV luminescence and optical absorption spectra differed from those seen in previously examined yellow synthetics: all but one fluoresced under LWUV and the HP/HT treated stones fluoresced stronger under LWUV than under SWUV (as would be seen with fluorescing natural diamonds). HP/HT treated stones showed a moderate to strong yellow phosphorescence under UV and most of the samples showed a green luminescence under visible light.

Press releases from the American Company Nova Diamond state that the brownish colour of Type Ia diamonds is altered to yellowish-green by HP/HT treatment without the use of irradiation. The pressure may go up to 70 000 atm and the temperatures up to or above 2000°C. The Antwerp-based Diamond High Council examined seven Type Ia diamonds, all with the characteristic yellow–green post-treatment colour.

Black cracks surrounding inclusions could be seen in some stones. With the use of UV-VIS-NIR spectroscopy a strong H3 fluorescence could be seen with a maximum at 529 nm and this could also be seen under the stimulus of LWUV. This is the cause of the yellow–green to green colour. NIR spectroscopy figures show that there is a very strong N2 absorption, this causing the green colour to become more pronounced. With FTIR spectroscopy all samples showed considerable absorption due to A centres and B centres. SWUV topographic fluorescence shows that green H3 fluorescence correlates with strain patterns observed through crossed polars using the microscope. These can sometimes be seen in unpolarized light without the microscope.

As the H3 and H2 defects can be found in natural untreated diamonds the presence of the H2 defect is not on its own an indication of HP/HT treatment but the strong to very strong H3 fluorescence/absorption and the H2 absorption come close to proving this treatment. The presence of A, B and C centres together in the IR spectrum with H2 absorption in the NIR spectrum is a definite sign of this treatment.

This can also be said of the presence of a combination of A, B and H2 centres with other C centre-related defects (giving absorption at 637 and 575 nm): with strong H3 absorption/fluorescence and H2 absorption these are again a proof of HP/HT treatment.

The girdles of treated stones are significantly beaded and frosted naturals are sometimes seen.

Reports of interesting and unusual examples from the literature

Items in this section have been chosen to illustrate points made in the chapter and to bring one-off items to your notice.

Throughout the book, and not just in the case of diamond, reports have been chosen which deal with synthetic versions and serious imitations of the major species. In general they are arranged in chronological order. Readers should note that many of the examples given may be one-offs and remarks should not be taken to apply to all examples of a particular stone.

By literature we refer to papers in the major gemmological and mineralogical journals rather than to previously published monographs.

We should not forget that diamond rough is sold in some parts of the world, though some major producing countries allow citizens to hold rough only under licence. It is not surprising that efforts are occasionally made to simulate rough diamond: some synthetic diamond rough was misrepresented as natural diamond from Canada at the time of the early development of the Ekati mines. The crystals, however, contained characteristic metallic flux inclusions (*Gems & Gemology*, 31(2), 1995).

Five imitation diamond crystals turned out to be CZ cut to give a diamond shape. They were also given triangular markings, engraved parallel lines and frosted surfaces. The heft (hand-estimated SG) of 5.80 compared to diamond's 3.52 gave them away.

Coating of diamonds is sometimes used in an attempt to raise the colour grade. One example reported in 1984 described stones whose colour grade rose to G from H or I after coating. The coating seemed, however, to give an overall greyish appearance. Under special lighting conditions a faint band can be seen on one side of the girdle and some diamonds have been found to show this band both on crown and pavilion sides of the girdle.

The presence of metallic inclusions does not always make a diamond subject to the pull of a magnet. GIA reported a heavily included black diamond in which black inclusions were arranged in bands and were probably graphite.

A very light green marquise diamond contained a row of crystals which appeared to be mostly diamond even though one of them, from its colour and form, could have been enstatite or a pale green diopside. This inclusion was found to be the sole cause of the colour: a band at 741 nm showed that the diamond had not been irradiated.

Interesting flash colours in a filled diamond, reported in 1995, were seen one at a time, green in bright-field and red in dark-field lighting. The fracture filling showed small bubbles and flow lines under high magnification.

Several reports during 2001 identified the new name Bellataire diamond as a continuation of the original GE POL stones. The Bellataire diamonds were set to be marketed in New York City and it has been proposed that the letters bd will be lasered on them.

At the 28th International Gemmological Conference, George Bosshart from the Gübelin Gem Laboratory, Lucerne, Switzerland presented a summary of natural and HPHT pink and blue diamonds, the paper being summarized in *Gemmology Queensland* for January 2002.

The summary stated that natural pinks may be either Type Ia or Type IIa but that a wider variety of colours existed in the first group. Type Ia diamonds may be pure pink to red to purple but pink and red are more frequently combined with orange and brown modifiers. Type IIa diamonds are limited to light pink to pink colours which may be modified by secondary orange or brown. How the 560 nm absorption band (the cause of the colour) arises is not fully understood at present but pink Type Ia stones especially tend to show unevenness of colour. If the 560 nm band is particularly strong the diamond will be purple rather than pink. The colour cannot be caused by either nitrogen or hydrogen impurities.

In Type Ia diamonds red is a possible colour and Argyle stones from Western Australia commonly show brownish to purplish pink and even red. Rarely encountered pure orange colour is caused by an absorption band at 480 nm.

Blue diamond is modified chiefly by a grey colour. Natural blue diamonds are always Type IIb, a few stones showing grey–violet rather than blue and belonging to the hydrogen-rich Type IaB. Type IIb diamonds are electro-conductive from substitution of carbon by boron in ppb amounts. Examination of blue diamonds cannot easily be carried out in the visible region of the absorption spectrum, the strongest absorption band being at 2802 wavenumbers in the infra-red. This band is related to the boron content.

HP/HT methods have been used by General Electric to grow the crystals now known as Bellataire and have achieved colourless crystals. Today pink and blue diamonds have also been grown. It would seem that the 'improvable' original rough must have been Type IIa brown or Type IIb grey to brown with fairly high clarity grades. Strong colours have not yet been observed. Testing cannot be achieved with normal gemmological methods. Raman photoluminescence may be more useful as the colour-enhanced diamonds show fewer photoluminescence bands than naturally coloured diamonds.

It is possible that HP/HT methods may be used in the future to alter the Cape yellowish colour to colourless.

Black opaque cubic zirconia has been offered as black diamond. Heat-treated black diamond (most diamonds altered to black have been irradiated) shows graphitization of surface-reaching fractures as shown in *Gems & Gemology* (Fall, 2001).

The most prominent feature in all fracture-filled diamonds is the flash effect, though material containing lead could be detected by X-rays and EDXRF spectroscopy. Diamonds as small as 0.2 ct have been filled. Yehuda/Diascience, Koss and Schechter (Genesis II) and Clarity Enhanced Diamond House, a subsidiary of Goldman Oved Diamond Company, appear to be the major players in the enhancement game. In dark-field illumination orange and pink flashes predominate and in bright-field blue and green flashes. Enhancement could cause colour grades to be lowered. Goldman Oved appeared to gain clarity from treatment but did not drop a colour grade: this happened with the Koss stones. Flash colours in Goldman Oved stones are bluish-green and yellow (bright-field illumination), violet, purple and pink (dark-field).

Yehuda treated diamonds had previously contained a light-brown to brownish-yellow or orange–red coloured filler, though these colours were not seen in later examples of their treated diamonds. More recent products contained a filler with a noticeably yellow body colour: the filling lowered the colour grade in some cases. Some stones contained a white cloud which lowered transparency. The fillers in all three examples could be damaged by direct heating or by common cleaning methods.

De Beers have recently developed the DiamondSure to assist in the identification of synthetic from natural diamonds. The instrument depends upon the 415.5 nm absorption line not being present in synthetic diamonds, though it is present in almost all natural diamonds apart from about 5 per cent of colourless natural specimens, some fancy yellows, blues and pinks of Type IIb.

Chapter 3

Ruby and sapphire: the corundum gemstones

Ruby and the various colours of sapphire are varieties of the same mineral, corundum (Al_2O_3), and apart from colour differences their properties are similar: both, naturally, are extensively imitated, both are easily synthesized by a number of different methods and both may be colour and clarity enhanced. Their place in the gemstone trade is established and some dealers specialize in ruby and fine blue sapphire. The other colours of sapphire are also extensively traded but are less likely to have been improved: of these yellow sapphire is encountered the most and green sapphire the least. Pink sapphire is a distinct and desirable colour, nothing like a pale ruby (and has been considered worth synthesizing by one of the slower and more expensive methods).

Some varieties of fancy sapphires are considered highly desirable and worth arguing about and a familiarity with some of the names used by Sri Lankan dealers can be useful! No one agrees on the exact colour meant by the Sinhala terms *padparadschah*, *padmaraga* and others but many fancy sapphires are very attractive and collected for their own sake rather than for any particular ornamental use. Ruby and blue sapphire may show six-rayed stars when cabochon cut and such phenomenal stones have their own following.

Testing corundum varieties ought to be straightforward: is the stone corundum; is it natural; is it synthetic; has its colour been improved and does it matter? In the case of the finest ruby and blue sapphire more needs to be known. Examination of current saleroom catalogues shows that for all important stones and even some moderately important ones locality information is now specified; ruby and blue sapphire from Burma, blue sapphire from Kashmir and either variety from Sri Lanka (in exceptional cases) are citations commonly found in catalogue descriptions.

The addition of locality information to sale catalogue entries may not on its own be considered a very troublesome matter but major coloured stones have long been certificated and today locality information appears on the certificate too, thus accompanying the stone on its travels between dealers and customers.

Ruby

Introduction

Colour

Ruby is the best-looking red stone and other red species and varieties are sometimes judged by how close to the ruby colour they appear to be (of course this is a superficial judgement as many red species are very beautiful too). Why does the red of ruby attract so very much? As always colour perception varies with the individual and that single person can feel differently about the same colour according to their psychological state at any particular moment. Nonetheless there is a general agreement on what constitutes the most desirable red. The red of ruby arises from a small quantity of chromium (Cr) acting as an impurity: when certain types of energy are directed at the ruby the chromium is encouraged (it has no choice) to behave in such a way that the observer perceives not only red but a red glow too because the stone fluoresces under the stimulus of the energies, one of which, conveniently, is daylight/natural light. Ruby will also fluoresce when excited by UV or X-rays but wearers of rubies will not wish for their assistance. So it's lucky that rubies contain chromium which behaves in just this way? Yes, but not all follow the trend exactly – there are individual variations.

It is interesting to see in the natural world that for a species or variety to be successful several factors have to operate together – the reverse is also the case. Rubies from Burma are universally considered the finest of all – they tend to be the largest (always a help), they are not usually very heavily occupied by solid inclusions, the particular way in which the colour is zoned (subtle shifts in intensity in different parts of the stone) is notably attractive, they may show the star effect and, best of all, they contain less iron (Fe). While this common element turns up in gemstones of almost any colour, it tends to give a brownish overtone to stones which owe their red to chromium and if it is absent so much the better. Iron tends to diminish or even prevent fluorescence so that a ruby with an appreciable iron content will not give the red glow as well as the red colour. If you find a ruby coloured by Cr and if Fe is there too, giving a brownish cast to the red, and if the iron could be got rid of in some way ...? Read on!

Interfering with colour

Rubies from Thailand (Siam rubies) do contain some Fe and thus appear subtly different in colour from Burma rubies. The effect of the Fe can be diminished by heating so that the stone will appear much more like a Burma stone as far as its colour goes. This is one of the most debated problems in gem identification. Today, if a stone at the point of sale is known to have been altered from its original colour at some time between recovery and sale does the customer have to be told?

The question of disclosure depends on whether or not the seller knows about the treatment, whatever it might have been. The seller's knowledge depends on the information acquired as a buyer. Initially the knowledge has to come from gem identification by an individual or laboratory and it has to be reasonably simple to obtain. Obscure knowledge or mere suspicion is easier to challenge later on – facts are not so easily threatened. We shall constantly be meeting the disclosure question (it includes clarity as well as colour) when we are considering the major gemstones.

Synthesis

Ruby and blue sapphire were obvious targets for synthesis when this became possible in the nineteenth century. The first rubies were manufactured simply and cheaply though today's synthetic stones, grown over a longer period (months rather than hours), cost much more than many customers expect. Identification of these products was quite easy to begin with but today it is desirable for a laboratory to look at important suspected examples.

Imitations

Ruby's importance has inevitably led to a variety of imitations, red glass being the commonest. Some natural stones could, with some effort of the imagination, be said to resemble ruby so that making the distinction would need careful investigation and undoubtedly some dealers act in ignorance when other red stones are actually mined with ruby (red spinel is the best example).

Identification

All the corundum varieties fit quite closely into RI 1.765–1.773, DR 0.008 (these are typical values for Burma ruby), SG 3.9–4.00, hardness 9. Pleochroic colours are usually pale yellowish-red and a deeper crimson. The absorption spectrum is distinguished by a sharp line in the red which is in fact two lines close together (doublet) at 694.2 and 692.8 nm and which in some lighting conditions shows red (emission) rather than dark (absorption). The bright red emission line stands out against a background which is also red and while it does not on its own show that the specimen is ruby it proves the presence of chromium. Note that emission lines in any part of the spectrum *other than the red* arise from room or other extraneous lighting, never from the specimen. *This applies to all minerals and gemstones.* The other features in the absorption spectrum of ruby are two absorption (occasionally emission) lines in the orange and a strong absorption region obscuring the bulk of the yellow and the green and centred at approximately 550 nm: important absorption bands in the blue, two closely spaced at 476.5 and 475 nm, are never present in the somewhat similar spectrum of red spinel and are accompanied a little further off by another absorption band at 468.5 nm and a general absorption of the violet.

Before concluding absorption spectrum notes it is worth knowing that in some synthetic rubies the emission doublet may turn to absorption on transmitted light and that a weaker emission line indicates that iron may well be affecting fluorescence. Even some blue and near-colourless sapphires occasionally show the Cr emission line. Observers may find a blue filter useful when investigating the emission lines which are then more prominent against a dark background. The broad absorption region extends further over the spectrum when the ordinary ray is observed, thus giving a better red (since the green is absorbed) than in the spectrum of the extraordinary ray which is yellower since the absorption region for this ray allows some yellow to pass unabsorbed.

A red stone which shows the Cr emission line and the pair of absorption lines in the blue must be either a natural or a synthetic ruby.

The luminescence of Cr-rich ruby at least can be seen through the Chelsea filter when stones show an unmistakable fluorescent red. Viewed through a simple red filter with the specimen illuminated by monochromatic blue light,

Cr-rich and synthetic rubies in particular glow against a dark background – this technique is called *crossed filters*. Under LWUV Cr-rich Fe-poor ruby glows a bright red (again, synthetic stones in particular) and there is at least some glow when the same type of specimen is illuminated by strong green, blue or violet light.

Synthetic ruby

Distinguishing natural from synthetic ruby needs the microscope since none of the small instruments can give a satisfactory diagnosis: correct interpretation of the nature of the inclusions and to some extent their arrangement is essential when doubts arise about a specimen. The microscope too is the first-line instrument for the preliminary establishment of locality information. The gemmologist presented with a fine, large, apparently flawless ruby does not necessarily have to jump to the conclusion 'synthetic' but if presented with a parcel of fine large stones matching in colour (or small stones closely matching in colour) such a conclusion is more likely but testing is still needed. This point is made because like many other words 'flawless' has several meanings which comfortably co-exist: in this context it means 'free from obvious and unusually unsightly inclusions' rather than showing no faults of cutting or wear.

When inclusions are to be examined it is much easier if the specimen is immersed: in the past a number of liquids could be called into service but most of them have been found with later knowledge to be unsuitable for use on health and safety grounds and need laboratory conditions and supervision rather than the conditions under which much gem testing usually takes place. Nonetheless even water is better than nothing, since the aim of immersion is to diminish reflections from the surface of the specimen – water in a glass cell at the end of a horizontal tube microscope can be very effective, though try not to disturb the liquid and specimen while someone is looking along the microscope!

Verneuil flame-fusion rubies

The earliest synthetic rubies were made by processes rather like the flux-growth method today, the much larger and facetable *flame-fusion (Verneuil) rubies*, sapphires and spinels coming onto the market only in the later nineteenth century. Verneuil stones show large, random, strongly outlined gas bubbles (gas bubbles are rarely if ever seen in natural rubies except as part of a multi-phase inclusion), curved growth zoning resembling the lines on a vinyl disc and curved colour zoning.

The lines are not easy to see if you simply look through the top of the stone, especially if a very bright light source is being used. These curved lines are most commonly seen in the plane of the girdle and when reflected rather than transmitted light is used. A fibre-optic light source placed adjacent to the microscope can be moved around to get optimum illumination or dark-field lighting (available on the stereo binocular microscopes commonly used for gem testing) can be used instead of or in conjunction with the fibre-optics. In certain circumstances curved lines can be seen more easily when the specimen is immersed in liquid of high refractive index (di-iodomethane is really the only possible one today – RI 1.74, quite close to that of corundum). Light from a photographic enlarger stopped down to *f*22 will usually show the lines in the immersed stone – they can sometimes be seen in colourless synthetic corundum.

Another useful test for synthetic ruby takes a little setting up: the specimen is placed between crossed polars on the polariscope and the instrument lamp switched on. As the specimen is rotated and correctly positioned (remember, if you have nothing better to do, rotate your specimen) systems of straight lines can be seen with the lines of the second and third systems seen at 60° to those of the first system. In colourless corundum the lines may be glimpsed but soon disappear as the specimen is rotated. When the optic axis of the corundum specimen is at right angles to the polaroids (a group of spectrum colours inside the stone shows you when you have found the axis) look for the interference figure which shows as a black cross with slightly splayed arms and with coloured concentric circles at its centre. Sight of the axis will prove that you have finally found the optic axis direction. Then, preferably with the aid of a microscope, investigate the system of lines. Their presence usually indicates that the specimen is synthetic. The effect is known as the Plato effect.

During photography included gas bubbles usually catch the reflection of the camera lamp and many bubbles seen together will reflect it in exactly the same way, ruling out the possibility that they are crystallites.

The property of pleochroism (different colour seen in different directions) can be useful in distinguishing the Verneuil ruby from natural ruby or from the more elaborately grown synthetics. The boule which results from the growth process is a single crystal and behaves like a natural ruby in that pleochroism is present. It so happens that since for economy the cutter needs to obtain as many stones as possible from a boule pleochroism can be seen through the table facets – this is much less likely in natural ruby whose crystals make it easier to fashion stones in a direction at right angles to the pleochroic one. Of course any natural crystal may be of facet-grade quality in colour but physically distorted so that the cutter cannot completely avoid placing the pleochroic direction so that it is at right angles to the table facet of the finished stone when some pleochroism may be seen.

Why should this matter? While the presence of pleochroism in ruby or any other crystal means to the gem tester that the specimen is crystalline and cannot be glass, the lapidary needs to be able to produce the best available red through the table facet of the stone and so has to ensure that the red is not diluted by any other colour. In the case of ruby the table facet should be placed at right angles to the vertical axis of the original crystal. This is a direction in which pleochroism does not occur.

It is interesting to note that some very early rubies grown before the flame-fusion process was developed also show curved but sub-parallel lines. Modern products grown by slower and more sophisticated (and expensive) methods may also show straight, angular growth lines.

While the best indication of a synthetic ruby is the absence of those mineral inclusions which are virtually inseparable from natural specimens, we have to know which these may be – in any case some mineral inclusions assist the compilation of locality information which as we have seen can greatly affect prices.

Note that neither the locality of origin nor whether a ruby is natural or synthetic can be determined by measurement of refractive index or specific gravity alone, nor will the absorption spectrum provide a definite answer.

Rubies from different localities

Burma ruby

While no single mineral species inclusion is peculiar to Burma ruby, corundum in general is a relatively fast crystallizer and so encloses a number of species that form at the same time (syngenetic inclusions}. Burma stones contain crystals of apatite as hexagonal prisms, rhombs of calcite showing twinning bands as vestiges of the rock during whose metamorphism the corundum was formed, octahedra of spinel and shortish needles of titanium dioxide in the form of the mineral rutile. This mineral is of great importance to those identifying gemstones: it occurs in many species and in the case of ruby and blue sapphire it may be star-forming. Rutile needles in Burma ruby tend to be shorter and more stubby than their Sri Lankan ruby counterparts; groups of needles intersecting at 60°/120° form a characteristic network known as 'silk' which is more densely packed in Burma rubies that in Sri Lankan ones. Rutile crystals in Burma ruby often show re-entrant angles in the broader end. Other mineral inclusions found are yellow crystals of sphene (titanite), a calcium-titanium silicate and deep yellow crystals of the zinc sulphide sphalerite. Liquid inclusions are comparatively uncommon in Burma ruby though they are prominent in the corundum varieties in general.

Without magnification the observer can see Burma ruby's highly characteristic swirly colour mixing, long known as 'roiling', a term quite difficult to define although the effect is easy to recognize once seen. Roiling is not the only proof of Burma origin. Inclusions showing how the original crystal grew can also be seen in the form of polysynthetic twin lamellae which appear as parallel lines – while corundum has no easy cleavage, parting can take place parallel to the twin lamellae.

Mong Hsu ruby

In recent years fine rubies have been found at Mong Hsu in the Shan state in eastern Burma. Some crystals show a pale blue to violet core but this is not the only reason for the heat treatment of the stones. Heated stones are usually placed in borax to fill any fissures. Low temperature treatment may be detected by IR analysis. Unidentified whitish particles are common, plagioclase, fluorite and dolomite are fairly easy to identify but rutile and spinel inclusions less so. Water-bearing fluid inclusions show a variety of contents. At least 75 per cent of Mong Hsu rubies contain inclusions of diaspore or structural (OH) but information obtained from mid-IR spectral features should not be used alone as locality information.

Electron spin resonance (ESR) has been used to distinguish Burma from Thai ruby, the former showing the Cr^{3+} absorption line (resonance field 7500e) and the latter the Fe^{3+} line (resonance field 8500e).

Sri Lanka

Ruby from Sri Lanka does generally appear a softer red than the Burma stones: rutile needles are longer, more slender and more (sometimes less) densely packed than those in Burma ruby. In addition Sri Lankan rubies contain books of biotite or phlogopite (both mica group species) minerals forming before the ruby, and crystals of zircon, often surrounded by characteristic markings known as 'haloes': these markings are in fact stress cracks formed by the gradual change in mass of the zircon crystal after incorporation into the ruby.

The change arises from a slow breakdown of the crystalline structure due to radioactivity from its own uranium and thorium atoms. Other included minerals may be apatite and pyrrhotite.

Sri Lankan ruby contains more liquid inclusions than Burma ones and liquid droplets forming flattish planes like fingerprints or feathers (and often called by these names) are common. The presence of silk and feathers together strongly indicates a Sri Lankan origin.

Thailand

Rubies from Thailand contain very little if any rutile (hence no star stones) but can be fairly easily identified by the presence of plagioclase, one of the feldspar group of minerals, very small crystals of almandine (garnet) showing a brown colour and apatite showing yellow, and pyrrhotite which is black with a metallic lustre. All or any of these inclusions when combined with a lack of rutile and certainly when associated with fine parallel lines which could be mistaken for rutile but which are really polysynthetic twinning lamellae, often with long fine needles of the iron oxide, böhmite – these are always arranged in three directions almost at right angles to one another – indicate Thai ruby. Another strong indication is the presence of what have been called decrepitation haloes in successive planes – these arise from the alteration of inclusions during geological events. Round melt-residues are sometimes left after heating to eliminate blue sections in the ruby.

When working on any unknown it is always wise to consider several types of inclusion together before coming to a decision.

Vietnam

Northern Vietnam and the ruby-producing areas of Burma have similar geology and rubies from both areas are similar.

East Africa

Stones from the Tunduru-Songea area of Tanzania are notably 'hot pink' in appearance though purplish and red colours are also produced. Most appear to have been heat treated. A generally turbid appearance in ruby from this area may arise from the exsolution of rutile or böhmite. Apatite and rutile crystals are fairly common and signs of lamellar twinning have also been reported.

Pakistan

Pakistan ruby from the Hunza valley in the Northern Areas is clear only in small pieces. Larger ones, of very attractive colour, show turbidity arising from cracks, parting planes, polysynthetic twin planes and swirling (roiling) of colour similar to the effect seen in Burma stones. The commonest mineral inclusion is calcite which can be very prominent but dolomite may also be present along with reddish-brown flakes of phlogopite mica. Though rutile has been found in the rubies it appears to form neither silk nor stars: other minerals reported are spinel, apatite, pyrrhotite and pyrite. RI 1.762 and 1.770, DR 0.008, SG 3.995. Inclusions are the only distinctive features.

Afghanistan

Ruby from Afghanistan has RI 1.762, 1.770, DR 0.008, SG about 4.00. The classic Afghan ruby from Jagdalek (many Romanizations of the name) may contain rhombs of calcite and laths of phlogopite mica, rutile prisms, some forming geniculate (knee-shaped) twins, garnet, pyrite, spinel, graphite, hornblende, apatite, zircon and dolomite, primary negative crystals, secondary healed fractures, straight angular growth zoning and irregular treacle-like swirls – these stones show a notable prominent growth zoning. Inclusions are the only distinctive features.

Madagascar

Madagascar is now coming on stream as an important producer of ruby. So far specimens from the Toamasina Province in north-eastern Madagascar have been found to contain clusters of zircon inclusions very like those found in ruby from the Umba valley, Tanzania and from Ilakaka in south-western Madagascar. These rubies, purplish-red in the rough, are routinely heated for the market. Ruby from the Vatomandry area of eastern Madagascar may be comparable on occasion to the best Burma stones. They can be distinguished from Burma and Thai rubies by the lack of growth structures, the presence of short oriented rutile needles and profuse, colourless, transparent or white birefringent zircon crystals. Stones have a high content of iron.

Trapiche rubies resemble trapiche emeralds in that there are six sections delineated by six non-transparent arms intersecting at a small point in the centre of the crystal. This gives the effect of a six-rayed star. Specimens have also been reported from Burma, Vietnam and Mae Sai in northern Thailand. The specimens will no doubt be imitated and turn up at mineral shows.

Tajikistan

Ruby and pink sapphire from the Pamirs in Tajikistan have RI 1.761–1.762 and 1.770, with SG 3.99–4.02. There is a very strong orange–red to red fluorescence under LWUV with a much weaker medium red under SWUV. Very fine undiagnosed clouds, short rutile needles and pinpoint particles can be seen inside the stones and there are faint to moderately distinct growth structures, faint colour zoning, twin lamellae and crystals of zircon, plagioclase, calcite, sphene and rutile.

China

China is now producing some ruby, though at the time of writing in August 2001 complete details of the product are not available. However, ruby of Burma or Vietnam colour is reported to have been found in the Ailao Mountains of Yunnan province: potential star material is reported from deposits at Qinghai.

India

A survey of Indian rubies showed a mean SG of 3.93 with some specimens from Orissa giving 3.75 and some ruby from Karnataka giving 4.13. In general Indian rubies are more often opaque or translucent than transparent though star stones can be attractive.

Natural stones which may be confused with ruby: glass

Red spinel: RI 1.718 (singly refractive). SG 3.60, H8. Bright red fluorescence similar to that shown by ruby but the absorption spectrum does not show absorption bands in the blue.

Pyrope-almandine garnet series: Cr-bearing 'Bohemian garnets' may resemble ruby but certainly when seen en masse look oddly harsh, glassy and match in colour: single RI close to 1.745, SG 3.7–3.9, H7.5. Some of the darker red (i.e. more Fe-rich) red garnets may look like Thai rubies and may have RI around the ruby area but will show no DR. Iron absorption spectrum with three strong broad bands in the yellow, green and green–blue areas. No emission lines or fluorescence. No pleochroism.

Red tourmaline: DR of around 0.018 makes doubling of back facets and inclusions easy to see: pleochroism two shades of red and could be mistaken for Burma or synthetic ruby. No diagnostic absorption spectrum, no fluorescence.

Glass: single RI often in the range 1.50–1.70, characteristic of most glass. No diagnostic absorption spectrum, no fluorescence. Well-rounded, bold-outlined gas bubbles not forming part of a multiphase inclusion, suspicious matching of colour when several specimens are shown together, warmer to the touch than crystalline materials. Between crossed polars glass shows no regular change from light to dark, nor does it remain completely dark. Instead a striped effect popularly known as 'tabby extinction' can be seen which is characteristic and soon recognized. Fashioned glass will often show conchoidal (shell-like) facet edges as the result of local fracturing.

Synthetic ruby rough has been offered as natural in South-East Asian markets: curved and diffused colour banding can be seen with rounded, sometimes elongated gas bubbles. Masses of whitish alumina from part of the growth process are also seen.

Natural star corundum has been dyed to simulate star ruby, the dye, confined to cracks and irregular fissures, sometimes showing an intense orange fluorescence under LWUV. Other examples of dyed corundum have not shown fluorescence from the dye even though it is the cause of the colour.

Both ruby and blue sapphire are regularly imitated by various kinds of composite. One type has a crown of natural sapphire with a base of synthetic blue sapphire.

Crystals for the collector

It is extremely rare for crystals of gem quality to be offered to representatives of gem dealers other than at the place of recovery, but occasionally an old collection may come up for sale so the possibility of a fine ruby crystal coming onto the market is not entirely ruled out. Apart from this happy accident, there are plenty of ruby crystals which are too opaque to be faceted and too included on the surface to make cabochon material. In recent years ruby crystals from Vietnam have been seen for sale. Corundum crystals are usually either bipyramids or flat with hexagonal outline but with triangular markings on the larger faces and small faces recurring three times in a complete rotation to show that a three-fold rather than a six-fold symmetry is operating.

Sapphire

Introduction

While ruby's place among the corundum varieties is unique, blue sapphire from Burma, Kashmir and Sri Lanka is also considered sufficiently important to have locality information given on certificates and in sale catalogues. As in the case of ruby, inclusions provide the best guide to the origin of a blue sapphire and synthetic blue sapphires provide the most convincing substitute. The standard gemmological measuring tests will not in general assist very greatly, as sapphire and ruby share constants so close that only corundum/not corundum can be ascertained without magnification at least.

Magnification is certainly needed to identify synthetic sapphire and to spot colour/clarity enhancement: it is also needed for the establishment of locality where possible.

Inclusions

In general, colour in natural blue sapphire is seen as straight swathes: curved growth zoning very strongly suggests a flame-fusion synthetic stone.

Note that neither the locality of origin nor whether a sapphire is natural or synthetic can be determined by measurement of refractive index or specific gravity alone, nor will the absorption spectrum provide a definite answer.

Burma

Burma sapphires contain highly characteristic healing cracks like folded fingerprints or crumpled flags, crystals of pyrrhotite and rutile needles, the individual crystals appearing sectioned: the whole assemblage does not resemble the silk found in Burma ruby. Other mineral inclusions which may be present are apatite, fergusonite, monazite, phlogopite, zircon, dolomite and brookite: pyrrhotite is uncommon.

Kashmir

Kashmir sapphires have a soft, hazy appearance arising from minute fissures but are generally lacking in distinctive inclusions. Identification has to fall back on a close examination of the fissures which can be seen to run in three directions intersecting at 120°. An overall picture resembling webs or brush-strokes strongly suggests Kashmir origin. While mineral inclusions are rare, tourmaline sometimes makes an appearance.

Thailand

Sapphire from Thailand contains no rutile as acicular (needle-like) crystals and so no star stones are found. Crystals of plagioclase, orange–red garnet and black niobite, so far unreported from sapphires from other locations and often surrounded by a liquid pool, should be sought. Twin lamellae may be found.

Sri Lanka

Sri Lankan sapphire does contain rutile-forming silk (in general composed of longer and more slender rutile crystals than in Burma sapphire – though the 'arrowheads' from re-entrant angles at the broader ends of the crystals can still be seen).

Star stones are available: specimens additionally show a wide range of liquid inclusions. Other mineral inclusions are golden to brown biotite and red phlogopite (mica minerals are not often seen in Burma sapphire), red garnet, plate-like crystals of hematite, pyrite, chalcopyrite, red prismatic rutile crystals and zircon – these crystals show in high relief and are often surrounded by stress cracks, long known familiarly as 'haloes'. Crystals of apatite, pyrrhotite and zircon also occur. Negative (hollow) crystals are common. Healing fissures are very common and are almost an indication of Sri Lankan origin on their own.

Sri Lankan lapidaries make use of the smallest speck of colour, often placing it in the base of a faceted stone: the stone will face up blue but will appear colourless from every other direction. This is an effect which never fails to please.

It was from Sri Lanka that the whitish, oily, translucent sapphire known as geuda was first noted. Some of this material, once disregarded by gem dealers, was found to turn a very fine stable blue when heated and apart from solid inclusions which appear to have 'exploded' under the stress of the heating to around 1900°C there appears to be no method of ascertaining whether or not a particular sapphire was originally geuda.

Whether or not any particular inclusion has altered under stress is not easily discovered by general gem testing as very close magnification is required along with a considerable familiarity with the geology and mineral associations of the host stone. Heating to these high temperatures (corundum melts at 2037°C) also causes disintegration of liquid-containing negative or hollow crystals (cavities echoing the crystal structure of the original crystal and bounded by plane facets): their liquid contents leak and may heal neighbouring cracks formed under stress. Generally solid inclusions surrounded by stress cracks indicate that heating has taken place at some time – but this could be during geological time and long pre-dating any attempt by artificial means to effect some desired alteration. This is why familiarity with the nature of solid inclusions is best left to professional gemmologists.

Montana

Sapphires from Montana, USA, show characteristic hexagonal mineral grains with accompanying liquid films: these are the Yogo sapphires which are mined from hard rock (rather than from alluvial gravels). There are similarities between Yogo sapphires and Thai rubies. Biotite, though rare, calcite, reddish-brown rutile, pyrite and spinel may be found, analcime has been reported and growth-zoning is prominent. Montana sapphires from old river gravel deposits almost always show signs of heating since colour is routinely enhanced and the practice is accepted. Apart from signs of treatment, gravel-hosted stones contain negative crystals whose contents include natural glass and one or more gas bubbles. The pit formed by the bubbles is black while the surrounding glass is medium grey. The glass has a different surface lustre from the host.

Pailin, Kampuchea

Pailin sapphires from Kampuchea may contain arresting red crystallites of pyrochlore together with plagioclase and thorite crystals. Healing cracks are common and form hexagonal patterns. Twin lamellae can be found.

Australia
Australian sapphires from the Anakie area of Queensland may contain crystals of feldspar: the well-known black star sapphires contain mixed acicular crystals of hematite and ilmenite.

Colombia
Metallic greyish-faced rutile crystals with apatite and pronounced growth-zoning are common in blue sapphires from Colombia, which have occasionally reached the collectors' market.

As we note elsewhere, magnifications of around 40–50× would be needed to focus on solid inclusions. While the 10× lens and gemmological microscope can show a good deal of the internal furniture of a specimen, satisfactory work depends on the specimen being reasonably large in the first place. Mineral inclusions do not often show defined crystal forms but are often corroded or otherwise deformed so that comparing those features which you may be fortunate enough to see with a mineral book may not get you very far – but an acquaintance with mineral formation helps!

East Africa
Sapphires from locations in East Africa (for example, the Umba valley of Tanzania) characteristically show dislocation lamellae with fringes arranged at an angle to a lamella which penetrate into neighbouring lamellae. Böhmite occurring on the edges of the twin lamellae, and graphite, zircon, pyrrhotite and apatite (this mineral is very common in corundum) have been reported.

Synthetic blue sapphire

Synthetic blue sapphire is more easily grown by flame-fusion than by any of the more demanding and expensive methods. Those flux-grown specimens which have appeared from time to time have shown very marked hexagonal patterning which, with the lack of natural solid inclusions, show the observer what the specimen is. Growers find that expensive investment in time and apparatus is justified only for the most easily sold species.

Flame-fusion-grown synthetic sapphires can be identified only by their inclusions: as with ruby, this type of blue sapphire shows curved colour banding, best seen when the stone is immersed and viewed against a light background. Curved growth zoning is observed in the same way and strongly suggests a synthetic product; however, some of the sapphires manufactured by slower and more sophisticated methods may show straight growth zoning.

Many blue sapphires on the market today owe their colour to diffusion treatment: this gives a near-colourless or pale specimen a deeper colour. The treatment gives only the most shallow colour penetration and when examined from the side this can be seen as a very thin layer, the remainder of the stone being its original pale colour. Furthermore there is usually some damage to facet or girdle edges – all faceted stones, of whatever species, with pitted or fractured edges are very likely to have been heated. Diffusion treatment has also been tried with ruby and other colours of sapphire but appears not to have been widespread.

Identification

Blue and the other colours of sapphire do not differ from ruby in their measurable properties but blue sapphire owes its colour to an interaction between trace amounts of iron and titanium and its absorption spectrum naturally differs from that of ruby. Though titanium does not absorb usefully in the visible, iron in blue sapphire gives at its strongest three broad bands at 471, 460 and 450 nm (the last band is the strongest) but the three are usually observed only in the deepest blue specimens: they may best be seen in dark Australian stones whereas pale Sri Lankan stones may show only a faint absorption at 450 nm. Absorption in blue sapphire needs to be checked by direction (as it does in all coloured stones) since a band may be revealed in one direction while appearing to be absent in another. A Polaroid filter can easily be used or the light by which the stone is to be viewed can first be passed through almost any kind of blue filter which will keep out the otherwise intrusive rays in the brighter red–yellow part of the spectrum.

The absorption spectrum of blue Verneuil-grown synthetic sapphires usually shows no absorption bands, but if one is seen it will be close to the 450 nm position: reports of a three-band spectrum with shadowy bands in the blue in Verneuil stones suggest that gemmologists could be confused if they rely on an absorption spectrum or lack of it when testing a blue unknown: it is wise always to turn to the microscope. Reported spectra of the slower-grown Chatham synthetic blue sapphires show some likeness to that of Sri Lankan blue sapphire (450 band only in darker specimens).

Natural stones which may be confused with blue sapphire

Blue spinel has a colour which could be mistaken for that of sapphire but is in general darker. The refractometer shows a single shadow edge at 1.718 and the absorption spectrum immediately appears complex with bands extending as far as the orange: this band is accompanied by others in the yellow, green and green–blue with the main broad band in the blue centred at 459 nm and accompanied by another narrower band at 480 nm. Blue spinel's RI and SG may rise with the addition of zinc to maxima of 1.75 and in excess of 4.0.

Blue tourmaline shows a noticeably higher DR than sapphire (0.018 compared to 0.008) and the RI is lower at 1.62–1.64: SG is usually near 3.10. A narrow absorption band may be seen at 497 nm.

The pale blue *topaz* of nature is now routinely irradiated and heated to give a much darker blue but the RI and SG at 1.612–1.622 and 3.56 come nowhere near that of sapphire.

Tanzanite as the transparent blue variety of the mineral zoisite could be taken for sapphire even though the colour is not quite so deep and rich. This is a subjective judgement and so tests should be made. Even after heating, tanzanite shows very pronounced pleochroism with violet–red and deep blue colours. RI is 1.692–1.701, DR 0.009, SG 3.35, hardness 5.5–7 but specimens are very brittle and unable often to withstand ultrasonic cleaning.

Though *aquamarine* could hardly be mistaken for blue sapphire, an unusual dark blue beryl whose colour is unstable could be confused with it in unsuitable viewing conditions. This *Maxixe* or *Maxixe-type* beryl will be described further under beryl but it has RI and SG in the beryl range, 1.57–1.58 and 2.70.

The bright and highly dispersive *benitoite* could be mistaken for Yogo sapphire (also from the United States). It fluoresces a bright blue under SWUV (blue sapphire shows no fluorescence due to its iron content): the lens will show its very large birefringence of 0.047.

Synthetic stones which may be confused with blue sapphire

Deep blue spinel which contains cobalt shows red flashes of colour in any bright light and through the Chelsea filter shows an all-over fine deep red. The RI of 1.728 is well below that of any variety of corundum. Many blue synthetic spinels will give a chalky bluish or greenish fluorescence under SWUV.

Doublets with a thin layer of natural greenish-blue Australian sapphire at the girdle and the top and bottom consisting of blue synthetic sapphire caused some disquiet in the trade at the time of their introduction since in the upper natural section straight colour banding could be seen and the absorption spectrum appeared to contain natural sapphire elements. 'Think doublet' is the advice we give to students – they are not too difficult to catch out (the stones, not the students).

The more expensive types of synthetic ruby

While the Verneuil flame-fusion method produces by far the highest percentage of synthetic rubies, growth by other methods gives stones of better quality and, as growth can take up to one year, at higher prices. In fact prices of many synthetic stones often seem surprisingly high and many are collectors' pieces just as much as their natural counterparts.

Rubies grown by the flux-melt method show no curved growth lines or gas bubbles: most important of all, they (like the Verneuil products) contain no natural mineral inclusions. Pleochroism through the table facet is not necessarily a strong indication of synthesis as it is in Verneuil ruby. Plato lines are not seen in flux-grown rubies.

When rubies and some other species are grown by flux-melt, some of the flux usually remains in the finished gemstone and although it usually shows a metallic or silvery lustre by reflected light the flux could be mistaken for a natural solid inclusion if care is not taken. Even more likely to cause confusion are the flat planes of flux in small particles: these can quite easily be mistaken for the feathers/fingerprints of liquid inclusions which are particularly characteristic of natural corundum. But there is one redeeming feature: the planes of flux particles are almost always twisted and have been variously likened to smoke in still air or to twisted net curtains blowing in the wind. A good technique is to move the light from a fibre-optic source over the specimen while observing it though the microscope – this is a good technique in fact for observations in general and the microscope light sources are not always needed. As always, the investigator should experiment with a variety of different lighting arrays.

Flux inclusions can also take the form of 'paint splashes' or 'breadcrumbs', both terms describing small whitish spots or streaks. These smaller inclusions can be placed by the lapidary at the sides of the stone or in some other inconspicuous place so that the stone overall appears virtually inclusion-free. This would be exceptionally rare for any natural gemstone and should always be regarded with suspicion.

Flux growth demands the use of an open crucible made from a substance that will not form compounds with the growing crystal. The noble metals platinum and iridium are generally used and fragments may break from the crucible wall and become incorporated in the crystal. The fragments appear notably as angular dark shapes and do not resemble any natural mineral inclusion. The 10× lens will pick them out quite easily.

When a crystal is grown artificially the grower is able to control the chemical composition and produce colours and colour varieties at will. The ability to exclude iron enables very bright red rubies to be grown with chromium alone acting as a colour-causing impurity. The hand spectroscope, while able to show the presence of chromium, cannot distinguish between natural and synthetic rubies although a very sharp chromium spectrum might well be regarded with suspicion as it would not be characteristic of natural ruby. The hand spectroscope can distinguish between ruby and red glass, garnet or tourmaline, however.

Chrome-rich ruby will fluoresce very strongly under LWUV and will glow stronger as the iron content is diminished. A ruby with a very sharp clear Cr absorption spectrum and showing a very bright red fluorescence should be examined with the lens and microscope. On the other hand a ruby with a significant iron content will glow less strongly and show a less distinct absorption spectrum.

X-rays will help here as a synthetic ruby will not only glow a very strong red but phosphoresce as well so that the possibility of iron being present is virtually excluded. Iron-rich stones will not transmit UV rays so easily as stones with little or no iron content.

We have seen that the gemmologist can do quite a lot with a suspected polished ruby: but what about ruby crystals? Rough ruby of gem quality is very rarely seen outside the producing area or in the workshops of leading lapidaries, but you may encounter a well-formed and beautiful crystal – obviously of ruby (you can test crystals with the spectroscope). The specimen may well be accompanied by a convincing yarn often including such phrases as ‘life savings’, ‘sold to me by an old miner’ (surely young ones would be more likely to chance their arm – old ones would know the score and not sell disadvantageously to a passing stranger).

Verneuil synthetics are formed as boules with no outward crystal form – although someone could take the trouble to cut one to resemble a crystal the usual Verneuil features would soon be spotted. The late Professor Paul Otto Knischka grew superb rubies and also sold a few of the crystals which were in general far more magnificent than the natural rubies which are eventually polished. The crystals display more forms than natural ruby crystals and can be recognized after a few encounters. Anyone faced with a ruby crystal of obvious gem quality should look carefully at the section on Knischka rubies below.

Chatham and Kashan rubies

Carroll F. Chatham of Chatham emerald fame was selling Chatham products with a good colour, showing flux remnants in twisted veils and an absence of mineral inclusions. Other properties, as with all synthetic rubies, are in the normal range for corundum. Our impression is that the quantities produced were not very large and this also is typical for flux-grown ruby.

In the mid-1960s the first Kashan rubies were grown and marketed. They were made by the firm of Ardon Associates Inc. of Dallas, Texas. In the early days it was claimed that the rubies could not be distinguished from natural rubies (quite a normal claim, one imagines) and at one time sets of comparison stones were put on the market or at least advertised but we have never seen one. In 1984 production ceased but was later taken up again by a different grower.

From the outset Kashan stones showed traces of the flux used for growth: in these stones 'paint splash' markings, smaller dust-like particles ('breadcrumbs') were all characteristic. In early crystals the 'paint splash' markings showed a distinctive overall moccasin shape and were arranged in parallel groups. Examples of Kashan rubies were reported in the Bangkok gem markets in the 1970s.

Ardon Associates also offered crystals for sale: these were probably aimed at the very large American amateur lapidary market. Many Kashan stones contain flux arranged in the highly characteristic smoke in a still room effect seen in many gem species grown by this method. An effect which has been called heat shimmer has also been observed in Kashan rubies and resembles the quiver in the air often seen above a heated surface. Some observers have said that this particular effect can be confused with the twinning lines seen in many natural rubies.

Kashan rubies are strongly dichroic, one of the colours sometimes showing as a strong orange or brown: the two colours are sometimes seen concentrated in distinct parts of the specimen. While many rubies are transparent to SWUV some of the Kashan stones are not and this may be due to added iron: some stones give iron elements in the visible absorption spectrum.

Knischka rubies

The late Professor Paul Otto Knischka of the University of Steyr, Austria, began growing beautiful ruby crystals in the early 1980s, the transparent crystals themselves being collectors' items with many crystal forms, some not observed in natural ruby, and a magnificent colour. The faceted stones are equally fine and have always been rare and expensive. The bladed habit seen in ruby crystals from other growers is not seen in the Knischka productions.

The earlier faceted Knischka rubies showed a violet tinge along with the red: this feature and the notably strong pleochroism are insufficient to distinguish this ruby from natural ones. Under LWUV the stones fluoresce a strong, clear carmine red and there is a marked phosphorescence after subjection to X-rays. This is characteristic of many synthetic rubies since growers are able to exclude iron which otherwise would dull the luminescence. The RI is in the range 1.760–1.761, 1.768–1.769 with a birefringence of 0.008. The SG averages 3.986. These figures are well within the corundum range and are unremarkable. Inside the stones are liquid feathers, colour swirls, negative crystals and black metallic platelets, with two-phase inclusions. The combination of all these features is sufficient to show that the Knischka ruby is an artificial product while, for faceted stones at least, not absolutely distinguishing it from Kashan or Chatham rubies.

With the appropriate equipment closer examination shows that the negative crystals (which are hollow tubes of crystalline shape) are either isolated or form groups on the ends of long crystalline tubes and these may be the best way to single out the Knischka ruby from other flux-grown examples. In 1982 Professor Knischka promised that he would be able to eliminate the hexagonal metallic platinum platelets which mark his early products. The two-phase inclusions which contain large gas bubbles, easy to see under low magnification, are hard to find because their outlines are very fine, indicating that they are composed of material whose RI closely matches that of the ruby host. It is the crystals and the beautiful and rare cut stones which remain in the mind.

Ramaura rubies

Fine ruby crystals began to appear in the early 1980s from the Ramaura Division of Overland Gems Inc. of Los Angeles (which later became the J.O. Crystal Company, named after its founder, and pioneer crystal grower Judith Osmer who announced her retirement in 2001). The stated intention of the company from the beginning was to market faceted stones together with lower quality rough for cutting into cabochons. They were also to grow single crystals and crystal groups, presumably for collectors (they have certainly been keenly collected).

The method of manufacture, described in *Gems & Gemology* (Fall, 1993) employed high temperature flux growth with spontaneous nucleation (this latter would give the single crystals and crystal groups). Most gem-quality flux-grown rubies grow on seeds within the noble metal crucible, this method giving the grower firm control of the melting and cooling process to obtain just the form and size of crystal required. It was reported that the Ramaura rubies were not grown on seeds or on seed plates, so that in general the products contained fewer inclusions. Spontaneous nucleation can begin from irregularities in the crucible wall or even from specks of dust.

The colour of Ramaura rubies ranges from a pure red through orange–red to a slightly purplish-red. Some faceted stones contain lighter red areas and when these are extensive the whole stone appears pink. On colour grounds, like most other synthetic rubies, there is no way of distinguishing the artificial from the natural. The absence of a seed decreases the number of inclusions (with a seed a certain amount of forced growth is needed and impurities are more likely to be trapped). While some Ramaura rubies appear inclusion-free, others contain traces of flux – there are of course no natural solid inclusions. The best way to spot any inclusions with this product is to tilt the stones, this attitude also showing up colour zoning where it exists.

The use of spontaneous nucleation gives crystals with different orientations, so that pleochroism may or may not show through the table as it does with Verneuil specimens. The RI for the Ramaura ruby was measured at 1.762–1.770 and 1.760–1.768 with a birefringence of 0.008, all figures consistent with either natural or synthetic ruby. The SG was 3.96–4.00, also in the usual ruby range. Under LWUV the Ramaura stones responded with variable intensity with a range extending from moderate to extremely strong dull chalky red to orange–red, with some small zones showing chalky yellow. There was no observable phosphorescence. Under SW the fluorescence is more or less the same with additional chalky, slightly bluish-white zones in a few specimens: again no phosphorescence was observed.

Under X-rays some areas showed no fluorescence and no phosphorescence was seen (Verneuil rubies phosphoresce strongly in general under X-rays). The chalky yellow and bluish-white areas, provided the lapidary does not polish them away, are good indications of a Ramaura product. The absorption spectrum gives nothing away and the stones do not transmit SWUV in any noteworthy way, though in general synthetic rubies do this quite effectively in the absence of iron.

The best way to distinguish the Ramaura ruby is by a careful examination of the inclusions. In this product black metallic flakes from the crucible metal are not seen as in many other flux-grown rubies. Large flux inclusions are easy to spot, however: some show a characteristic orange–yellow colour while others are colourless. The size range is quite large, some inclusions are particulate while others are large, sometimes drop-shaped. Some flux-filled negative crystals may be seen, angular or rounded, while some of the coloured flux inclusions may contain near-transparent or opaque whitish areas. Some have a crackled appearance; this is best seen in voids or channels partly filled with flux material. Flux particles may form fingerprints – where they do the flux is white rather than yellow. Features superficially resembling two-phase inclusions are in fact solid: early Ramaura stones at least show white wispy flux veils.

A suspected Ramaura stone (any stone) should be examined with a variety of lighting. Twinning and parting evidence shows as parallel lines which may be angular but other shapes are reported. Portions of the stone containing no flux inclusions will be seen to be colour-zoned. Some Ramaura rubies are notably large and attractive so it is essential that tests are carefully carried out. The parallel twinning lines seen in many natural rubies can be seen from any angle while similar colour zoning in the Ramaura product appears and disappears as the stone is brought in or out of focus. This effect is seen in the Kashan ruby already described. GIA has described as 'comet tails' the minute undissolved flux particles which sometimes appear in the stones. In natural rubies this phenomenon (not comprising flux particles of course but natural solids) seems to traverse a crystal randomly whereas in the Ramaura ruby the 'tails' appear to originate in an included crystal.

Magnification is the clue to identifying Ramaura rubies just as much as it is for all other artificial products. Before or coincident with the appearance of the stones the manufacturer stated in publicity material that an additive to the ruby composition would in some way show that the rubies were artificial but this has so far not been reported. It would seem that such an additive would probably stimulate a particularly recognizable fluorescence (a rare earth element might well have provided this and also shown an absorption spectrum which would immediately alert the gemmologist or jeweller). Faceting might well have polished a surface layer away but it would still be present in unpolished crystals.

The Douros synthetic ruby

This new product has been manufactured in Greece by unseeded flux growth using controlled spontaneous nucleation and slow cooling techniques. The Douros ruby resembles the Ramaura product and contains yellow flux inclusions often associated with rounded bubbles and minute unidentified solid inclusions. EDXRF and/or UV-visible spectrophotometry may be needed for a precise assignment of name though their artificial nature can be established by gemmological tests: the immersion microscope will be found very useful here.

Lechleitner rubies and sapphires

In the case of Johann Lechleitner's product it is convenient to deal with ruby and blue sapphire together. Lechleitner had for many years produced emerald overgrowth on colourless beryl which we shall meet in the emerald chapter: in the mid-1980s he began to produce complete rubies and blue sapphires which are rapidly becoming collectors' pieces as they do not seem to have penetrated the market very far – there was already sufficient competition from other synthetic corundum growers.

In a letter to GIA dated 1985 Lechleitner stated that he had been growing synthetic ruby and blue sapphire since 1983, all his crystals going to Gebrüder Bank of Idar-Oberstein, Germany. He is also reported to have grown colourless, padparadschah (pink with some orange), yellow, green pink and alexandrite-like corundum with some sales taking place in Japan. I have not examined these more exotic colours.

A transparent ruby examined by GIA showed a strongly saturated purplish-red with some haziness. Flux inclusions could be seen under magnification but not easily with the unaided eye. Dichroism could be seen through the table as in most Verneuil rubies. RI 1.760–1.768 with birefringence 0.008, all normal figures for ruby though the SG at 4.00 is a little higher than in some synthetic rubies. Under LWUV specimens fluoresced a strong red with slightly chalky white overtones and no phosphorescence. No phosphorescence was visible after X-ray irradiation. The absorption spectrum was normal for ruby. Flux inclusions as wispy veils and fingerprint traces were easy to see as well as curved growth striae as in Verneuil rubies. In a paper GIA wondered whether a seed might have been used to initiate growth – this would explain the curved striae provided the seed was not cut away from the grown crystal. It was also possible that a larger Verneuil ruby had been used as a depository for a ruby coating grown by the flux method and some of the flux could have entered this rather large 'seed'.

This experiment has in fact been tried but presumably did not have sufficient saving in growth costs to displace rubies grown by more straightforward methods. Lechleitner has produced pink sapphire over colourless corundum, synthetic ruby over Verneuil ruby and synthetic ruby over natural corundum, the latter presumably colourless. Interestingly the 'crazy-paving' effect seen in Lechleitner's emerald overgrowth on beryl has not been reported in his corundum products.

Lechleitner blue sapphires and other colours of corundum

Lechleitner blue sapphire examined by GIA was in the form of a brilliant cut stone of 0.69 ct. Parallel to the main crystallographic axis (c-axis) the colour was a strong violet–blue – the table facet was placed at an angle of approximately 20–30° to this direction. A pale greenish-grey–blue colour was seen in a direction at right angles to this. No response was seen with LWUV and there was no phosphorescence (as blue sapphires need iron as part of their colour-causing mechanism this is hardly surprising). Under SWUV a very weak chalky whitish-blue fluorescence could be seen though phosphorescence was absent. A similar effect as seen under X-ray irradiation.

No absorption bands could be seen with the hand spectroscope but a broad absorption region covered some of the far red and some of the violet. The RI was measured at 1.760–1.768 with a birefringence of 0.008 and SG 4.00.

The internal pattern was similar to that shown by the Lechleitner ruby though flux seemed more profuse. Curved colour banding was visible. The flux inclusions serve to distinguish this product from a Verneuil stone.

Blue and orange sapphires grown by Chatham

While these two productions had a brief vogue among gemmologists and collectors they did not seem to make a great impression on the jewellery trade (unusual colours sell very slowly). Crystal groups of both colours were grown, the small crystals held together by a ceramic glaze found to have an SG of 3.08 and to contain prominent gas bubbles. One of the authors (MO'D) examined a faceted Chatham orange sapphire some years ago and noted that the flux inclusions formed an obvious and recognizable hexagonal pattern.

The faceted blue sapphires themselves show marked colour zoning with a range of near colourless to light to very dark blue. Inclusions cause many areas to appear whitish: crystals in the groups range from colourless to dark and from translucent to opaque.

Orange faceted Chatham sapphires show pleochroic colours of strong pink–orange and brownish-yellow. A variable response to LWUV has been reported with strong to very strong orange with some stones showing a chalky yellow fluorescence. Similar but weaker colours can be seen under SWUV. No phosphorescence has been noted. Under X-rays the colour response varies with some areas reddish-orange and others inert: there is no reported phosphorescence. The spectroscope gives the emission lines of chromium and also some iron characteristics: this pattern echoes the natural orange sapphire which is quite uncommon.

As well as the hexagonal flux patterns (which are very easy to see if the stone is immersed) stones often contain platinum fragments, colour zoning and healed fractures. In the blue sapphires thin whitish needles have been seen.

Synthetic star rubies and sapphires

Virtually all star rubies and blue sapphires (stars in other colours of corundum do not occur naturally in the same way as in blue and red specimens though sometimes a similar pattern can arise from conveniently and accidentally placed inclusions). Virtually all synthetic star corundum is made by the cheap and easy Verneuil process and the characteristic curved growth lines and gas bubbles can usually be seen.

A star stone of notably attractive colour with a well-centred star with rays of equal length and reaching the edge of the cabochon has to be viewed with suspicion. Most natural star stones either show an indifferent colour or a rather weak, sometimes broken star: fine colours and good stars make a very valuable stone. Lapidaries in countries where star stones of good colour are found (Burma and Sri Lanka) traditionally leave quite a lot of weight on the base of the finished cabochon: the resulting rough surface contrasts well with the flat surface of the synthetic stone.

Linde star stones

The Linde Division of Union Carbide began to produce star rubies in 1947: two distinct types have been noted. The earlier Linde stars produced up to 1952 are quite notably transparent compared to later examples and look much more like high quality natural star stones. Growth lines were prominent in the pre-1952 stones and the stars were somewhat faint. The firm at one time made a few star stones in other colours which included purple, green, pink, yellow and brown.

Kyocera and Inamori star stones

In 1988 the Japanese firm Kyocera produced star rubies which were marketed as Inamori. The stars were intense and whitish and the inclusions could be deceptive. Some specimens showed pitted surfaces and some of the stars appeared broken or wavy. The colour resembles natural ruby with a purplish-red colour but the backs are usually flat and semipolished.

Under LWUV the Kyocera star rubies fluoresce a very strong red and the response to SWUV is also a strong red, with a chalky blue–white overtone. Very fine rutile needles have been observed as inclusions, the fineness exceeding that of similar needles in natural rubies. Round and distorted gas bubbles are found and when stones are closely examined with a fibre-optic light source a fine consistency whitish material can be seen forming bluish-white swirling veils randomly intersecting the rays of the star. This effect has been noted in rubies grown by the Czochralski pulling method so it is possible that the Kyocera star stones have been made by a similar method instead of by the Verneuil flame-fusion process. The absence of curved striae rules the Verneuil process out.

A Kyocera star ruby tested by GIA showed a roughly hexagonal pattern through the top of the stone and this could have been the remains of a seed crystal. Gas bubbles and swirls, which are seen as dark-edged wavy bands in shadowed transmitted light, are especially characteristic of Kyocera star stones.

There are no solid inclusions: at the time of first production the company stated that there would be two grades, A and B. Grade A stones would be 'perfect' (presumably inclusion-free) while B stones would contain some inclusions. Physical and optical properties of these star stones are unremarkable apart from what has been already mentioned.

Synthetic pink sapphire

Pink sapphire is a very attractive gemstone and it is not surprising that various syntheses have been attempted. While some pink sapphire has been and continues to be grown by the Verneuil process, the most successful examples have been grown by crystal pulling.

Union Carbide pink sapphire

In 1995 a commercial pink sapphire was produced by Union Carbide under the name of 'pink T-sapphire'. Growers in Russia and in Japan (Kyocera) have also produced good quality material.

While there is no particular problem with growth, the marketing of pink sapphire has always posed problems of nomenclature – is it merely a paler version of ruby (not very easy to sell) or a distinct colour in its own right? This is one for the advertisers. Judging by sales, pink sapphire comes and goes rather than being ever-present in the trade. The 'T' in 'T-sapphire' probably refers to titanium, which is added to crystals originally grown for laser use, which would explain the escape of slightly imperfect ones into the gemstone trade.

Faceted Union Carbide pink sapphire shows orange–pink and purplish-pink with the more saturated colour said to arise from annealing after growth in an oxygen-free (reducing) atmosphere. The titanium-doped and annealed stones show only a faint orange fluorescence under LWUV and a weak to moderate blue colour with slight or strong chalkiness under SWUV. This response is not seen in natural or synthetic pink sapphires which are coloured by added chromium.

The absorption spectrum of the annealed stones is weak and none is shown by the as-grown stones. Natural pink sapphires show a chromium absorption spectrum: Verneuil synthetic and pink flux-grown and pulled crystals also show a Cr spectrum: there have been no reports of natural pink Ti-coloured sapphires.

The Union Carbide sapphires contain minute bubbles in high relief and pinpoint inclusions. The pulled stones in general lay show elongated gas bubbles. On immersion some of the annealed sapphires were found to show very elusive colour banding. The best immersion medium (about the only available one today) is di-iodomethane. Examination of the stones under polarized light was also useful but not all specimens showed the zoning. No curved colour bands are reported although some pulled sapphires are said to show them.

These sapphires can be quite difficult to identify and the unusual fluorescence probably gives the best clue, along with the absence of natural inclusions. This being seen, the lack of a chromium absorption spectrum and the presence of tiny bubbles should increase suspicion that the specimen may be a Union Carbide synthetic.

It is interesting that with the importance of corundum the hydrothermal method of growth should not have progressed beyond the experimental stage save in a few cases. The method used is more costly than the Verneuil or flux-growth techniques and since pressure is involved greater care and time would have to be taken. Nonetheless there have been some attempts at hydrothermal growth of ruby. I was once shown a fine large dark red crystal with platinum wires protruding from it: the seed crystal showed as a white area within: the grower was Pierre Gilson of synthetic emerald fame.

Synthetic yellow sapphire

If diamond presented the problems that yellow sapphires can pose gemmologists would be in for a hard time! Nassau (1994) identifies seven types of yellow sapphire (natural and synthetic) in some of which the colour is unstable. In general yellow stones are not quite so keenly sought as bright reds and greens but fine yellow sapphires can be magnificent.

The problem with yellow sapphire is not so much one of identification but of colour stability: irrespective of original appearance some stones will fade in bright sunlight or when heated, as in the setting process. Gemmological testing cannot establish whether or not a particular specimen's colour will be stable: the only test possible is a fade test, in which the stone is subjected to strong light or heat over a period. This is not part of normal gemstone identification.

The finest natural yellow sapphires, mainly those from Sri Lanka, Thailand and Australia, in general do not fade. Most synthetics (virtually always the cheap Verneuil stones) are colour-stable.

The colour-fading problem usually affects those yellow sapphires which owe their colour to the irradiation of material which began life as colourless. Though there have been occasional reports of the fading of heated yellow sapphires they are likely to be misleading or incorrect since the mechanism of both the inducement or the diminution of colour is well understood. Details of how this works can be found in the more advanced gemmological textbooks and in Nassau (1994). The fade test in sunlight takes a few hours only but who would submit their sapphire to it? It has been argued that the colour may be restored by further irradiation but the specimen will inevitably fade once more and in any case who has access to the type of equipment needed? Most of the cases cited are from laboratories.

Other colours of sapphire are usually known as 'fancies' as in the case of diamond. Many if not most fancies come from Sri Lanka or East Africa – if they turn up anywhere else the crystals may be discarded. The chief fancy colours are orange, purple to mauve (no two people agree which is which!) and green finds a place here though it is much more common.

Taking green first: this is one of the colours whose crystals may not make it to the lapidary. Iron is the cause of the colour which is dark but not without attraction. It is usually darker and less bright than the green of peridot or tourmaline. It is not synthesized, save by the Verneuil method, since the colour is too dark to be really popular. It may be worth mentioning that faced with a boule section specimen of Verneuil-grown apparently dark green sapphire it should be quickly checked with a fibre-optic lamp to ensure that no red flashes can be seen since a dark green synthetic spinel which looks quite like dark green sapphire has been made as a (quite successful) alexandrite imitation. The physical and optical properties of green sapphire do not vary from those of the natural mineral and the strong iron absorption spectrum (the classic iron spectrum in corundum) is present in the Verneuil stones.

The heat treatment of ruby and sapphire

Ruby and blue sapphire are important enough for any improving treatment to be tried out to bring indifferently coloured material up to a more acceptable level. We shall not examine the fine details of how this might be done – there are many different techniques and those who enhance stones do not publish their methods too readily – but we certainly do need to know how to recognize enhanced material since the customer will want to know about it: details should in any case be given on any accompanying certificate.

First we need to know whether or not the 'new' colour is stable. In the case of ruby and blue sapphire the answer is 'yes' though as we have seen this may not be true of some yellow sapphires. Stones treated by diffusion of colour may have the shallow layer produced by this method polished away by the lapidary though repolishing an already faceted stone would result in a virtually colourless specimen – what the lapidary does will only affect rough material though some sapphires are given colour-diffusion treatment in their finished state.

What is the treater aiming at? The ideal would be to bring a ruby up to 'Burma' level of colour and Burma colour would be the aim of those working with blue sapphires. A good place to begin may be with ruby from Thailand, even now known as 'Siam' ruby to many. This ruby has an appreciable iron content which, with the existing chromium-caused body colour, gives a faint but perceptible brownish cast to the stone. Heat treatment can eliminate this by altering the nature of the iron, leaving the chromium red in sole possession. The darker red 'Siam' rubies are not seen so often now, except of course in old stock or jewellery, since heating is now a matter of routine. It is possible that the original stone will be recognized from its inclusions which have already been described, but heating can alter their appearance so that considerable familiarity with inclusions in general is certainly needed by those dealing with corundum gemstones.

It is worth recapping the main differences between Burma and Thai ruby: Thai stones contain no rutile needles but do show sets of long parallel lines with crystals on them, this effect being absent from Burma rubies. Thai rubies also contain plagioclase crystals. A ruby advertised as 'Burma' but which appears to contain no rutile should be very carefully tested. As a beginning the girdle and facet edges should be examined for pitting: only then should the interior be investigated, remembering that rutile needles and the parallel twinning lines might superficially resemble each other. Heating 'explodes' some inclusions but the effect is hard to distinguish and the tester needs to be familiar with the effect.

The white, rather oily-looking geuda sapphire from Sri Lanka was around for many years before its potential for gem use was discovered when some was heated, reputedly by Thai workers. While the exact nature of the heat treatment is not readily disclosed, the end result can be spectacularly beautiful. The unheated material is translucent and must contain some titanium and iron for a blue colour to be possible. It is reported or at least rumoured that temperatures up to 1600°C are needed for the blue colour to develop. Nassau (1994) suggests that stones are preformed or otherwise worked on so that as many inclusions as possible are removed before heating. Nassau (1994) believes that a reducing atmosphere is needed. Heating may be repeated if the first round fails to give a good enough colour.

In general these are the conditions used for sapphire heat treatment, such as that given to many Montana sapphires: the fine Yogo material apart, Montana river-gravel deposit sapphires occur in many colours, most of which are routinely and successfully enhanced by heating. Crystals with a central orange spot are especially well handled as the final product after heating comes out a fine bright overall orange: yellow and green heated crystals are of good gem quality when heated and there are also some fancy colours.

Heating may lighten the often rather dark colour (they can appear virtually black) of some blue Australian sapphires.

It is possible for heating to remove or at least modify the curved growth lines in Verneuil-grown corundum. A perhaps cruder practice which also involves heating is the inducement of fractures by thermal shock, quenching heated material (commonly ruby) in water or liquid nitrogen, the surface-reaching cracks being filled with coloured dye or covered with an overgrowth: the idea is to simulate natural inclusions or fractures. The colour of the dye is never entirely natural looking but some of the stones are quite convincing. Similar practices take place with colourless rock crystal as we shall see later.

Treatment and enhancement: diffusion of colour

Both cabochon and faceted blue sapphires have been described in which colour has been diffused to form a thin surface layer which can be seen when the stones are immersed. High temperatures are needed for the process to be successful even though the penetration of colour into the surface is no more than tens of micrometres. One disadvantage of the method (which seems to persist in popularity despite obvious snags) is that the layer can easily be polished away if the stone needs to be repaired.

Diffusion can also produce a star effect providing that titanium dioxide is present. These stones are less common. While blue is the easiest colour to diffuse into corundum, orange diffused stones have been reported, the colour arising from chromium. Nassau (1994) reports that while it takes one-hundredth of a per cent of iron and titanium impurities to give a good blue throughout a stone, 1 per cent of chromium would be needed to make ruby – this is why few diffusion-treated rubies are reported.

Both heat- and diffusion-treated corundum can be identified having first made sure that the specimen is corundum and not an entirely different species. The treatment does not affect physical or optical properties. Pitted facet edges and damaged girdles usually result from the treated stone having to be repolished to conceal or minimize damage caused by treatment at the high temperatures needed. Exploded inclusions have already been mentioned: if you look for stress fractures which look rather like haloes or discs surrounding what are clearly solid inclusions you are probably on the right track. Koivula (1986) states that if inclusions containing carbon dioxide are identified (this is a laboratory test) then the specimen cannot have been heat treated.

For the gemmologist, the absorption band at 450 nm is not usually seen in enhanced-colour blue sapphires (but it is not always easy to spot anyway): more likely to catch the eye is the unusual pleochroism shown by some examples. Violet–blue and greenish-blue seem to replace the more usual dark and light blue. Heat-treated blue sapphires may show a chalky green fluorescence under SWUV and magnification will reveal no silk (students can work out why this may be!). Diffuse or patchy colour distribution is another feature which should command further attention for the specimen.

In cases where profuse rutile has been eliminated from sapphire by heating, or the brownish component removed from Thai ruby the investigator has to fall back on whatever inclusions there may be. These will almost always be sufficient to identify a Burma or Thai ruby despite treatment. We have already suggested that a good test for diffusion-induced colours is immersion but if facilities for this are not available the stone can be viewed over a white background when thin colour layers at the surface may be seen and the colour in any case appears uneven.

Diffusion-treated star stones need high temperatures for the asterism to be induced so that damage is more likely – as there will be no facet edges to show pitting the gemmologist will want to study the surface and base carefully. Stars are usually strongly defined.

Some diffusion-treated stones show a characteristic loss of colour in some areas, particularly from the girdle area or in some sets of facets. Surface pits when examined may show colour at the bottom as it could not be polished out at that depth. Finally Nassau (1994) mentions that heat-treated corundum gives a 'plink'-like sound when dropped onto a hard surface: unheated ones 'plonk'!

Imitations of ruby and sapphire

Glass as always has to come first as the colours of ruby and blue sapphire are not hard to equal – we repeat that glass will almost certainly contain swirls and gas bubbles and may show signs of a process known as devitrification which produces small crystals which could well be mistaken for natural solid inclusions (which in a sense they are!). The RI and SG will be well below those for corundum: corundum has an RI 1.76–1.77 and SG 3.99, the glass RI is most commonly between 1.50 and 1.70 and facet and girdle edges may well show distinct conchoidal (shell-like) fractures. Between crossed polars glass does not show the clear alternation between dark and light shown by corundum and glass shows no corundum absorption spectrum.

Garnet-topped doublets very often imitate ruby or blue sapphire: they are glass with a thin slice of almandine garnet fused to the table facet – this when examined with a strong light gives an unmistakable red flash which would immediately rule out blue sapphire. In the case of ruby clearly other tests are needed: examination of the red almandine slice will usually reveal rutile needles in and confined to it – in the appropriate orientation an almandine absorption spectrum may even be seen. As always with composites their existence has constantly to be remembered. The difference between the red of almandine and that of ruby is best seen when the stone is immersed.

While in theory many different types of composite could be made to resemble ruby, in practice the availability of the cheap Verneuil ruby makes the devising of such composites rather a waste of time. Far more garnet-topped doublets imitate emerald. Star corundum may be imitated by glass cabochons with a flat back on which star-like rays have been engraved: more ingeniously some imitations have an engraved mirror placed at the back of the stone. At one time corundum could be coated with some form of plastic to improve its colour but in most cases the layer does not survive intact. Coating of corundum, mostly blue sapphire, by a process similar to the 'Aqua Aura' blue coated quartz has been tried but is not convincing since the coating often shows signs of tarnishing.

Corundum with a *star engraved on the back* is yet another possibility. It is likely that few such imitations are made today but they may well turn up in valuations.

Reports of interesting and unusual examples from the literature

Items in this section have been chosen to illustrate points made in the chapter and to bring one-off items to your notice.

Near-surface flux inclusions in a Chatham pink sapphire glowed green and ones deeper inside the stone yellow when under either type of UV light. The inclusions formed a grid-like structure and the stone gave a strong orange–red fluorescence under both types of UV light.

Lasering of girdles is normally seen only in diamond but an American firm, True Gem Company, at one time at least lasered a registration number on the girdle of their rubies and sapphires. They were said to have been grown by the crushing of natural rough which was subsequently purified and finally melted and pulled. Trace elements were said to be added to assist identification.

A dark blue cabochon of more than 40 ct into which a star effect had been diffused was found on the basis of its inclusions to be natural corundum. It responded chalky greenish-white under SWUV and did not show the iron absorption line at 450 nm. The needles from which the star was formed were seen under immersion to be confined to the surface. A similar effect was observed in a light red 6 ct star ruby. The ruby turned out to be flux grown and the needles forming the star were much coarser than those found in most synthetic star corundum.

Occasionally very small beads, ruby in this case, can be coloured by dyeing the interior of the drill-hole. Suspicion is the only safeguard.

In 1994 the *Journal of Gemmology* described a large inclusion in a cabochon ruby: occupying a large area of the base of the stone the inclusion showed a whitish cast unlike the usual filling material. The inclusion was found to be calcite (by observing its reaction to a small drop of dilute hydrochloric acid). It is possible to be so guarded against fillers that natural inclusions can be the deceivers – familiarity with the forms of natural inclusions is essential.

A heat-treated blue sapphire examined by GIA showed a zone of brown, not normally seen in heat-treated stones. The absorption spectrum showed no distinctive bands: this suggested Sri Lanka rather than Australia as the place of origin since Australian sapphires are iron-rich and absorb strongly. It was thought that the sapphire was originally a geuda as these sometimes show brown stains on heating.

Brown synthetic star sapphire is rare but GIA found a specimen which showed no absorption in the visible and which was inert to both types of UV light. The stone was opaque except for some semi-transparent areas near to the surface and showed widely separated growth bands.

A fine orange–yellow sapphire of 7.6 ct was found to have been heat treated. It showed three zones of chalky blue fluorescence close to the girdle when observed under SWUV. Immersion in di-iodomethane showed straight angular yellow-coloured zoning alternating with near-colourless areas (these characteristic of natural corundum), and three straight blue zones close to the girdle, coinciding with the fluorescent areas. GIA believed that these features arose from heat treatment.

An update on the Knischka rubies published in *Gems & Gemology* (Fall, 1990) stated that faceted stones up to 11 ct were being grown as well as preforms which may have exceeded 25 ct on occasion. Crystals were being offered as 'macroclusters, plates and microclusters', presumably for collectors. Colours ranged from 'Thai' red through light, medium and dark 'Burmese' pink. A crystal which had been sawn in half showed a dark purplish-red with accordion-like deep growth steps perpendicular to and along the length. Glassy two-phase inclusions and platinum platelets were found.

Two pieces of synthetic corundum were offered as water-worn rough: the larger piece, weighing 56.04 ct was purple in daylight and purplish-red in incandescent light. The smaller crystal of 21.8 ct was medium blue and resembled a broken, worn pebble of sapphire. Both crystals showed characteristic flame-fusion-growth signs with concentric colour banding. The larger crystal had 'steps' on the surface, the smaller one showing step-like elevations and depressions. These markings were from a grinding wheel: the crystals had been purchased in Sri Lanka.

As we have said several times throughout the text, fine quality gem rough virtually never enters the normal gemstone trade in Europe or the USA. Fine rough is sold right at the mine to dealers who have bases not too far away. Mining of most of the classic gemstones is and has always been family based and chances of an outsider getting in on the act are nil. We have explained this over 30 years to hundreds of keen students who very much want to mine or purchase at source for themselves: this is an excellent ambition but you have to know the score! Curiosity is among the highest of the virtues and without it not much would get done in the sciences so we commend all such ambitions whatever their outcome!

Some Verneuil-grown ruby, blue and orange sapphire has been found to contain needle-like inclusions and sometimes straight twinning. Triangular cavities containing gas bubbles have also been reported. An unusual flux-grown orange–red ruby of 3.5 ct showed near-colourless straight parallel bands between areas of orange–red. This effect was reported to look natural.

Some dealers long in the trade still refer to 'Geneva ruby', the term denoting early Verneuil stones with very tightly curved striae, prominent gas bubbles, colourless areas and strain cracks, with some black inclusions said to be undissolved alumina. The products are known to date back to 1886 at least when the French Syndicate of Diamonds and Precious Stones ruled that they had to be described as man-made.

Very good imitations of the pink–orange padparadschah variety of sapphire has been made from both YAG and the much rarer lithium fluoride. In both cases the colour is obtained by doping with the rare earth element erbium: fortunately this gives a strong rare earth spectrum with many sharp lines and bands and it doesn't matter if you do not recognize what the imitation really is – it certainly is not sapphire. The padparadschah, which can be most beautiful and is certainly rare, is only one of the fancy sapphires which from time to time come onto the Sri Lankan gem market: there are other varieties also with Sinhalese names but no two dealers agree what the colours of this group really should be. Recognition of the inclusions as well as the presence of an anomalous absorption spectrum will make distinction of any imitation quite simple.

Signs of twinning are not always as easy to interpret as we should like. Some blue and orange stones tested by GIA and reported in 1991 showed twinning lines but fortunately gas bubbles and curved growth lines were also present. However, crystals of böhmite were present between the twinning lines as so often seen in Thai rubies, so it is perhaps as well that the same effect has not been tried with ruby.

The curved growth lines of the Verneuil product are not always easy to see, being most obvious in the 'alexandrite' imitation synthetic corundum (this material looking more like amethyst and showing a diagnostic absorption line at 475 nm). In the imitation the lines are usually best seen not through the table but at right angles to this direction. In paler synthetic corundum the lines can in fact be very hard to distinguish. Immersion can certainly be a help but the choice of the appropriate liquid is getting limited and gemmologists to all intents and purposes are left with di-iodomethane. However, there are some aids around: after trying bright-field and dark-field illumination and moving a fibre-optic source placed close to the microscope (this can take the place of the built-in microscope lighting – just because it is there you don't have to stick to it), further assistance can be given by filters. GIA recommends using a green plastic filter placed between objective and specimen to give a colour contrast when looking at some corundum – a blue filter certainly helps with synthetic yellow sapphires which are very hard to test, apart from the absence of natural mineral inclusions.

Distinguishing coated from diffusion-treated blue sapphires is not always easy as some specimens can deceive. A stated colour-enhanced blue sapphire reported in the Spring 1994 issue of *Gems & Gemology* showed a large broken surface with a colourless stone beneath but the expected dark outlining of facet junctions of diffusion treatment was not apparent: the surface showed dimpling and a gentle repolishing removed the coloured layer, this proving a coating rather than diffusion.

As if yellow sapphires were not trouble enough, material grown by pulling was on sale at the 1994 Tucson Gem & Mineral Show. The colour was bright greenish-yellow and stones were said to resemble golden beryl rather than yellow sapphire. While some nickel was found to be present there were no other chromophores in any significant amounts. A chromium emission line at 690 nm was detected with the hand spectroscope and three specimens tested gave a faint orange glow under SWUV, one showing a similar effect under LWUV. Very small inclusions which may have been gas bubbles and curved growth lines were seen when a blue filter, bright-field lighting and immersion were used.

A green star sapphire was found to be coloured by cobalt, this element giving an absorption line at 670 nm. The specimen contained a thin, cloudy mottled area below the dome and this, rather than the customary rutile needles, was responsible for the star effect. The stone from the photograph in *Gems & Gemology* was a quite attractive light green. Though most natural green sapphires owe their colour to iron there have been reports of colouring by trivalent vanadium or trivalent cobalt.

In *Corundum* (Butterworth-Heinemann, 1990) Richard Hughes suggests that the combination of low magnification and SWUV can be used to spot growth details in synthetic colourless sapphires, with the added use of an SWUV blocking filter. A white diffusion filter can be used to resolve colour banding in blue synthetic sapphires and a blue filter for yellow ones. Growth sectors in synthetic diamonds can also be seen better with filters.

Sapphires from Sri Lanka, East Africa and very likely from other sources may show a colour change from bluish-green in daylight or fluorescent light to pinkish-purple in incandescent light. Some specimens showing this behaviour but with no natural mineral inclusions (and no gas bubbles or negative crystals) but showing apparently natural laminated twin planes were in fact Verneuil synthetic corundum. Through the pavilion facets curved striae could be seen. This is a lesson to anyone testing a stone known from its RI to be corundum – examine it in every direction. It is quite likely that with the increasing interest in fancy sapphires (which do not seem likely to be in exceptionally short supply) more cheap synthetics will be grown.

Some blue sapphire grown by pulling was found in gem shows around the middle of the 1990s. GIA reported that some of the crystals were unevenly coloured with the earlier-grown portions lighter than the later. Verneuil boules show the opposite effect with darker colour in the centre and lighter on the outside. Some faceted examples with normal corundum properties gave a faint pink through the Chelsea filter and a weak red fluorescence under strong light from a fibre-optic source (this reaction has been reported from some natural Sri Lankan blue to violet sapphires and in some similarly coloured Verneuil stones). It proved hard to find slightly curved blue banding until specimens were immersed in di-iodomethane.

Curved growth banding and striae were also seen in 'recrystallized ruby and pink sapphire' products reported to have been manufactured by the TrueGem Company of Las Vegas. These stones had identification numbers lasered on the girdles and showed corundum properties. Gallium, titanium and vanadium were detected in specimens using X-ray fluorescence, as well as chromium and iron which presumably were the chromophores.

A deep purple colour seen by transmitted light was characteristic of GGG crystals offered as water-worn natural ruby. The crystals had been ground to give a rough surface and showed parallel striations.

Rubies from the Mong Hsu deposit in Burma (see above) often show a blue central zone which has been ingeniously imitated by the insertion of a blue waxy material into a cavity within a synthetic ruby.

Finally yet another yellow sapphire – or so it seemed. The stone in question was near-colourless, weighed 4.57 ct and showed curved growth lines characteristic of Verneuil material when examined under SWUV. After irradiation for about 3 minutes the stone had turned a medium brownish-yellow. After 6 hours in a solar simulator (fade test) most of this colour had disappeared but the flame of an alcohol lamp returned it to the yellow colour within a few minutes. Even the faintest yellow colour in a specimen otherwise known to be sapphire should be very carefully examined.

Three types of fluid inclusion have been identified in sapphire from Thailand and suggest that the stones are of magmatic origin. One type consists of vapour-rich CO_2, another has multi-phase inclusions with several associated minerals, hypersaline brine and a CO_2-rich vapour phase. The third type consists of possible silicate-melt inclusions with immobile bubbles of vapour in an isotropic or weakly anisotropic phase of low relief.

One way of distinguishing natural from synthetic colourless sapphires is by examining their trace-element composition and their transparency to SWUV. EDXRF shows higher Fe, Ti, Ca and Ga in natural sapphires, these impurities causing a diminution of transparency to SWUV. UV-visible spectrometry detects this as a total absorption in the UV below 300–280 nm which is not seen in the synthetic material. A SWUV transparency tester exists which can rapidly identify the nature of parcels of colourless sapphires (and spot them when they are acting as a diamond imitation).

Diffusion-treated blue sapphires of 1.86 and 3.02 ct examined by the GAGTL London laboratory were clearly natural from their inclusions but the edges were lighter in comparison with the rest of the stone – an effect the reverse of that normally seen in diffusion-treated sapphires. Blue blotches could be seen on the surface and were similar to those reported in blue diffusion-treated blue–green topaz. The specimens gave a cobalt absorption spectrum showing that a cobalt-doped treatment had been used. The topaz-type diffusion process stops short of pit edges but can be found in the pits themselves. The specimen of 3.02 ct showed a red colour under magnification due to the presence of chromium.

Synthetic yellow sapphire with no obvious inclusions can be difficult to test: one procedure makes use of the Nelson M17 Gemstone Cooling Unit which is filled with liquid nitrogen. When the specimen is cooled it is irradiated by UV: if it proves to be a flame-fusion product coloured by nickel and chromium a fluorescent line at 693.5 nm is produced – this cannot be seen at room temperatures.

Colour-change sapphire from a new source in the far south-west of Tanzania shows bluish-green in daylight and reddish-brown under incandescent light. Strong colour zoning made pleochroism hard to describe: RI and SG were in the normal range for sapphire. Specimens showed red through the Chelsea filter but were inert to both forms of UV. No single element could be responsible for the effects shown though both chromium and vanadium were present. Nine of the eleven stones tested showed a true alexandrite effect.

Sapphires from Mingxi and Shandong, China, can be distinguished from one another by their IR absorption spectra.

Kashmir sapphires are reported to contain apatite, zircon and biotite as well as tourmaline, crystals of which were identified by X-ray powder diffraction. Kashmir sapphires have also been found to contain inclusions of liquid and gaseous CO_2.

Blue sapphires from Laos show prominent growth structures but no rhombohedral glide planes: negative crystals in basal orientation are surrounded by iridescent fluid rosettes similar to those reported from Thai rubies.

There has always been dispute over what constitutes the padparadschah variety of sapphire: in testing a large number of orange–pink sapphires a report concluded that a chromium content of between 0.4 to 0.8 per cent was needed to produce the colour as well as the presence of undefined colour centres.

Some milky geuda sapphires from the Mogok stone tract, Myanmar, turn to a fine blue when heated to around 1600°C. In *The Heat Treatment of Rubies and Sapphires* (1992) and in the *Journal of Gemmology*, 26(5), 1998, Themelis describes how the original material appears milky or contains profuse silk: it also may occur as wax-white to colourless, yellowish or bluish. Some of this material was heated from room temperature to 1200°C and back to room temperature over an 8–10 hour period. The milky sapphires turned transparent to translucent colourless stones (with the single exception of a very light blue milky stone with a zonal structure – this turned white).

The temperature was raised to 1300°C and held for 2 hours but with no difference in the results. When the temperature was raised to 1500°C the single stone turned to blue. Heating to 1600°C with an extended period produced a beautiful dark from a medium blue: increasing the temperature to 1700°C turned the beautiful blue to a medium blue. Oxidizing conditions were used.

Fluid inclusions in sapphire from Myanmar, Kashmir, Sri Lanka, Madagascar and Rwanda have been found to consist of pure CO_2 while daughter crystals of diaspore, AlO(OH), suggest that water was once present in the fluid phase. Unequal densities of the primary fluid inclusions in sapphire from Sri Lanka and Kashmir allow these stones to be distinguished from specimens originating from the other countries listed.

Sapphires from an alluvial basalt in south-west Rwanda are mainly deep blue with some hematite or rutile inclusions giving silk or with a slight milkiness perhaps due to submicroscopic exsolution of these minerals. RI is reported to be 1.770–1.762, DR 0.009, SG 3.98–4.00. Specimens are inert to both forms of UV and show the characteristic blue sapphire absorption spectrum.

Transparent blue sapphires with RI 1.760–1.771, DR 0.008, SG 3.99–4.02 are reported from a location near Indaia, Minas Gerais, Brazil. Rutile and mica with healing cracks and distinct growth and colour zoning can be seen.

Blue to blue–green gem-quality sapphires from alkali basalt sources in southern Vietnam show prominent growth structures and colour zoning in geometrical patterns. Sharply bordered blue bands alternate with narrow colourless or yellowish to brownish bands, many stones showing a colourless core. The commonest inclusions are various types of cloud with cross-hatch or lath-like patterns not so far recorded from other sapphires.

Mineral inclusions identified so far have been plagioclase, ilmenite, uranpyrochlor and spinel group species. Spectral curves in the 880–280 nm region might be used for country of origin determination but these results have to be interpreted with care. Southern Vietnam sapphires have a high iron content, too high to be from a metamorphic source: the colourless core together with the cross-hatch and lath-like cloud patterns with the distinctive colour zoning are good clues to locality.

Synthetic blue hydrothermally grown sapphires have usually suffered from uneven colour distribution but a process developed by Tairus in Russia uses Ni^{2+} as a dopant, giving a sky blue colour. By varying the concentration of Ni^{2+}, Ni^{3+} and Cr^{3+} and varying the oxidation-reduction environment, differently coloured sapphires have been produced. Swirl-like patterns are common: greenish-blue Cr-doped specimens fluoresce red under UV. There are crystalline copper inclusions and the IR spectrum shows five small peaks between 2500 and 2000^{-1}. The Ni-doped greenish-blue sapphire absorption spectrum includes three intense bands at 970, 599 and 377 nm and two weak bands at 556 and 435 nm.

Some hydrothermally grown blue sapphires made at Novosibirsk, Russia, have RI 1.760–1.770, DR 0.008, SG 3.98–4.06. Specimens showed swirl-like growth marks and black residues resembling breadcrumbs.

An interesting example of a mineral inclusion in sapphire was given by a blue commercial-quality stone of Sri Lanka-type colour. Under the microscope and viewed by transmitted light two different types of inclusions were identified. One was found to be an idiomorphic spinel with appreciable Cr^{3+} content, the second was a metamict zircon.

Two natural sapphires of treated blue colour were reported by GIA in 1996. One had an incised design on the back and contained fluid-filled fingerprints and unidentified crystals altered in such a way as to suggest heat treatment. The origin of the blue was a dye which could be seen when the stone was immersed. The other stone was a quench-crackled diffusion-treated sapphire.

Blue sapphires from the Andranondambo area of south-eastern Madagascar have RI 1.761–1.771 with DR 0.008, SG 3.91–98. Stones show strong pleochroism with green–blue and violet–blue colours. Nearly all samples have been found to be inert to both forms of UV but a few gave a weak greenish-blue and one a weak and uneven orange under SWUV.

A broad absorption band with a maximum at 570 nm in the yellow part of the ordinary ray is ascribed to an intervalence charge transfer between Fe^{2+} and Ti^{4+} and a sharp band at 450 nm arises from Fe^{3+} absorption. There is strong colour zoning: mineral inclusions reported include fluorapatite, calcite, minerals of the spinel group, plagioclase, scapolite, thorium and uranium phases. Milky growth structures parallel to the basal pinacoid and exsolved rutile needles are also reported.

Many of these particular Madagascan sapphires are heat treated in Bangkok though it is not difficult to separate treated from untreated specimens. Greyish-white or greyish-brown fine- medium- or coarse-grained bands (heat-treatment bands) are present. Mineral inclusions become turbid, translucent or opaque on treatment and stress fissures may develop around them. In most unheated specimens the most pronounced spectral minimum is near 490 nm.

Green flame-fusion sapphire has been grown to give faceted stones of about 2–3 ct. The original boules are usually no more than 50 mm long and 20 mm wide. They are said to grow initially in a dark blue cobalt colour but this changes to green shortly after crystallization. One end of each boule retains a small area of blue. Studies of green synthetic sapphire have cited Co^{3+} or a combination of Co^{3+} and V^{3+} as the cause of colour.

Yellow to blue sapphires from Antsiranana province, northern Madagascar, have a low TiO_2 content but are relatively high in Fe_2O_3. Virtually all colours are heat treated to reduce milkiness and give a better colour.

A type of blue sapphire found in basaltic rocks at Amboudrohefeha in the north of Madagascar shows a typical broad absorption band in the near-IR with a maximum at 850 nm, due to an Fe^{2+}/Ti^{3+} charge transfer. A parcel of Madagascan blue sapphires contained all three types.

Yellow, orange and brownish-red Madagascar sapphires have absorption patterns similar to those shown by sapphires from the Umba and Tunduru areas of Tanzania. Yellow stones show absorption maxima at 450, 387 and 377 nm. Orange sapphires are coloured by iron and chromium. In addition to the Fe^{3+} bands producing the yellow, broad pleochroic bands of Cr^{3+} can be seen, these providing the red component. Absorption maxima are at 555 and 410 nm and the characteristic chromium line is seen at 693 nm. Increased Fe gives a brownish tone and such orange to brownish-orange stones have been called African padparadschah. Some Madagascan specimens, though, show the more usual pinkish-orange padparadschah colour.

Distinction between natural and synthetic (pulled) pink sapphire can be made on the basis of the synthetic's lack of luminescence in the LWUV region, chalky violet–blue fluorescence under SWUV and pinpoint inclusions, some of which can be resolved as gas bubbles if sufficient magnification is available. The colour is caused by Ti^{3+} rather than by chromium as in natural pink sapphire. The absence of the expected chromium absorption spectrum should alert the investigator.

Pink flux-grown Chatham sapphires show RI 1.759–1.768 with DR 0.009, SG 3.99–4.00. The most common inclusions are flux particles and many minute droplets are arranged in rows along growth steps. Two faceted stones reported in the *Journal of Gemmology* in 1994 showed presumed flux inclusions arranged in neat intersecting rows: secondary flux inclusions showed as partially healed fingerprints, the majority with an undulating form. Angular metallic inclusions were tentatively identified as platinum-occurring least commonly as large single platelets. Unusually, the reaction to both forms of UV was of similar intensity.

Sapphires in colours ranging from dark blue through blue, greenish blue to yellow have been found in a basalt in the Changle area of China. Properties fall in the normal range for sapphire but there is little pleochroism. Red zircon, feldspar, ilmenite and spinel are found as inclusions.

Some Kashmir blue sapphires have been found to contain rutile, green tourmaline, zircon and plagioclase crystals. Uraninite (U-rich) and allanite-(Ce) (Ce-rich) as well as fine lamellae with exsolved material and fissures are also reported. Another report stated that some Kashmir stones have been found to contain chromium and that these showed an emission line at 693 nm. This report (*Revue de gemmologie*, 121, 1994) tentatively ascribes the turbidity to rutile rather than to minute fissures.

An interesting synthetic sapphire with strong colour zoning showed a uniform yellow face up but from the side all but a small orange area appeared pink. Curved banding in the orange zone proved artificial origin.

In 2000 sapphires were reported from the Municipio de Manhuacu area near Sacramento, Minas Gerais, Brazil. Colours include pink, violet–red, lilac and blue–lilac with some bicolours, mainly yellow–blue and yellow–violet. RI and SG were in the normal corundum range. Minerals found as inclusions were zircon, apatite, mica and oriented rutile needles. Healing cracks, twinning lamellae and growth zoning were also reported.

Transparent blue, blue–green and some multi-coloured sapphires from the Primorye placers in Russia contain inclusions of hercynite and ferrocolumbite. Reddish-orange gem-quality sapphire is reported from the Kalalani area of Tanga province, Tanzania – the stones closely resemble sapphires from the Umba area which is about 3 miles north.

A large percentage of pale blue, pale green or near-colourless sapphires can be converted into saturated blue and yellow stable colours. The strong yellow colours are attributed to a broad absorption band extending from 600 nm to the violet end of the spectrum. Blue coloration is due to the dissolution of rutile in the presence of dissolved iron followed by reduction of some of the iron.

An interesting chatoyant greenish-blue sapphire contained no rutile but a series of nearly planar, parallel, liquid-filled fingerprints. The presence of diffuse colour haloes surrounding the inclusions indicated that the sapphire had been heat treated. Some diffusion-treated blue sapphires when immersed show uneven coloration from one facet to another.

Spherical blue clouds surrounding dark blue crystal inclusions have been reported in some heat-treated sapphires. The heating temperature probably approached the melting-point of corundum.

A sapphire advertised as ‘pink geuda’ has been on sale in Vietnam: specimens are reported to be opaque and cloudy with a pronounced milkiness. Heat treatment raised some of the stones to cabochon quality.

Some pulled sapphires with a slightly saturated greenish-yellow colour owed this to nickel as their main chromophore. Curved growth lines could be seen under the microscope. Some green synthetic sapphires owe their colour to cobalt – one reported example showed curved growth lines. Chatham pink sapphire with less than 0.2 per cent chromium gave an orange fluorescence under SWUV and a similar and equally powerful response to LWUV. General characteristics were otherwise in the normal range for corundum.

Generally speaking, locality determination needs a variety of chemical and physical properties to be considered, including growth zoning, chemical composition and spectrum analysis as well as a close examination of inclusions. To accomplish this needs sophisticated instrumentation; this may include UV-VIS-IR spectrophotometers, energy-dispersive X-ray fluorescence (EDXRF), the scanning electron microscope (SEM) and optical microscopy. A large reference collection is essential.

A hydrothermally grown yellow sapphire from Guilin, China, appears to be grown on a seed of colourless synthetic corundum and to owe its colour to oxides of nickel, cobalt and chromium. Darker greenish-brown colours are obtained by increasing the nickel content. A pinkish near-padparadschah colour was made by further altering the chemical composition. Variously shaped two-phase inclusions, planes of gas bubbles, branching acicular shapes, presumably crystals, wedge-shaped intersecting growth banding and colourless seed plates have been observed in the stones. The main paper, in *Guilin Journal of Gems and Gemmology*, 3(1), 2001, was abstracted in *Gemmology Queensland* in 2001.

Blue sapphire star cabochons reported by GIA showed stars which appeared to float over the surface. No trace of oriented needle-like inclusions could be seen under magnification and it appeared that the stars had been diffused. With transmitted light opaque and transparent areas could be seen. Fibre-optic lighting showed a very shallow and hazy whitish surface layer: this effect has previously been noted with diffusion-induced stars.

A good imitation of ruby with quartz constants is made by plunging heated rock crystal into a cold red dye. The dye enters the stones though the cracks induced by thermal shock.

'Reconstructed' ruby is a Verneuil single crystal product.

Chapter 4

Emerald, aquamarine and the beryl group of gemstones

Introduction

The beryl family of gemstones includes not only emerald but aquamarine, red beryl, golden beryl (sometimes called heliodor), yellow beryl, pink beryl (morganite) and a green beryl which can sometimes approach the colour of a bright emerald but which has its own considerable charm – it is coloured by vanadium and is known to the trade as green beryl (gemmologists often call it vanadium beryl). Colourless beryl is sometimes faceted.

As with the finest rubies and blue sapphires, the best emeralds need locality information for certification purposes: there are lots of imitations and synthetics of many types exist and are still grown. Colour is now routinely enhanced by several different methods and the question of disclosure is just as important with emerald as with ruby and blue sapphire. There are no true star or cat's-eye emeralds although effects resembling both types may show up more or less by accident.

As with ruby, gem-quality crystals of emerald very rarely come onto the market, though at the time of writing in September 2001 a crystal of approximately 655.4 ct appeared in Christie's New York Magnificent Jewels sale on 22/23 October 2001. The upper estimate is given as US$200 000. The crystal is said to have come from the Muzo mine and to have undergone a moderate degree of clarity enhancement. This shows how important crystals, which feature occasionally at high profile gem and mineral shows such as the annual show at Tucson, Arizona, can cost quite as much and more than fashioned stones of the same material.

Again like ruby, emerald has one classic country as a source which exceeds all others in esteem. As we have seen for ruby the country is Burma: for emerald, Colombia. Stones are judged by their nearness in colour to the finest Colombian material. Colombian stones are sometimes equalled in colour by Pakistan or Zimbabwe specimens but these never attain large sizes. Stones from some sources, notably the Sandawana stones of Zimbabwe, hold their colour down to the smallest sizes and this is a desirable feature. It should be remembered that compared to ruby of the highest class emerald of high quality is nothing like so rare geologically and examples can always be obtained if sufficient resources are available.

We have seen that country of origin evidence is fairly easy to obtain with ruby and sapphire provided that the investigator becomes familiar with the inclusion patterns peculiar to each area. Unlike the corundum gemstone specimens from the different areas of production, emeralds can vary perceptibly by locality in refractive index and specific gravity, though gemmologists would do well to remember that such figures cannot be completely relied upon for locality certification purposes for which inclusion evidence is required.

All the beryl minerals are beryllium aluminium silicates with the composition $Be_3Al_2(SiO_3)_6$. Emerald is coloured by a chromium impurity, aquamarine and yellow beryls by a trace of iron: red and pink beryl is coloured by manganese. The general hardness, not a testing feature to rely upon but of course worth knowing, is 7–7.5 and a mean RI would be something like 1.56–1.59, DR 0.004–0.009 (exceptions will be given below): mean SG 2.71.

A very simple instrument for testing emeralds was devised many years ago with the aim of distinguishing natural from synthetic stones. This is the Chelsea colour filter, named after the Chelsea Polytechnic in London where pioneer gemmology classes were held from the 1930s: successor classes are held under the supervision of one of the authors (MO'D) at London Metropolitan University. The Chelsea filter tells the observer whether or not chromium is present in the specimen being examined and it is only after that is established that the nature of the specimen can be investigated. In the early days of the filter synthetic emeralds glowed a notably bright red when they were viewed under a strong light: in general many natural emeralds contain sufficient iron (as well as the chromium) to show a strong but less brilliant red than the synthetics.

Many anomalous results are found when natural specimens are being tested and it would be highly unwise to rely completely upon the filter but it does distinguish emerald from, for example, green sapphire, green tourmaline and green glass, which show green. Some natural emeralds, notably from India and the Transvaal mines of South Africa, show pale red rather than green.

We have mentioned vanadium green beryl above but it needs further comment at this stage: many vanadium beryls look like emerald and are, of course, members of the beryl family of gemstones. Nonetheless, only green beryl in which chromium can be proved to be present can officially be called emerald – this is a rule of CIBJO, the international organization which represents gemstone dealers and the jewellery trade. The hand spectroscope cannot easily detect the presence of vanadium but chromium lines and bands are generally easy to see so that there is not a particular problem. There are no synthetic vanadium emeralds presently on the market though some were grown experimentally in the 1960s.

Emerald

Identification

Colombian emeralds are traditionally reported to be from either the Muzo or Chivor fields: other producing-area names come up occasionally (Gachalá, for example, seems to have been worked out though stones with a characteristic banded appearance but good colour were on the market briefly in the late 1960s). Both Chivor and Muzo stones, if they are important enough, are named as such in catalogues so that the gemmologist or auctioneer needs to know what distinguishes stones from one field from stones from the other. Clearly such information has to come from the inclusions though measurable properties can give some guide: Chivor emeralds usually fall in the RI range 1.571–1.574, 1.577–1.580 with a birefringence of 0.006. Muzo emeralds have RI in the range 1.576–1.580, 1.582–1.586 with a birefringence of 0.006. The respective specific gravities are 2.69–2.71 and 2.71–2.72.

Emeralds from Chivor show a number of interesting and quite recognizable inclusions of which the most celebrated is the three-phase inclusion (meaning the three phases of matter).

The three-phase inclusions, which can adopt shapes to which fanciful names could all too easily be given, contain a cubic crystal of halite (sodium chloride or common salt), a liquid (brine) and a gas bubble (carbon dioxide). The edges of the liquid are often jagged and the whole inclusion can resemble hanging banners with tattered edges: on the other hand they can look like the gaping jaws of a crocodile (the liquid), with the animal's eye (the gas bubble) and the cube of halite representing whatever you fancy. Such inclusions also occur in emeralds from Muzo.

Chivor emeralds may also contain very small plates of albite (these are also seen in Gachalá stones) and recognizable brassy-yellow pyrite crystals. Muzo emeralds may contain small brownish-yellow crystals of the mineral parisite. With some good fortune and careful arrangement of the lighting it is sometimes possible to obtain a rare earth absorption spectrum from the larger parisite inclusions. Rhombs of calcite also identify Muzo stones.

On the whole these inclusions are probably sufficient to distinguish crystals from one main Colombian emerald-producing locality from those of the other.

Both Chivor and Muzo produce hexagonal 'trapiche' emeralds in which crystals have a central core of emerald from which arms or rays, also of emerald, radiate spoke-like to the crystal edge. The emerald is usually of very fine colour and the whole crystal can be attractively polished and set.

Liquids, apart from those making up one part of a multiphase Inclusion, do not play a great part in Colombian emeralds but in Brazilian stones, which account for a high proportion of the cheaper emeralds on general sale, they are far more common. This is not to say that there are not fine Brazilian emeralds but merely that emerald is relatively common and the lower qualities can be afforded by most customers. Brazilian stones contain two-phase inclusions, biotite, talc and dolomite crystals as well as liquid in films.

To some extent Brazilian mines can be identified, with crystals from the Santa Terezinha mine (often said to produce the best Brazilian stones) showing very small crystals of dolomite, pyrite and chromite: emerald from the Carnaiba mine contains brown mica laths and stones from the Itabira/Belmont area show biotite, pyrite, chromite and multiphase inclusions. This information is interesting for research purposes but will not be particularly relevant to the gem trade.

In 1956 emeralds were found in south-eastern Zimbabwe, the field becoming known as Sandawana. These stones, which do not reach very large sizes but retain their colour however small, are a particularly fine bright green. Crystals of the amphibole mineral tremolite are prominent and recognizable as thin grass blade-like shapes. The occurrence is in a tremolite schist.

Emerald from the Ural Mountains of Russia is formed in a mica schist and thus contains brown laths of biotite, one of the mica group minerals. Emeralds from this source also contain crystals of calcite and bamboo stalk-shaped crystals of the amphibole mineral actinolite. Emerald from Austria (Habachtal emerald) also occurs in a mica schist and has a rather clouded interior with some biotite flakes and thin actinolite crystals. This material can sometimes be purchased as quite large blocks of schist, studded with dark actinolite with occasional tips of emerald crystals showing: on one occasion such an emerald crystal, of facetable quality and about 3 cm long, was extracted from what at first appeared to be a mere speck!

Emerald from Egypt is known from ancient jewellery: it also occurs in schists and contains mica and actinolite. There should be no particular difficulty in identifying examples as natural emerald as the quality is not usually outstanding.

Emerald from Zambia contains tremolite, tourmaline, phlogopite–biotite mica (looking like black dots), fibres of an asbestiform mineral and two-phase inclusions: these, with mica and actinolite, can also be found in the rather bright and attractive emeralds from Lake Manyara, Tanzania, together with liquid-filled cavities.

Emerald from Pakistan can be of very fine quality but up to now it has not been found in the largest sizes: nonetheless some published price lists, which also continually assess colour qualities, have placed Pakistan material on a level with that from Colombia. Stones occur in a soft talc-carbonate schist and contain wavy liquid feathers with some colour zoning, healing cracks, rhombohedra of dolomite and jagged two-phase inclusions. In the 1980s a deep green emerald was reported from the Mohmand Agency of Pakistan: these stones contained fluid inclusions, talc fibres and mica plates.

Emerald from the Panjshir Valley of Afghanistan is a fine dark green with two- and three-phase inclusions. Crystals are said to be found in quartz–albite veins.

Indian emeralds are well known from Mughal jewellery. They are found in biotite mica schists and often contain rectangular cavities containing a gas bubble – these two-phase inclusions have been called 'hockey sticks' (or 'commas') from their resemblance to that equipment – the cavities lie parallel to the vertical axis of the crystal. Biotite is a common inclusion in South African emerald and quartz crystals have been found in the fairly uncommon emerald from North Carolina, USA.

This is a convenient point to give representative RI and SG for emerald from the different localities quoted. Readers should bear in mind that the figures will often overlap those of emerald from other places and so they are to be taken as a guide only. Colombian figures, the most important, have already been given.

	RI	*DR*	*SG*
Brazil	1.566–1.593	0.006–0.008	2.67–2.76
Zimbabwe	1.583–1.596	0.004–0.006	2.74–2.75
Russia	1.581–1.588	0.007	2.74
Austria	1.584–1.591	0.007	2.72–2.76
Egypt	1.580–1.585	0.006	2.75
Zambia	1.581–1.591	0.007–0.009	2.68–2.74
Tanzania	1.578–1.585	0.007	2.74
Pakistan	1.588–1.600	0.007	2.75–2.78
Afghanistan	1.578–1.585	0.005	2.71
India	1.585–1.595	0.007–0.010	2.72–2.74
South Africa	1.578–1.585	0.006	2.74

Synthetic emerald (constants)

While the RI and SG figures help a good deal with distinguishing emerald from its many imitations they do not entirely provide the only means of identifying the many types of synthetic emerald that have been on the market for years. As we have seen, the Chelsea filter was designed to make a start with the spotting of the earlier (Chatham) synthetic emeralds but certainly the earlier products in general had a lower RI and SG. This is not true of later products so close behind the table of properties for natural emerald are some general figures for the main synthetic ones: the first group consists of flux-grown products by Chatham, Zerfass, Gilson, Gilson 'N', Lennix, Russian products and Inamori (Japan). These have a general RI range of 1.558–1.567 with birefringence 0.003-0.005 and SG 2.60–2.65.

Emeralds grown under pressure by the hydrothermal method can be separated more clearly:

	RI	*DR*	*SG*
Lechleitner	1.566–1.587	0.005–0.006	2.68–2.80
Linde, later Regency	1.566–1.578	0.005–0.006	2.67–2.70
Biron (Pool)	1.569–1.573	0.004–0.005	2.68–2.71
Russian	1.573–1.586	0.006–0.007	2.68–2.70

The reader may have noticed that in general and not merely with emerald a high SG means a high RI so that a specimen found to have a low SG and a high RI should be looked at again. There are plenty of exceptions, as always.

In the case of beryl the higher figures usually mean a higher iron content or even the presence of heavier metals: caesium is an example though it does not affect emerald.

Identification of emerald

Natural stones

We have already looked at inclusions, RI, DR and SG so we now have to consider those other properties of both natural and synthetic products, including composites, which are frequently the cause of confusion.

Among natural stones, in no particular order, are chrome tourmaline, which can be quite like emerald: demantoid garnet, rather bright but confusion can arise: chrome- or vanadium-green grossular garnet: green jadeite when opacity/translucency is appropriate. Other green stones include peridot, green sapphire, green zircon, emerald-green fluorite, chrome chalcedony, vanadium kornerupine, chrome diopside, lighter colours of nephrite, a rare green variety of andalusite, alexandrite, green diamond, dioptase, chrome enstatite, common (nickel-bearing) opal: sinhalite, sphene, the hiddenite variety of spodumene. Many readers may think such confusion is highly unlikely to arise, at least under trade conditions, but we have to remember that there are many collectors of gemstones for whom these species will not be unfamiliar.

Chrome tourmaline can look quite like emerald but can be distinguished under magnification by its much higher birefringence, which when the specimen is examined in different directions, will cause facet edges to appear doubled. *It is always vital to examine specimens in different directions*. The comparative birefringence figures are 0.007 (emerald), 0.018 (tourmaline). *Demantoid garnet* shows no birefringence and has a notably high dispersion (0.057 compared to emerald's 0.014). Should a spectroscope be available a strong absorption band can be seen at 443 nm, this showing as an 'early cut-off' of the visible spectrum. Under magnification demantoid almost invariably shows the well-known 'horsetail' fibres of crocidolite. Even a single fibre is enough to place the unknown as demantoid.

The two transparent green varieties of *grossular garnet, tsavolite* (one coloured by vanadium, the other by chromium) have higher RIs than beryl (1.73–1.74) with no birefringence. Inclusions are quite unlike those of emerald and much less prominent and complex, showing most commonly as lines of whitish dots: crystals of apatite and actinolite have also been reported. The variety *hydrogrossular*, which is opaque to translucent, can in some instances resemble emerald of similar diaphaneity: it fluoresces orange under X-rays but there is usually no need for so expensive a test since the RI of around 1.728 is not difficult to obtain as much hydrogrossular is fashioned into cabochons with flat backs which can be tested on the refractometer.

Green jadeite, when carved or in cabochons, can quite closely resemble emerald: when a flat surface is available the RI of about 1.66 is well above that of emerald. With luck jadeite of sufficiently deep (emerald) colour will show the 437 nm absorption band which is diagnostic: other bands may be seen but this is the one to look for. Chromium-rich green jadeite will also show characteristic chromium absorption in the red with the strongest line at 691.5 nm.

Peridot is not really like emerald and no gem dealer would be taken in by a possible masquerade or mistake. Under magnification peridot shows strong birefringence (around 0.036) with tourmaline-like doubling of facets when the stone is appropriately orientated. The refractive index in the area of 1.65–1.69 is well away from that of emerald: the specific gravity is also higher at around 3.34. These figures vary since peridot is a member of an isomorphous series.

Should the spectroscope be handiest, peridot gives a recognizable and characteristic iron spectrum with three broad bands in the blue at 493 nm, 473 nm and 453 nm.

The properties of *green sapphire*, when tested, rule out emerald with ease. The SG is higher at 3.99 and the RI is 1.75–1.77. Green sapphire, which owes its colour to iron, does not look anything like emerald but if you have not seen many emeralds.... The spectroscope will show the iron bands (which are best seen in this colour of corundum) at 471 nm, 460 nm and 450 nm, the two last bands often coalescing while the first stands a little apart. You should not need to look for inclusions but if you start at that end of the testing process note that there will be no beryl-type three-phase inclusions while those solid inclusions notable in corundum should be present.

Neither does *green zircon*, though most interesting in its own right, look anything like emerald and it will show no chromium elements in its absorption spectrum. It may in some cases give a refractive index reading at about 1.78 at its lowest but many examples will give no reading on the refractometer. The specific gravity is usually in the range 3.95–4.20 with many specimens reading 4.00.

While zircon as a species shows very strong birefringence (0.059), green stones often show none – this will be discussed further in the zircon chapter – so that other tests will be necessary. Most interestingly, some green zircons, which generally have a cloudy interior, show bright fissures or 'angles': these indicate the positions of certain crystal forms dating from the time of the crystal's original formation. Emerald shows no such marks. The absorption spectrum of green zircon may show a band at 653.5 nm but this is generally diffuse: a band at 520 nm in the green is sometimes visible.

Fluorite is quite often found in an emerald-green colour and even those green specimens whose colour could not be equated with the colour of emerald may be mistaken for that species by accident or design at a gem show. Since we have not come across this topic for a while, let us remember that fluorite has a very easy octahedral cleavage and that crystals (which will turn up more frequently than faceted stones) will usually show interference (rainbow-like) colours at the pointed edges which is where the cleavage is initiated by a blow or by attempted fashioning. If the cleavage continued the original cube would become an octahedron. The refractive index at 1.43 is nowhere near the beryl range: the specific gravity at 3.18 is higher than that of the beryls. Nonetheless confusion with a faceted emerald could arise when a three-phase inclusion is found in an emerald-green Namibian fluorite in which they have occasionally been reported. Green fluorite will usually give a sky blue or violet–blue fluorescence under LWUV light.

Chrome chalcedony, sometimes called mtorolite, looks nothing like emerald to the practised eye but could be unwittingly or deliberately sold as low quality beryl. The refractive index will be that of quartz and will read around 1.55: the specific gravity will be lower than any *natural* emerald at about 2.65. Here care might need to be taken as some synthetic emeralds may have a similar SG and while there is no real resemblance between the two materials it depends what the customer is told.

Kornerupine is a quite rare and often attractive gemstone, though it resembles emerald only when coloured by vanadium and this material has been reported so far only from Kenya. Stones are notable bright and very attractive – they could just be mistaken for the lighter-coloured emeralds. The refractive index can be easily read, at 1.665–1.680, a good deal higher than that of emerald. The specific gravity is about 3.3. These tests should be enough without having to resort to the microscope. Some Kenyan kornerupine from a different mining area shows a different, bluer green but there should still be no confusion with emerald when stones are tested. Occasionally an emerald-green stone will show a weak blue in one direction with an intense emerald green in another. This particular pleochroic effect is not seen in emerald. It is possible that some East African emerald-green kornerupines may give a yellow fluorescence under both types of UV light.

Chrome diopside could be mistaken for a dark emerald but has a higher RI and SG at 1.664–1.721 and 3.22–3.38. Some light green diopside, which can be chatoyant, resembles the occasional cat's-eye emerald. Chrome-rich diopside shows chromium elements in its absorption spectrum.

It is possible that the occasional specimen of *nephrite* may be mistaken for a dark opaque emerald but nephrite's higher SG of around 3.0 and its RI of close to 1.62 should make testing reasonably simple. Some light green nephrite, in particular some specimens from New Zealand, could cause confusion in the mineral collecting arenas.

The uncommon green, low pleochroic variety of *andalusite* does in fact look quite like emerald at first glance and specimens are usually small. Any suspected small darkish emerald which shows fine absorption lines grouped in the yellow and the green with general absorption of the blue and violet may turn out to be this type of andalusite which is a collectors' stone.

The majority of *alexandrites* show green as the predominating colour and although the colour is due to chromium the green is not very like that of emerald. Nonetheless in certain conditions confusion could arise though testing easily gives the correct answer. Both species may give a red through the Chelsea filter but the RI and SG of alexandrite (chrysoberyl) are much higher than those of beryl. Fibre-optic illumination will show a trace of red or at least pink in alexandrite.

It may seem absurd to cite possible confusion between *green diamond* and emerald but in poor lighting and specimens of small size (plus customer unfamiliarity) there just might be a problem. Of course, green diamond shows a much greater dispersion than emerald and it will show no chromium absorption spectrum.

Crystals of the copper silicate *dioptase* are more collected for their mineral specimen value but the very bright green could be mistaken for that of emerald if the buyer has not seen many faceted dioptases: since dioptase has an easy cleavage faceted specimens are rare and usually very small. The blue and violet areas of the spectrum are strongly absorbed and there is also an absorption band centred at 550 nm. If an RI test is possible the values, at 1.644 1.709, are well above those of beryl: SG is 3.28–3.35.

Chrome green *enstatite* is more familiar (though not very familiar) to gemmologists as one of the mineral inclusions sometimes found in diamond but some South African diamond-producing areas, Kimberley, for example, produce small very attractive bright-green stones which have been fashioned. Most enstatite shows a strong absorption band at 506 nm but this may not show strongly in iron-poor chrome-rich material. Enstatite's RI is higher than that of emerald, at 1.65–1.66, with SG around 3.2–3.3.

The term *common opal* is used for opal showing no play of colour. While most common opal has no particular colour, nickel-bearing stones often show an attractive green (which resembles chrysoprase more closely than emerald). Specimens have the relatively low RI and SG of 1.45 and 2.10.

Sinhalite very occasionally shows a green which might be thought of in the emerald context but has the absorption spectrum with strong bands at 493, 475, 463 and 452 nm: RI is 1.66–1.71 with SG about 3.5.

The high dispersion and birefringence of *sphene* (titanite) both far exceed those of beryl group minerals. Nonetheless the rare and very beautiful chrome sphene is dark enough for both properties to be diminished or hard to see and confusion could certainly arise. The RI (1.84–2.1) cannot be read on the refractometer: SG of sphene is 3.54.

The *hiddenite variety of spodumene* is extremely rare in either crystal or faceted form. The green is very like that of the brighter shades of emerald: hiddenite shows a chromium absorption spectrum but an RI determination is less likely to lead to confusion: spodumene has an RI higher than that of beryl, at 1.65–1.68: SG near 3.18.

Synthetic emerald

Flux growth

We said previously that the earlier synthetic emeralds at least showed rather lower RI and SG than some of their natural counterparts. They also show a notably bright red under the Chelsea colour filter but it would be unwise to rely completely on either of these two contexts. It is vital, rather, to study photographs and actual examples of the inclusions in as many natural and synthetic emeralds as possible since these are the key to successful identification.

As a silicate, emerald cannot be grown by the Verneuil flame-fusion method like the oxides corundum and spinel. Gem-quality emerald is either flux or hydrothermally grown. Two features at least distinguish flux-grown emeralds: they take up to a year to grow into crystals from which reasonably sized fashioned stones can be cut and prices are correspondingly much higher than those asked for Verneuil crystals of other species. The prices asked for good quality synthetic emeralds can easily reach £150 a carat or more. The stones look very attractive and it is worth making the point that emerald, like all transparent coloured stones (except diamond), would look glassy were it not for the scattering of incident light by the variety of solid and liquid inclusions contained in them and so the few flux inclusions in the synthetic emerald in fact play a large part in the attractiveness of the finished stone to the prospective purchaser. Manufacturers do obviously try to obtain crystals with as few inclusions as possible but it is very likely that the pursuit of near-absolute clarity is consciously not taken too far (it would cost more in any case). The economics of emerald growth are affected by the relative ease of obtaining gem-quality natural material, compared, for example, to ruby.

Compared to the corundum gemstones, emerald has not been grown by a very large number of firms or institutions and flux growth has always been preferred to hydrothermal methods. The first experimental growth took place in the middle to late nineteenth century and by the 1930s the arrival of the Chelsea filter was welcomed as by then the early Chatham flux-grown emeralds were coming onto the market.

As with the corundum gemstones, synthetic and natural emerald have identical compositions – the SG can be as low as 2.65 and the RI 1.56–1.58, these figures should be seen against those for the majority of natural emeralds (2.66–2.80 and 1.56–1.60). There is plenty of room for mistakes to be made if these are the only tests employed. If a source of UV light is available then you will find that synthetic emerald transmits down to 230 nm at least, where natural emerald's transmission stops around 300 nm. Like all other properties, they may not be identical for all specimens and it is possible that some emeralds grown in the past few years may behave in this respect more like natural stones.

The crystal grower can decide what elements shall enter the growing crystal and iron can be excluded (with some care).

It is possible for both natural and synthetic emerald to show a red fluorescence under LWUV, from chromium. In practice a red fluorescence is often hard to see when a natural emerald is tested – the inevitable trace of iron will be the reason. The synthetic emerald will show a much brighter red, as with the Chelsea filter: another simple, useful and attractive test illuminates the emerald with monochromatic blue light, the specimen then being viewed through a red filter (as it passes some red light the Chelsea filter will do). If the specimen now appears red, then chromium must be present. This test is called 'crossed filters'. With a strong glow the specimen is more likely to be synthetic than natural, though other tests, particularly of inclusions, are necessary before a final assessment is reached.

The earlier synthetic emeralds are definitely collectors' pieces and would attract substantial interest were they more plentiful. During the early years of the last century work in Germany at IG Farbenindustrie produced some gem-quality emeralds which were sent in presentation cases to selected recipients worldwide (I (MO'D) have seen only one of these in over 35 years of studying man-made crystals).

Igmerald

The name 'Igmerald' was given to this product which showed striae parallel to the basal plane of the crystal and stones show a weaker pleochroism than most natural emerald. After the Second World War the German firm of Zerfass continued production of emerald with the same properties – this is described below. The Igmerald showed characteristic absorption bands at 606 and 594 nm which were additional to the customary emerald absorption features. The SG was unusually low in the range 2.49–2.70.

Inside the stones the observer could see wisp-like inclusions with very small bubbles in each individual liquid patch. The inclusions appear to be grouped in swarm-like lines, these crossing the stone in slightly curved directions.

The work of Richard Nacken

Emeralds were grown in Germany by Nacken in the 1920s. The material was thought for years to be hydrothermally grown but later work by Nassau showed that crystals were flux grown. Like the Igmerald, Nacken emeralds are highly collectable. One major feature of the investigations was that the Nacken stones showed no signs of water being present (this can be established by IR spectroscopy). Water is found only in natural and hydrothermally grown beryl: however, some specimens known to have been grown by Nacken do show the

IR spectrum of water, this arising from the natural beryl used as a seed upon which the emerald grew. The story shows the need for careful testing and literature consultation before final judgements are made. Neither Igmerald nor the Nacken emerald ever entered the gem market. Chatham entered the market after the Second World War, and today Chatham Created Gems Inc. grows emerald single crystals and crystal groups, taking months to grow each crystal and perhaps a year to harvest crystals large enough to provide fashioned gemstones of several carats.

Chatham

Chatham's emeralds are grown by the flux-reaction method though there is some doubt whether or not the seed crystal employed is natural or synthetic beryl (some seeds at least do not show the IR spectrum of water). Nassau (1980) has suggested that the Chatham growth process is similar to that used by Nacken as the inclusion patterns are alike.

Chatham emeralds are a rich green with RI 1.560–1.566, DR 0.003–0.005 and SG in the range 2.65–2.67. Inside can be seen twisted veils of flux, phenakite crystals, two-phase inclusions and metallic fragments which have become detached from the crucible wall. There will be no natural solid inclusions.

Gilson

The pioneering work of Chatham was followed as far as commercial synthetic emerald was concerned by crystals grown in France by Pierre Gilson who was later to produce versions of opal, lapis-lazuli, coral and turquoise. The emeralds are grown on a seed of natural colourless beryl which when coated with emerald are then used as seeds for the final product. Rate of growth at least in the earlier years was in the region of 1 mm a month.

Readers may find in earlier textbooks references to a Gilson emerald to which iron was added (the N series). Examples are not very common so it would appear that they represent an experiment. Specimens show, in addition to the normal chromium absorption spectrum, a band at 427 nm. Gilson also produced very attractive crystal groups of emerald which were sold for quite high prices. Individual crystals have highly characteristic swirls on the flat ends, an effect not seen in natural emeralds which do not cluster in this fashion.

Gilson introduced his own quality grading, the highest classification (by number of stars) being reserved for the finest green and virtually complete absence of inclusions. Stones up to 18 ct have been reported. The RI, SG and DR are in the same range as the Chatham product: details and diagrams of the flux-transport method of growth can be found in O'Donoghue (1980) and elsewhere.

While it has been reported that Gilson and Chatham emeralds can be tested by immersing them in colourless transparent benzyl benzoate (RI 1.57), this liquid is dangerous and should be used only under laboratory conditions. It was said that facet angles grew brighter as the microscope focus was raised. Bromoform diluted with xylol gave a liquid in which synthetic emeralds were reported to float while natural emeralds sank but again such liquids have to be confined to the laboratory.

As always with a synthetic product examination under the microscope is the only reliable test. Both Gilson and Chatham emeralds will show twisted veils of flux, showing a metallic lustre under appropriate lighting conditions: we have described the veils as 'cigar smoke in a still room' though without specifying the origin of the cigar. Any emerald which at first sight seems to be inclusion-free should be examined with especial care in the girdle region or on the edges of the facets as this is where small traces of flux may have been left. The setting can easily hide them.

The Gilson process was sold some years ago to Nakazumi Earth Crystals Corporation of Japan but we have not seen their products, nor do lwe know at the time of writing whether there are any. In general, currently produced synthetic emeralds may show less brightly through the Chelsea filter than their ancestors did and they may be less transparent to SWUV light. The world of crystal growth is not at a standstill.

Zerfass emerald

We have said above that the Zerfass product was the direct descendant of the Igmerald of IG Farbenindustrie. Crystals were grown in Germany for a short period only during the 1950s and examples are rare and collectable, especially the crystals. One crystal examined by one of the authors (MO'D) showed a flat hexagonal form from the prism faces of which smaller prismatic emerald crystals protruded. Observed specimens have been full of flux inclusions, in some cases at least giving an overall subhexagonal pattern with pronounced twisting. The RI has been measured at 1.555–1.561 with DR 0.006 and SG 2.66. When immersed in benzyl benzoate (see above for cautionary note) the inclusions show up particularly well.

There is a weak red fluorescence under LWUV light (an effect not always easy to see with any emerald): the colour of the Zerfass emerald is very attractive and specimens would be well worth setting in jewellery.

Lennix emerald

The first Lennix emerald crystals appeared on the market in 1966 the name being taken from that of the grower, M. Lens of Cannes, France. Faceted stones were on the market at least by the early 1980s. Lens has stated that his emeralds are grown by the flux-melt process. Crystals of tabular hexagonal habit with the pinacoid as the dominant form were examined by GIA and reported on in 1987.

The colour of the fashioned specimen (1.30 ct) examined was said to be a homogeneous dark green though under magnification a more intense colour could be seen in areas parallel to the vertical axis of the original crystal. Some specimens were found to be heavily included while others were clear. The RI was in the range 1.556–1.568 with a birefringence of 0.003 (low for any emerald) and the SG was in the range 2.65–2.66.

GIA found that the darker areas of the emerald gave higher RI readings: the stones showed bright red through the Chelsea filter. Stones gave the expected chromium absorption spectrum and a bright red fluorescence under LWUV light with a weaker red seen under SW light. As usual, fluorescence responses cannot be treated as diagnostic. A cathodoluminescence test (laboratories only) showed some Lennix emeralds responding purple or bright violet–blue and this has not been reported from any other type of synthetic emerald.

Inclusions found by GIA in the Lennix emerald confirmed that flux growth had been used. Opaque tube-like structures were preferentially oriented parallel to the vertical axis of the original crystal and other inclusions were found in clusters along the borders of successive growth zones following the edges of the basal pinacoid (the flat bottom of the hexagonal crystal). Also noted were thin crystals of both beryl and phenakite as well as healed fractures lined with flux, giving the familiar twisted veil/cigar-smoke pattern. Two-phase inclusions along the edges of the basal pinacoid and in a direction at right angles to this were observed.

This is quite a complicated interior which could without careful examination be mistaken for the *jardin* of natural emerald. Some Lennix emeralds have been found to contain opaque black material, no doubt associated with the flux.

Inamori and Seiko

Two types of synthetic emerald grown in Japan, the *Inamori* and *Seiko* products, are sold as bright and attractive stones. In the first instance the Inamori emerald, produced by the Kyoto Ceramics firm (Kyocera), was sold only in a range of jewellery (Quintessa) where the emeralds carried the trade name Crescent Vert (Inamori Created Emerald in the United States). This was a flux-grown product. The emerald produced by the Seiko firm was also flux grown. It can be recognized by planes of radiating phenakite crystals occurring in groups between growth layers: isolated phenakite crystals have also been observed. Flux particles, many near-rectangular, appear to lie in a single direction in a plane between colour zones – the zones are alternating green and colourless. RI is reported at 1.560–1.564, SG 2.65.

Russian synthetic emerald

Emerald from Russia first appeared on the market in the early 1980s. *Gems & Gemology* described clusters of hexagonal crystals very like those produced by Chatham and Gilson, described above: the crystals had pinacoidal terminations and radiated from a polycrystalline crust. Specimens up to 3 cm in length and 4.2 cm in diameter have been recorded.

The emeralds fall into the customary synthetic emerald range of properties. A lead vanadate flux was used rather than the lead molybdate used by Chatham and Gilson with natural beryl as a nutrient. A crystal cluster tested by GIA found RI 1.559–1.563, SG 2.65. A near-transparent included substance found at the surface proved to be phenakite and there were also groupings of alexandrite crystals and silvery material from the crucible walls. Further flux and crucible particles could be seen inside the crystals so that their artificial origin could be clearly established.

Emerald grown by the hydrothermal process

Among the major gemstones only emerald is regularly grown by both flux and hydrothermal techniques although the former is much the most used method. Hydrothermal growth involves a pressure vessel and crystals are grown on seeds. There is no spontaneous nucleation at multiple sites so that crystal clusters cannot be grown. Hydrothermal emeralds show much fewer recognizable inclusions and many stones would be quite difficult to test were it not for the ever-suspicious absence of natural solid inclusions. Growth rate is around 0.3 mm per day.

Linde: Regency

In the mid-1960s the Linde Division of Union Carbide Corporation in the United States produced hydrothermally grown emeralds which were set in their own jewellery product, Quintessa. It appears that by 1970 the stockpile of emeralds had grown too large for satisfactory disposal and the process was sold to Vacuum Ventures Inc. of New Jersey. They continued to produce hydrothermal emerald, presumably along the same lines, under the name Regency Created Emerald. By then the crystal growth rate had increased to 0.8 mm per day.

The Linde emeralds had RI 1.566–1.578, DR 005–0.006, SG 2.67–2.69. These properties on their own could indicate a natural emerald so that careful examination of the inclusions is necessary. There are some phenakite crystals and very fine two-phase inclusions: some hydrothermal emeralds show a red flash in strong white light and some may contain tapering growth tubes extending from phenakite crystals. Very small white breadcrumb-like inclusions may also be seen and many hydrothermal emeralds show arrow or chevron markings – of course the absence of natural mineral inclusions should alert the investigator.

Biron emerald

Hydrothermal emerald marketed as Biron was grown in Western Australia from 1977. In 1985 GIA described the product in the Fall issue of *Gems & Gemology* from 150 faceted stones and two crystals. The colour was found to vary from green to slightly bluish-green and the faceted stones were notably transparent with some specimens apparently free from inclusions (others showed prominent inclusions). The IR absorption spectrum showed that water was present but this only tells the investigator that the specimen cannot be flux grown. No clue to artificial origin was found in the visible absorption spectrum – this would not be expected from emerald in any case. The RI was measured at 1.569–1.573, DR 0.004–0.005, SG 2.68–2.71, placing the stones well inside the range for natural emerald. No response was observed to either type of UV light and it has been suggested that a high content of vanadium may be the cause. The general response of emerald to UV light is in any case less definite than many textbooks state.

Under the microscope the Biron emeralds show some unique features and others which have been found in similar hydrothermal products. Two-phase inclusions consisting of a fluid and a gas bubble form shapes resembling fingerprints and curved veils; large irregular voids containing one or more gas bubbles have been noted and nail-head spicules, a very characteristic feature of hydrothermal emeralds, are also present.

While the fingerprints and veils seen in flux-grown stones are growth defects indicating healed fractures, in hydrothermal crystals they are found to be made up of many minute two-phase inclusions concentrated at curved and flat interfaces. Similar-appearing flux inclusions are solid. This means that in the hydrothermal emeralds the veils strongly resemble structures in the natural stones. Looking more closely at the nail-head spicules it can be seen that they are cone-shaped voids containing a fluid and a gas bubble. The head of the nail is a single crystal or a group of phenakite crystals. The spicules often form groups arranged parallel to the vertical crystal axis.

Interestingly, small crystals of gold have been observed in some Biron emeralds: the pressure vessel (autoclave) used in hydrothermal crystal manufacture is often lined with gold to prevent the formation of undesirable compounds during growth. The gold crystals may occur as flat plates or as angular grains. It is worth mentioning that the apparently ubiquitous phenakite crystals should be carefully studied since there is the possibility of confusion with crystals of natural inclusions (calcite, for example). Phenakite crystals are usually well formed and show prismatic form rather than the rhomb form taken by calcite. As phenakite has a similar composition to beryl its presence within a synthetic beryl crystal is usually due to a variation of temperature during growth. It has been suggested that if crossed polars are available and the inclusion is sufficiently large the observer will note that the interference colours shown by calcite will be a higher order than those shown by phenakite. It is also worth noting that calcite and dolomite crystal inclusions in natural emeralds do not occur in the same concentrations as in the synthetic emerald.

Sometimes a characteristic 'Venetian blind' or shutter effect can be seen in the Biron emeralds and there are several different zoning patterns.

Some Biron emeralds were found to show near-colourless seed plates flanked by planes of gold inclusions. Chlorine has been found in the stones but detecting it is not in the gemmologist's repertoire. Nonetheless it has not been reported from other synthetic emeralds nor from the natural material.

There has been confusion between the Biron emerald and the Pool emerald. At one time they were thought to be quite different products but this is not the case. The Kimberley emerald, also produced in Australia, may be the same as the Biron type. The Emerald Pool Mining Company at one time said that their emerald was recrystallized natural emerald. This was untrue as it was Biron-type synthetic emerald. The Biron trade name was reinstated having once been dropped.

Hydrothermal emerald grown in Russia

Hydrothermal emerald was grown by the Laboratory for Hydrothermal Growth at the Institute of Geology and Geophysics, Siberian branch of the Russian Academy of Sciences at Novosibirsk. GIA found that features of this emerald distinguished it from other hydrothermal emeralds though its artificial origin was not hard to prove.

The RI was measured falling in the range 1.572–1.584 with DR 0.006–0.007. The SG was 2.67–2.73. Through the Chelsea filter stones examined by GIA showed only a weak red glow when viewed at a low angle to the light source. There was no response, from the eight stones tested, from either type of UV light.

No chevron-like patterns, so often seen in hydrothermal emeralds, could be seen in this Russian product. Dense clouds of very small reddish-brown particles, unidentified at the time of the report and layers of minute white particles were observed in all the stones tested. Fibre-optic lighting shows these inclusions most adequately. A phenakite crystal was found in one sample and in another two opaque black hexagonal plates showing silvery-grey in reflected light. The reddish and white particles would seem to be particularly characteristic of the Russian stones, nothing particular being visible with the spectroscope or dichroscope.

Water was found to be present and also traces of nickel and copper: sets of parallel lines with step formation and sometimes associated with colour zoning also seem characteristic. GIA have found that fragments or slices of what must have originally been larger crystals show colourless or slightly greenish residual portions of the seed. Parallel growth planes at about 45° to the optic axis (direction of single refraction) could be seen in some specimens. Stones appear to have been cut with some care so that characteristic growth patterns could be avoided or at least be not too obvious.

Lechleitner emerald

Emerald grown on a seed of faceted natural beryl was the first production of Johann Lechleitner of Innsbruck, Austria, who later made a number of other ingenious stones. After the faceted seed had been coated with emerald the crystals were polished, sometimes with back facets left in their original rough state to increase weight. They were marketed in the United States by the Linde Division of Union Carbide Corporation under the names Emerita or Symerald. There were also outlets in Germany.

Despite the thin emerald layer the complete stones show a particularly attractive green. A *Gems & Gemology* report of 1981 suggests that if the high chromium content of this layer was to be found in a complete emerald the colour would be too dark to make a satisfactory gemstone.

On immersion the junction between seed and overgrowth is very easy to recognize – the name 'crazy paving' has been appropriately given to it. Very minute almost dust-like crystals of euclase or phenakite also occur at the junctions.

Complete hydrothermal emeralds have also been grown by Lechleitner but there are also sandwich emeralds in which a seed plate of colourless beryl is coated with emerald using the hydrothermal process. This being accomplished the specimen is then coated hydrothermally to give a colourless layer. The stones would need to be shown in closed settings to conceal the different layers.

A similar coating method consisting of an epitaxial layer of flux-grown emerald onto opaque white beryl has been used for cabochons rather than faceted stones. The material was manufactured in France and was marketed under the trade name Emeraldolite. The growth of the coating takes place differentially, ruling out the manufacture of satisfactory spheres: also the surface of polished cabochons is irregular and uneven. The RI and SG are 1.56 and 2.66, in line with the flux-grown core. Neither core nor overgrowth responded to either type of UV light. A chromium absorption is reported and stones showed brownish-red under the Chelsea filter. The material is both hard (8) and tough.

The flux overgrowth layers showed groups of very small parallel crystal faces; where the coating was missing or broken the white beryl was visible. Emerald crystals in the overgrowth showed typical prisms and basal planes characteristic of beryl but there was no crazy paving as in the Lechleitner overgrowth emerald. Quite large flux inclusions could be seen in the emerald layer and near-spherical voids in high relief closely resembling gas bubbles seen in most glasses. Surface pitting shows opened bubbles.

Summary

Having surveyed natural emerald from various sources and its synthetic counterparts and placed ourselves in a position where we know what to look for in both types, we can take a further look at one or two classic testing features that have not yet been fully described. It was more important in the case of emerald to establish locality and natural/synthetic clues and concentrate upon the use of the microscope in this context.

When faced with an emerald-like green stone the spectroscope can tell us whether or not chromium is present, but not necessarily that the unknown is emerald. The absorption spectrum of emerald varies between the two main refracted rays. The spectrum of the ordinary ray gives two narrow lines in the red, forming a doublet and measured at 683 and 680 nm, followed by a sharp line at 637 nm. A broad and weak general absorption covers most of the yellow from 625 to 580 nm and specimens rich in chromium show narrow lines in the blue, one of them at 477.4 nm.

The spectrum of the extraordinary ray shows a bolder doublet in the red but there is no absorption at 637 nm. Two diffuse lines at 662 and 646 nm follow, the latter usually accompanied by 'windows', patches adjoining the absorption bands on the SW side which appear transparent in comparison with the rest of the spectrum. This effect seems only to occur in emerald. The broad absorption is less strong than in the ordinary ray and is closer to the red, thus allowing some yellow to pass. There is no absorption in the blue.

With the dichroscope emerald shows yellowish-green (ordinary ray) to blue–green (extraordinary ray) though other tests are more useful. As we have mentioned several times already, chrome-rich emeralds and in particular their synthetic counterparts glow red through the Chelsea filter though natural emeralds do not show as much response to either type of UV light as do synthetic emeralds.

Aquamarine

Aquamarine owes its colour to iron but when mined it is usually a much greener blue than that seen in modern jewellery today: this is because almost all aquamarine is heat treated after mining and before sale. The dichroscope is much more useful here than with emerald since one of the two colours seen is a definite blue while the other is nearly colourless. As always, examine the specimen in a number of different directions before coming to a judgement. Some aquamarines which have not been heated (as virtually all aquamarines are) will appear a greenish-blue to the eye and will show a deeper green than the unaided eye registers using the dichroscope, the other image in the dichroscope being colourless.

However, the dichroscope is not always easy to use with small cut stones but thc London dichroscope, developed by Ian Mercer and made from two pieces of Polaroid, resembling the Chelsea filter in overall appearance does work very well with aquamarine crystals which may show two shades of blue along their length. The spectroscope, directed at the 'modern' metallic blue aquamarine, will show two weakish bands in the blue and violet, at 456 and 427 nm. As is often found, the larger the specimen the easier it is to make out any absorption bands that may be present.

As gemstones go, aquamarine crystals often grow to large sizes and provide the lapidary not only with the possibility of a large fashioned stone but also with clear, virtually inclusion-free areas. When inclusions are seen they generally take the form of parallel channels (sometimes known as 'rain' – the tropical kind) which are in fact growth tubes (negative crystals) containing crystallites and liquid and gas vesicles. Groups of two-phase inclusions can also be found in aquamarine and these can appear in a mosaic pattern. Red hematite tables are occasionally found as are flakes of mica and skeletal crystals of ilmenite.

The RI of aquamarine does not vary significantly with location, and is usually within the range 1.567–1.590 with a birefringence of 0.005–0.007. The SG is in the range 2.68–2.80.

A very fine dark blue beryl comes onto the market from time to time: this is the Maxixe or Maxixe-type beryl whose colour is sadly likely to fade over time on exposure to strong light or heat. This material can be recognized by its beryl RI together with an absorption spectrum quite unlike that of aquamarine: it shows strong bands in the red at 697 and 657 nm with a weaker band in the orange at 628 nm. The convention is to call naturally occurring blue beryl of this type Maxixe beryl, using the term Maxixe-type beryl for a similar-looking beryl obtained by irradiating an originally pink beryl with UV for some weeks. The colour is magnificent. The dichroism is the reverse of that seen in aquamarine.

Aquamarine is quite easily imitated by a number of natural species and by some synthetic ones. *Blue topaz* when mined quite often resembles aquamarine though the crystals would be quite easily distinguished from one another. Even when blue topaz is irradiated to give the stronger colours now familiar in world markets it is not quite the same blue as aquamarine. But colour is an unreliable test and it is better and in this case easier to use the refractometer: topaz has a higher RI than aquamarine, in the range 1.62–1.64: the SG is also higher at 3.56. *Blue zircon* has even higher constants with the RI above the limits of the standard refractometer (at 1.926–1.985) and the SG 4.69. Despite the difficulty of obtaining an RI reading the DR is so strong, at 0.059, that doubling of facet edges and inclusions would be hard to miss.

Blue topaz does not show a distinctive absorption spectrum but blue zircon will usually show sharp absorption bands at 653.5 and 659 nm, most easily seen by reflected light – other zircon bands may lurk in the yellow and the green. In bright spotlights blue zircon, despite its body colour, will display far more dispersion than aquamarine.

Synthetic blue spinel is grown in some quantities by the cheap and easy Verneuil method. Close control of the growth process allows virtually any colour (though red is rare). A light blue Verneuil spinel will show a higher RI at 1.728 and SG 3.64. It will not show much in the way of inclusions, though there may be some very small idiosyncratically shaped bubbles – but this can also be true of aquamarine.

The Chelsea filter is useful here as blue synthetic spinel shows a bright orange–red, from added cobalt. While natural cobalt-bearing blue spinel has been found to some extent in recent years the darker stones do not resemble aquamarine.

Blue glass can also resemble aquamarine and on the refractometer its values (commonly in the range 1.50–1.70) can coincide with those of the beryl gemstones. Glass shows no birefringence and will give a striped effect between crossed polars. Facet edges are likely to be rounded rather than sharp and inside the stones there will be large, randomly spaced, well-rounded gas bubbles (these are never seen in natural minerals): the colour may be in swirls.

While many textbooks mention touching a stone with the tip of the tongue to get some idea of whether or not it is crystalline, crystals conduct heat better than non-crystalline (amorphous) materials. Remember that the specimens may just have come out of an immersion test and that the liquid used may well not have been water! Some of the chemicals used to test gemstones are dangerous to health. Wash the stone first!

The investigator should never forget that *garnet-topped and other doublets* can turn up in any colour and that it is quite possible that one could be mistaken for aquamarine, as the garnet layer is thin enough not to influence the colour of the rest of the stone.

A borosilicate glass once known as '*mass aqua*' and with a hardness close to 7 may have RI 1.50–1.51 and SG 2.37–2.42. The RI is of course too low for aquamarine but specimens look quite convincing. Fused beryl glasses, which can also be the colour of aquamarine, have RI 1.515–1.516 and SG 2.44–2.49. Emerald is also imitated by this type of glass.

Pink beryl (morganite)

The pink variety of beryl is called morganite and probably owes its colour to manganese though other elements may play a part. The spectroscope may show absorption bands at 540 and 495 nm in one ray and 555 and 355 nm in the other – the latter band could not be seen with a hand spectroscope and even the other bands are hard to detect. The RI is usually near 1.59, DR 0.008, SG 2.80.

Fortunately those natural species with which morganite could be confused all possess properties which can easily be tested so that distinction presents few problems. *Pink topaz* has a higher RI and SG at 1.62–1.64 and 3.53 and the spectroscope may often show at least the red emission line of chromium. *Kunzite*, the transparent pink variety of spodumene, has RI close to 1.67 with a notable DR of 0.015, SG 3.18. Kunzite may also show interesting fluorescent effects, orange with phosphorescence under X-rays and LWUV, for example, but if in testing for pink beryl X-rays have to be used something has gone wrong! The rainbow-like interference colours indicating cleavage are very likely to be found in kunzite but not in morganite.

Natural or synthetic *pink corundum* can be distinguished by its higher RI and should cause no trouble as a faint or pronounced chromium absorption spectrum ought to be visible. RI around 1.76, SG around 4.0. Pink synthetic *spinel* can also be distinguished by its RI of 1.728 and SG 3.64. Natural spinel is also found in pink but its lower RI and SG (1.718, 3.60) are still well above those for beryl.

The high birefringence of *pink tourmaline* at 0.018 immediately separates it from morganite and if the specimen is carefully examined in different directions doubling of facet edges and inclusions cannot be missed. It is possible that some *amethyst*, which sometimes shows crystal sector zoning of a light pink colour, could just be taken for morganite but it is unlikely: nonetheless distinction should be straightforward with the RI at 1.55 and SG 2.65.

Transparent pink *scapolite* resembles morganite quite closely: the RI, DR and SG are 1.54–1.57, 0.007–0.009 and 2.59–2.63. While the SG appears to be the best method of testing this is not so easy today (hydrostatic weighing takes too long or is fiddly and the heavy liquids contravene health and safety requirements). Scapolite does show a stronger pleochroism than that shown by morganite.

Yellow beryl

As sometimes happens in the gemstone world the preferred name for yellow beryl is heliodor which should be given only to truly golden specimens. Most yellow beryl is not golden and can be simulated or, more probably, merely mistaken for by other natural or synthetic species. Yellow beryl has RI 1.57, DR 0.006, SG 2.68 – all figures within the general beryl range. There is little pleochroism and no distinctive absorption spectrum. As it is coloured by iron yellow beryl shows no luminescence. There are no particularly significant inclusions.

Taking possible causes of confusion, *yellow topaz* has higher RI and SG, at 1.62–1.64 and 3.53: *yellow sapphire* also has higher RI, DR and SG at 1.76–1.77, DR 0.008, SG 3.99. Some yellow sapphire, especially material from Sri Lanka will fluoresce an apricot-yellow under LWUV light and this will quickly eliminate any suggestion of yellow beryl. Yellow sapphire will show the commoner inclusions of corundum, in particular the liquid fingerprints.

Yellow tourmaline will show prominent birefringence of 0.018 and also display distinct pleochroism. *Yellow orthoclase*, a beautiful soft yellow stone from Madagascar looks more like a pale yellow beryl than most other gemstones: it may be distinguished by its absorption spectrum with two bands in the blue and violet, at 448 and 420 nm – the latter is the stronger band of the two. *Danburite* does not feature in jewellery but is certainly collected: it can be distinguished from yellow beryl by its higher RI of 1.63 and SG of 3.00. If no instruments are available stones may show a sky blue fluorescence under LWUV light whereas yellow beryl is inert.

We should not forget that *chrysoberyl* can appear in attractive yellow colours but the greater hardness gives it a quite different lustre from yellow beryl. If you don't want to test an unknown faceted stone for hardness – and of course you won't – you can try the absorption spectrum. Chrysoberyl shows a strong band in the deep blue at 444 nm with some weaker bands: the stronger this band the deeper the colour of the stone. The mean RI, DR and SG are 1.74, 0.008 and 3.72. Chrysoberyl does not fluoresce and may not show any distinctive inclusions so it is worth getting to know the absorption spectrum. Though some varieties of chrysoberyl show distinct pleochroism this is not quite so obvious in those stones of a near-yellow beryl colour.

Citrine, the transparent yellow variety of quartz, does not always look very like yellow beryl. The RI, DR and SG are 1.55, 0.009 and 2.65. It should not come to this but quartz shows a distinctive interference figure of a black cross with splayed ends of the arms that do not meet in the centre, which is occupied instead by coloured concentric circles. This may be seen when the specimen is examined between crossed polars and the figure for the quartz varieties is diagnostic.

Yellow apatite quite closely resembles yellow beryl and would be hard to distinguish by colour alone. The spectroscope will show two sets of fine lines, one set in the yellow and one in the green, which indicate the presence of the rare earth elements neodymium and praseodymium. The spectroscope should be used as early as possible in gem testing with coloured stones! If a spectroscope is not available the mean RI is 1.64 with DR 0.002 and SG 3.20. Apatite may show a variable fluorescence where yellow beryl shows none: both yellow beryl and apatite may show the cat's-eye effect.

Zircon in a yellow colour shows the high birefringence and some of the dispersion shown by its colourless varieties, and is not very like yellow beryl. The DR is the best feature to look for since at 0.059 it can hardly be missed. The spectroscope will show a marked series of bands across the width of the spectrum, evidence of the presence of uranium which is not found in any other major gem species. The effect is strongest in stones from Sri Lanka.

Yellow spodumene with fine material from Afghanistan could be confused with yellow beryl but shows two narrow absorption bands in the violet at 438 and 432.5 nm. Spodumene has an RI in the range 1.660–1.675 with a DR of 0.015 and SG usually near 3.18.

The *yellow garnets* do not usually come close to yellow beryl in colour or are too small as in the case of yellow andradite which might more easily be confused with yellow diamond. On the other hand some varieties of grossular, especially those from East Africa, are sufficiently golden to cause confusion with the better varieties (true heliodor) of beryl. These garnets have a higher RI and SG than beryl, at 1.742–1.748 and 3.65.

While yellow diamond could hardly be mistaken for beryl as it is too brilliant and highly dispersive, *yellow glass* is always around. The properties of glass can be found in many other places in the book but a quick glance at a specimen placed between crossed polars will show a striped effect which should not be visible with yellow beryl.

Sphene (titanite) with high dispersion and birefringence (0.051 and 0.134) does not look much like yellow beryl though the even rarer *brazilianite* certainly does. This is softer than beryl and has RI and SG 1.603–1.623, 2.94–2.998. The DR at 0.020 is higher than that of beryl and may give doubling effects.

There has been no serious attempt to synthesize yellow beryl though it could be accidentally confused with Verneuil yellow corundum.

Red beryl

Red beryl, coloured by manganese, is a comparatively recent addition to the ornamental beryl range. While stones are almost always small, the colour is a very fine red with a good deal of life. The RI is in the range 1.564–1.574 with SG 2.66–2.70. Confusion with any material other than glass is unlikely. Synthetic red beryls have been grown in Japan and Russia; details are given below.

Colourless beryl

Fashioned examples are usually large and some specimens may show the merest hint of yellow. Properties are in the normal beryl range.

The colour enhancement of emerald

The reader can at any time take a walk round the jewellery quarters of large cities such as London and see considerable numbers of emeralds set in jewellery – there will be more emeralds than any other major gemstone except for diamond. Small sizes will predominate and colours will vary from a bright to a dark, perhaps oily green. 'Oily' is the appropriate adjective, since virtually all emeralds mined today are immersed in oil with the aim of filling surface-reaching cracks and minimizing the effect of inclusions so that light would be less scattered and the colour consequently enhanced by being deepened.

Oiling is not confined to emerald – any stone treated in this way will smell of oil, even more so when a large number are in a single parcel whose paper will show oil stains. A finger used to stir the stones around will emerge feeling tacky and, over time, persistent heat from most kinds of lamp (though not fibre-optic sources) will cause the oil to seep from the stone.

When this kind of treatment is detected the onus is on the dealer or other seller to disclose it to the customer. Whether or not this has been carried out faithfully since the practice first began is not easy to check since oiling goes back to classical (Greek and Roman) times. The oil has to be light rather than heavy and of course as near-colourless as possible. If an oil with an RI similar to that of the host can be introduced the result is more effective: Canada balsam with an RI of 1.53 which comes quite close is often chosen as the medium for emerald oiling. Oiling is not confined to finished stones but is also carried out on crystals too.

The question of oiling is considered of sufficient importance for the major auction houses to mention it in their jewellery sale catalogue entries (other methods of treatment are also mentioned). Only if one follows the path taken by a stone right from its first recovery at the mine to final cutting and sale can you be sure that it will not have been treated in some way.

Some fracture-filling substances used in Colombia may give a dull yellow fluorescence under LWUV light – the emerald host would not show this effect without the oil so it is quite a useful guide to treatment. The filled fractures will probably appear dull when light from a fibre-optic source is directed at them. To avoid confusion with natural liquid-filled inclusions assume that the supposedly filled fracture reaches the surface. Where there is a marked difference in RI between filler and host spectrum colours may be seen. Use of the thermal reaction tester (hotpoint) may bring a bead of oil out of the specimen but you should ask the owner first!

While many of the oils used with emerald are colourless there are reports that some coloured oils have been tried out. It was recognized that there would be a danger of concentrating colour in certain parts of the stone and that the colour produced would not quite match that of the host. Such concentrations can easily be seen if the specimen is placed on a white plastic or other background and examined while illuminated by an intense source of light placed underneath. When areas of deeper colour than the predominating colour of the host are seen the stone should be very carefully examined as unfilled fractures will tend to reflect green light from many parts of the host.

The filling of surface-reaching fractures by the introduction of glass or epoxy resins has occupied the gemstone trade for some years. Opticon was one of the first and probably best known of the resins (the name has passed to other resins). Studies made include not only identification but predictions of durability as they may degrade or fragment under certain conditions.

Details of some of the processes used can be found in O'Donoghue (1997) and in issues of *Gems & Gemology* and the *Journal of Gemmology* but for the purposes of the present text we need to know how the presence of fillers can be identified. Opticon may be colourless but a green version has been used for emerald filling.

Stones treated with fillers usually show an arresting flash of a colour quite alien to the colour of the host; orange and blue appear to be the commonest colours seen – they are caused by dispersion. In addition to the colour flashes emeralds treated in this way sometimes show a distorted appearance internally or will appear less transparent. The overall effect resembles that of heat-shimmer from a hot road. Some of the filled fractures may give a weak chalky white to blue fluorescence under LWUV light (only Opticon, apparently, shows it). There is no response to SWUV light.

The orange and blue dispersion flashes resemble those found in fracture-filled diamonds: GIA have noted that when some filled fractures were closely examined nearly edge-on some of them showed a slightly orange–yellow colour. Sometimes this could be seen over the whole fracture and when the stone was tilted the flash colour changed to blue. In other cases the orange colour did not change but disappeared and returned when the stone was tilted and moved back to its original position.

Examination with the microscope is advised as filled fractures have low relief whereas unfilled ones containing air will have a higher relief and will be easier to see. Today, a careful examination of surfaces should be practised with all stones until some proficiency at spotting fillings is arrived at. This should be done with the assistance of fibre-optic lighting which should be arranged to reflect light from the surface to be examined. The use of dark-field illumination is recommended and when the top of a fracture is found the portion of the stone beneath it should be examined with especial care.

Some stones known to have been treated with Canada balsam or with cedarwood oil have shown no sign of coloured flashes so their absence should not be taken as an indication that no treatment has taken place. The majority of fillers contain bubbles surrounded by cloudy areas and if the bubbles are closely examined they may be found to show interference (rainbow) colours when examined under overhead lighting. The bubbles may occur on their own or in groups and some of the fractures display a flow structure seen best in dark-field lighting. Some fractures are sealed at the surface using a hardener.

The durability of Opticon fracture fillings has not yet been determined though tests carried out by GIA seem to indicate that it is more lasting than Canada balsam or cedarwood oil. Repeated ultrasonic cleaning may have an effect and some evidence of filling deterioration may be visible at the surface. Steam cleaning was more harmful to fillers than ultrasonic cleaning and in general, when fillers are affected by a cleaning or setting/repair process, material is lost and evidence of this should be visible. Opticon has been found to take on a yellow tinge after heating of the host: the process in question was retipping of prongs, a familiar exercise in jewellery repair.

A low polymer epoxy resin, palm oil, has been used as a filler in emeralds imported from Colombia into Japan. This has an RI of 1.57 and shows weak bluish-green to orange–red flash colours; bubbles have also been found. Some other fillers reported from emerald include a cyano-acrylic material which gives a white 'brushmark' in fractures and a hardened epoxy resin with few bubbles and no colour flashes.

Beryl other than emerald is not often enhanced since the commercial gain would be small. Aquamarine is routinely heated but there has never been any dispute arising from this practice. The only really critical example of beryl treatment is the Maxixe-type deep blue beryl described above. It is worth mentioning that this material was at one time marketed under the name Halbanita – a further, personal account of the history of one of these stones can be found in O'Donoghue (1997). As a matter of interest and just in case the reader comes across a dark blue beryl of this kind, the early rumours that they would fade more or less at the drop of a hat turned out to be exaggerated: Nassau, in a paper of 1996 in the *Journal of Gemmology*, showed that the original tests for fading involved higher temperatures than would obtain under normal conditions of wear and that fading takes place more slowly – nonetheless such stones should not be subjected to strong sunlight or heat at any time if their colour – which is very beautiful – is to be preserved. Nassau in the same paper reminds us that both aquamarine and golden beryl could be affected by the higher temperatures sometimes encountered during the jewellery setting or repairing processes so care needs to be taken.

Composites

Composites imitating emerald can be very tricky and there have been many examples over the years though it is possible that the arrival of synthetic emeralds from a number of manufacturers has diminished the desire of composite makers (who are usually anonymous). There will be references to especially interesting (or dangerous) examples below but in the main emerald is imitated by composites often known as *soudé emeralds*. These are made primarily of glass with the colour being provided by green material at the junction between crown and pavilion. If the crown is of glass an RI measurement of the table may give the game away and if, as often happens with glass, the table is slightly rounded this will also be a clue. If there is an absorption spectrum the bands will be woolly rather than sharp.

Doublets which are beryl-on-beryl with a central coloured layer of glass or gelatine can be detected when the stone is immersed – if you think about doing so! Stones may be assembled in such a way that natural inclusions appear to be continuous between top and bottom. Beryl-on-beryl and emerald-on-emerald specimens have been put together in this way – this is one of the consequences of the general availability of emerald of moderate quality. If the colour in the junction area is not derived from a thin slice of emerald the absorption spectrum will not be a chromium one: in such cases there may be absorption bands in the red–orange.

A beryl and glass triplet emerald imitation reported by GIA in 2001 gave an RI of 1.582–1.591 when the crown was tested but the microscope showed a single plane of gas bubbles parallel to the table. When immersed the crown was seen to be colourless and that it was fixed by a colourless cement to a green pavilion. While the crown was birefringent the pavilion was not and showed an RI of 1.49. The colour was reflected from the green pavilion rather than from green cement.

Reports of interesting and unusual examples from the literature

Items in this section have been chosen to illustrate points made in the chapter and to bring one-off items to your notice.

This is not likely to be encountered but a specimen of emerald-green glass with RI 1.635, SG 3.75 was found to be dangerously radioactive, up to 12 times background levels.

A recent study compared oil, Canada balsam and artificial resins as emerald fillers. An artificial ageing test showed that the artificial resins turn yellow from colourless. Clove oil can be distinguished from artificial resins and carbon–hydrogen chemical bonds present in some of the inclusions in natural emerald can be detected by IR spectrometry. Soap and refractometer contact liquids can affect spectra from emerald fillers.

Even though filling materials in emerald can deteriorate over time it may be possible to replace them. Oils are the most volatile: artificial resins are harder to remove. IR and/or Raman spectroscopy can be used to tell the trade which filler has been used. After visual examination under the microscope and with LWUV light, which should show the presence and location of any filler, IR spectroscopy is used as a macro-method and Raman spectroscopy as a micro-method of assessing the nature of the filler. Results are compared with others in a databank.

New emerald-producing areas have been reported in the past few years: stones from *Somalia* have been found to be a mid-green colour and to contain frequent natural cracks. They also lack the depth and regularity of colour and clarity shown by the better Colombian stones. Thin platy crystals of phlogopite have been noted. RI and SG of emeralds tested so far are in the range 1.568–1.574 with DR 0.006 and SG 2.76. The emerald examined showed red through the Chelsea filter.

Emerald from the Kagem Kafubu mines in *Zambia* falls in the RI range of 1.581–1.597 with DR 0.008, SG 2.69–2.77. The commonest inclusions seem to be mica laths though rod-like inclusions, possibly amphiboles, have also been observed. Fluid inclusions are oriented, with long axes parallel, and are liquid-filled negative crystals. Some gaseous inclusions have also been found.

Emeralds found in the high Andes of *Chile* near the town of Ovalle, close to the border with Argentina, show strong zonal coloration which is best seen between crossed polars. Actinolite, dolomite, pyrrhotite and rutile with other minerals occur as inclusions. These Chilean emeralds closely resemble emeralds from the Austrian Habachtal.

The Delbegetey deposit in *Kazakhstan* has produced emeralds with a strong bluish-green colour with RI 1.558–1.570 and SG 2.65. The iron content was found to be low with primary fluid inclusions containing an unusual liquid-to-vapour ratio of 50:50 to 90:10.

Though aggregates of emerald crystals have so far apparently been confined to the artificial productions of Chatham and Gilson, such specimens have recently (1998) been reported from the Coscuez mine at Muzo, Colombia.

In a paper of 1999 30 fillers were investigated and placed into six substance categories (it may be useful to cite these as disclosure rules at some time in the future). 'Presumed natural' were essential oils, including cedar wood oil and other oils and waxes and three 'artificial resins' including epoxy and other prepolymers, including UV-setting adhesive and polymers. Whatever their composition, fillers with RI >1.54 show flash effects in emeralds. On the basis of IR and Raman spectroscopy the fillers are separated into five spectral groups A–E. Most but not all commonly used artificial resins have spectra distinct from that of cedar wood oil though the detection of one substance in a filler does not imply the absence of others.

Further help on locality information for emerald is given in a paper published in 1999 in the *Journal of Gemmology*. Specimens from Australia, Brazil, Mozambique, Russia, Tanzania (Lake Manyara) and Zambia belong to schist-type deposits and show inclusions of mica group minerals, talc, pyrophyllite, chlorite group minerals, wollastonite, chromite, calcite, dolomite, iron oxide, apatite, quartz, pyrite, fluorite and plagioclase. Emeralds from Colombia, Nigeria and Sumbawanga (Tanzania) give related assemblages of chlorine-rich Al-glauconite, sphalerite, Ti-rich mica, anhydrite and beryllium minerals.

Electron microprobe studies of 29 rough and fashioned emeralds from the same countries also showed that schist-type emeralds are characterized by high concentrations of MgO, FeO and Na_2O while Colombian and Nigerian emeralds had low concentrations of MgO and Na_2O. Hydrothermally grown emeralds had the lowest amount of Na_2O.

An update of Sandawana emeralds from Zimbabwe by Raman microscopy has shown hitherto unreported Cr-rich ilmenorutile as well as actinolite, albite, apatite, calcite, Mn–Fe-bearing dolomite and quartz. Most stones are less than 0.5 ct but are a vivid, even green with RI 1.584–1.594 and SG 2.74–2.77. The amphibole minerals found included are actinolite and cummingtonite. The amphibole group has been extensively revised in the past few years: the best up-to-date list is in *Fleischer's Glossary of Mineral Species*, 1999.

Absorption spectra can also help in placing the origin of emeralds, as shown in a study published in *Australian Gemmologist* during 1999 which examined specimens from Afghanistan, Australia, Brazil, Colombia, Mozambique, Nigeria, Russia, Tanzania amd Zambia. These showed that a mixed type of absorption spectrum with characteristics both of emerald (Cr^{3+} and V^{3+}) and of aquamarine (Fe^{2+} and Fe^{3+}) was very common but that the relative intensities of the Cr and Fe bands differed with different deposits. Optical spectra of emeralds from the same region were found to be identical.

Emerald (and green beryl) from central Nigeria show locality-specific solid inclusions, described in a paper in the *Journal of Gemmology*, 1996. The commonest are albite with fluorite, the albite forming colourless, transparent angular to irregularly shaped grains appearing translucent white in reflected light as they themselves are heavily included. Fluorite forms perfect octahedra but some cubes and rounded irregular shapes have been found. Mica inclusions are iron-rich and other included minerals are ilmenite, monazite, beryl and quartz. Fluid inclusions are also noted. RI and DR are 1.560–1.574 and near 0.005, SG 2.64–2.68. The absorption spectrum shows an emerald component (Cr/V) and an aquamarine component (Fe^{2+}/Fe^{3+}).

Emeralds from Malipo, south of Kunming, China, occur as heavily included long prismatic crystals with dark needle-shaped tourmaline crystals. Some of the crystals are of gem quality and have RI and DR 1.573–1.583, 0.007–0.008, with SG 2.68–2.71. Crystals show fine cracks and distinct growth zoning under magnification.

Emeralds from the Kala Guman (Kaliguman) area of India are dark to watery green and have RI 1.574–1.581, DR 0.007 and SG 2.75, values lower than those cited by some authorities. Chlorapatite and beryl inclusions have been reported along with some talc, albite and a few three-phase inclusions. Primary and secondary fluid inclusions are reported and also growth tubes of large dimensions.

Emerald from the Mananjary area of eastern Madagascar contains inclusions of plagioclase, quartz, pyrite, goethite/hematite, phlogopite in hexagonal or pseudo-hexagonal plates, allanite and barite. Fluid and two- and three-phase inclusions have also been observed.

Emerald from Tocantins state, Brazil, was on display at the 1998 Tucson Show. Stones were darker than most Colombian emeralds and showed signs of thermal treatment.

A further test for the origin of emerald is the laser-induced (time-resolved) photoluminescence technique (LPL). This showed that spectra of specimens from Afghanistan, Russia and Zambia (schist-type deposits) were less intense than spectra of hydrothermally grown emeralds which showed a 'bulge' structure in the luminescence curves.

The popularity of trapiche emeralds has led to their further study. In a treated specimen from Colombia the relatively broad dark area between the six emerald sectors surrounding the hexagonal emerald core was found to consist of a fine-grained substance with resin acting as a stabilizer. Light greyish-green trapiche has been reported from Madagascar.

Dominant absorption due to chromium can be seen in Colombian emeralds while absorption peaks from both Cr and Fe are recorded. Synthetic hydrothermal emeralds show absorption due to Cr and a high transmission of UV light. An absorption relating to Ni is found in the Russian synthetic emeralds. In flux-grown emeralds only Cr-related absorption can be seen and the transmission peak in the UV lies at a higher wavelength than that of natural Colombian emeralds. From NIR (1200–2700 nm) spectra the origin of emeralds and their growth conditions can be conjectured from the absorption peaks since they show the presence, absence and nature of H_2O.

The country of origin of emerald may be determined by the isotopic composition of oxygen of the various rock types and the emeralds associated with these rocks. This is particularly useful in the case of Colombian emeralds.

Synthetic emerald has been made in China using the hydrothermal process. Crystals were grown in an alkali-free chlorine-bearing solution at pressures higher than those usually employed in this type of growth. Seeds of natural or synthetic beryl could be seen in the emerald crystals together with characteristic growth zoning parallel to the seed. The zoning formed at angles between 20° and 40° to the vertical axis of the crystal and no crystal angles in natural emerald meet within this range.

Unidentified spicules oriented parallel to the main axis were commonly associated with minute crystals of chrysoberyl. Needle-like inclusions were oriented almost perpendicular to the seed on dominant growth faces. EDXRF analysis and IR spectroscopy showed that chlorine was a diagnostic element. Iron was not detected: RI and SG overlap with natural emerald.

An example of the rare star emerald has been described: the stone, of 10.03 ct, showed a well-defined six-rayed star and strong pleochroism with an intense blue–green and a light yellowish-green visible in different directions. The asterism was caused by three sets of unidentified small particles oriented perpendicular to the main crystal axis.

Triangular bodies of unknown composition have been found in some Colombian emeralds and might cause confusion with metallic crucible material inclusions seen in many synthetic emeralds.

An emerald-like substance marketed under the trade name Swarogem was found to be a silicate glass with a relatively high RI of 1.608–1.612. Specimens were virtually inclusion-free and had a medium dark bluish-green colour.

Further notes on Swarogem were published in the Summer 1994 issue of *Gems & Gemology*. Stones tested had an RI in the range 1.608–1.612 (the manufacturers gave 1.605–1.615). The SG was in the range 2.88–2.94, this was in accordance with the manufacturer's statement. Under SWUV light there was a faint green fluorescence and a weak to faint yellowish-green to greenish-yellow fluorescence under LWUV light. With the hand spectroscope absorption bands could be seen centred at about 590 nm with other bands at 483, 472 and 466 nm, these forming a triplet. A strong broad doublet could be seen at about 448 and 442 nm. The stones were inclusion-free and remained green when viewed through the Chelsea filter. EDXRF showed that Swarogem was a calcium aluminium silicate glass with some praseodymium also present, this explaining some elements of the absorption spectrum. The hardness and RI were higher than those found in most ornamental glass.

AGEE synthetic hydrothermal emerald, manufactured in Japan, is probably grown from low grade emerald feed material. It has RI 1.573–1.580, DR 0.004–0.007 and SG 2.67–2.72. Stones show a strong red through the Chelsea filter and either a variable weak red fluorescence under UV light or no response, partially spiral fingerprints and angular growth zoning are reported together with parallel cavities and widely spaced colour zoning. The producers, A.G. Japan Ltd, appear also to have grown emerald with the same properties as the Biron material and flux-grown stones, as well as a vanadium beryl.

Guilin hydrothermal emerald from China is characterized by a low content of alkalis, iron and chlorine. Nail-like inclusions are characteristic of this and other hydrothermal emeralds. Some orange–red colour has been shown, presumably with the colour filter or dichroscope. At the time of writing we have seen no reports of commercial exploitation.

Synthetic emeralds grown by Tairus in Novosibirsk, Russia, show two broad absorptions in the near-IR which have not been found in natural emeralds. Chevron-like patterns often seen in hydrothermal emeralds do not appear to be present. The emerald is grown on a seed in such a way that the usual prominent growth zoning is largely avoided. Metallic inclusions show that iron, chromium and titanium oxides are present.

A filled emerald with the unusual flash colours of orange to pinkish-purple in one direction and a blue–orange in the other may have produced two crossing dispersion curves in conjunction with a non-crystalline filler, the flashes occurring at the wavelength at which the refractive indexes match. Bright-field illumination gives blue while dark-field gives orange as the complementary colour. In the other orientation orange to pink forms the dark-field colour and the expected bright-field colour of green to blue green was probably masked by the stone's body colour.

In an interesting paper in *Australian Gemmologist*, 17(12), 1991, Lechleitner's production of emerald was classified as types A–F. Type A stones were emeralds flux grown onto seed plates of natural beryl, grown between 1958 and 1959 and sent only to gemmological laboratories: Type B stones were sold commercially and had cores of colourless or slightly greenish natural beryl with hydrothermally grown emerald overgrowth in which fissures and cracks can be seen. Type C emeralds used natural beryl seeds cut obliquely to the c-axis (main vertical axis of the beryl crystal) and were reported to have been grown in a single run, not entering the market. The seed showed dark green emerald layers on both sides of a colourless centre. Type D stones are complete hydrothermally grown emeralds using seed plates from Type C material. They have been grown since 1964 and can be seen to have a layered structure when examined under immersion. Type E stones were grown for research only and Type F emeralds, grown by the flux-melt method between the early 1970s and the 1980s, are with the B and D stones, and still available (i.e. in 1991).

A faceted natural emerald, probably Colombian, was overgrown by a layer of green artificial glass, the assemblage weighing 5.62 ct. The RI was 1.570–1.578 with one of the back facets giving a reading of 1.547. Spotting this could only have been achieved by a habitual back-facet tester – back facets are not usually very large!

Another emerald imitation reported was a crystal with a core of green artificial resin: the stone showed bubbles in the core and anomalous double refraction between crossed polars.

An emerald which may have been grown in Russia had been filled with a resin and gave very strong orange, blue, yellow and purple flash colours dependent upon the angle of viewing. A fracture fluoresced a moderate yellow to white with a very short phosphorescence under SWUV light. As there was no reaction to the hotpoint on the surface the filler may have been hardened.

An emerald whose inclusions indicated Zambia as its origin gave an RI of 1.48. When examined under the microscope using reflected light an iridescent coating could be seen. On removal of the coating with an ink eraser an RI of 1.579–1.588 was observed.

A emerald-imitating triplet made from synthetic spinel and glass was made in France by Jos Roland as long ago as 1951. The name *soudé sur spinelle* was used. Stones were made in a variety of colours. Crown and pavilion were colourless synthetic spinel. A specimen examined by GIA gave an RI from the table of both 1.724 and a weaker reading (unexplained) of 1.682. The glass layer was about 0.55 mm thick with a hardness of around 4 and showed swirls and small flattened, rounded and irregularly shaped bubbles when viewed from the side. Immersion showed the different components very clearly: no absorption could be seen with the hand spectroscope. The crown showed a strong chalky yellowish-white fluorescence under LWUV light viewed nearly at right angles to the girdle with the table closest to the UV source. While the glass layer was inert the pavilion gave a strong clear yellow fluorescence with no trace of chalkiness. When the culet was placed close to the source the effect was reversed. The stone was virtually inert under SWUV light and under X-rays showed only a very weak chalky green – there was no phosphorescence.

Emerald crystals which are probably the most obtainable of major gem rough and which can to some extent get onto the market have been imitated over the years. The 'crystals' are almost invariably glass usually with a coating of what the maker supposes to be appropriate – I have seen the 'wrong kind of mica' on a crystal from Central Africa. An ingenious imitation emerald crystal was reported in 1989: a pale green beryl crystal was broken in half and the cores drilled out, the resulting cavity being filled with a dyed green epoxy or similar substance. The two sections were cemented together. The absorption spectrum was characteristic of dyed materials.

A similar enterprise reported by Thomas Chatham to *Gems & Gemology* in 1990 also featured a hexagonal prismatic crystal of beryl. On sawing through the piece a green fluid leaked out (the tester must have been sure of his material!). In this case the crystal was green even without the filler. The crystal had been excavated and the hollows coated with a green substance: an apparently water-worn elongated subhedral 'crystal' had been inserted into the cavity: this was probably a plastic. A cap made of a soft grey metal, perhaps lead, covered by a mixture of ground mineral material set in a polymer, melted when the hotpoint was applied. It is reported that this composite was manufactured in Bogotà.

While coated stones intended to simulate emerald have to some extent been replaced by synthetic emerald as far as manufacture is concerned, the old ones are still around and likely to give trouble. An emerald-green stone of 4.39 ct examined by GIA and reported in 1993 had a green coating covering the entire pavilion: the stone showed red through the Chelsea filter. The properties were those of natural beryl and natural inclusions characteristic of beryl could be observed. A dye-related absorption band was seen between 690 and 660 nm with the hand spectroscope. The coating had worn through in several places.

GIA reported an unusual occurrence of a piece of jewellery with all stones consisting of synthetic emerald overgrowth on faceted near-colourless beryl. Such products are almost always first encountered as individual stones.

Some of the stones grown by the Kyocera Corporation have had different characters lasered onto a pavilion face just below the girdle. At the time of the report in the Summer 1995 issue of *Gems & Gemology* the characters were a stylized CV (for Crescent Vert) with the weight of the stone and its quality grade. Some 'recrystallized' corundum marketed by the TrueGem Company of Las Vegas has also been lasered, showing identification on the girdle.

Biron, manufacturers of hydrothermal emerald, have also grown a pink beryl with normal beryl properties but with no natural inclusions. The material shows a moderate apricot fluorescence under LWUV light.

An iridescent coating reported on aquamarine and some golden beryl is apparently quite easily removed by light polishing. Some examples have shown poor polishing on the coated surfaces.

A synthetic water-melon coloured beryl with a pink centre and green coating has been made by the Adachi Shin Industrial Company of Osaka, Japan. The RI was 1.559–1.564 and SG 2.66. According to the Spring 1986 issue of *Gems & Gemology* a new method of synthesis had been used, in which fluorine and oxygen were reacted at higher temperatures with crystalline or amorphous beryllium oxide, silicon dioxide and aluminium oxide: various dopants are added. Manganese gives the pink of the core and chromium the emerald-green outer colour. The company stated that crystals had been grown up to 1 cm in length and that brown, reddish-brown, pink, colourless, sky blue, yellowish-green and emerald-green colours were available.

Beryl crystals grown in Russia have been doped with cobalt to give a rich orange–red, with purple, from chromium and manganese, and blue (copper). Manganese on its own gave a red colour according to the Winter 1988 issue of *Gems & Gemology*.

A set of apparently aquamarine beads consisted of some pale-blue aquamarine and some colourless beryls, these being coloured blue from the coloured thread and from dye in the drill holes. On soaking with acetone (hopefully with the customer's permission!) most of the dye was removed.

A synthetic red beryl grown in Russia has RI 1.570–1.580, DR 0.008 and SG 2.63–2.65. Growth is hydrothermal and faceted samples showed a step-like growth structure and distinctly inhomogeneous growth was shown by irregularly changing sub-grain boundaries almost perpendicular to the step-like growth. These features are also found in Russian hydrothermal emeralds.

Planar and veil-like feathers consisting of liquid and two-phase inclusions have been reported as well as black platelets, perhaps of hematite. These beryls are of low alkali type with a low concentration of sodium. There are also significant amounts of cobalt and manganese oxides.

In what may be a different Russian hydrothermal red beryl a purplish-red pleochroic colour is seen parallel to the main crystal axis with an orange–red to orange–brown at right angles to this direction. Specimens are inert to both forms of UV light and colour distribution is generally even, though there is some brown banding along growth zones. Metal wires have been observed in a few crystals, some of which contained nailhead spicules not reported in faceted stones. The spectroscope gives absorption bands between 600 and 480 nm with superimposed bands at about 585, 570, 560, 545, 530 nm and broad absorption between 470 and 400 nm. There are weak bands at 410 and 370 nm. The IR spectrum shows the presence of water which is absent from natural red beryl.

Fading of colour in beryl was examined by Nassau and reported in the *Journal of Gemmology*, 25(2), 1996. In the study 43 large high quality faceted beryls recently mined in Brazil were examined: 29 were intense blue Maxixe-type beryl and the remaining five yellow and nine greenish beryls were found to be golden beryl and green aquamarine. If an accelerated fading test (heat and light exposure) is used the heat must not exceed a maximum of 50°C since both aquamarines and golden beryls which do not fade from light alone would lose their yellow component and perhaps be identified as Maxixe beryls which had gone some distance along the fading path. One recorded example of this happening stated that a temperature of 110°C had been used. In another reported experiment three blue and one light greenish-blue beryl were found to be Maxixe beryls. After 4 hours in a solar simulator the three blue stones had begun to fade.

A stone of a violet–blue colour examined by GIA and thought to be tanzanite was found to be beryl, showing sharp and closely spaced absorption lines between 690 and 575 nm. It was strongly pleochroic. This colour is unstable and prone to fading.

A faceted red beryl of 4.66 ct from Utah contained a cluster of opaque inclusions just beneath the girdle: the RI was 1.568–1.572 with DR 0.004. The spectroscope showed a wide absorption band at 590–500 nm with a weak band at 455 nm and a stronger one at 430 nm with general absorption from 420 to 400 nm. Most if not all of the inclusions were found to be of the manganese–iron oxide bixbyite.

Colourless beryls from the Badmal mines, Orissa, India, were found to change to green and greenish-yellow on irradiation. On heating to 300°C the colour of the irradiated beryl changes from greenish-yellow to yellow and at 500°C to colourless. The irradiation carried out employed a linear beam accelerator and samples were also gamma-irradiated from a ^{60}Co source. Electron spin resonance and optical absorption methods showed that the colour was due to the formation of a defect centre similar to the Maxixe type as well as to other phenomena, the combination giving the greenish-yellow colour.

A green beryl with a cabochon Australian opal set in the culet area showed the opal's play of colour through the table when the stone was viewed face up. The same type of assembly has been found with amethyst, aquamarine and golden beryl as the host. Gilson synthetic opal chips have been set in glass to form briolettes, hearts and spheres.

Golden beryls from the Laghman district of Afghanistan show etch figures and have RI 1.560–1.580, DR 0.008 and SG 2.7. The same area produces brownish-yellow tourmalines.

A reasonably convincing doublet imitation of emerald was made by adding a green plastic layer to a fully faceted beryl.

A faceted hydrothermally grown aquamarine of 0.85 ct showed RI 1.572–1.580, DR 0.008 and SG 2.71. Under the microscope the stone on immersion showed inhomogeneous growth patterns, characteristic of hydrothermally grown beryl.

Santa Maria aquamarine has RI 1.578–1.588, DR 0.007–0.008 and SG 2.68–2.70. Absorption bands have been recorded at 891, 833, 557, 426 and 370 nm. Disc-like inclusions are oriented parallel to one another and have a filmy appearance.

A green aquamarine showed an unusual absorption band at 537 nm as well as the more common ones at 467 and 427 nm. The 537 nm band is characteristic of untreated aquamarines and is also found in yellow and colourless beryls.

We should not forget the quite convincing emerald imitation made by heating rock crystal and immediately plunging it into a cold green dye. The dye enters the cracks induced by thermal shock. Specimens have quartz constants but the colour is good.

Chapter 5

Opal

Introduction

Opal occupies a unique position among the better-known gem species: not only does it give a completely different colour effect from other gemstones, but, unlike them, it is not crystalline and thus some of the methods of testing that might well be used for ruby or emerald cannot be applied. This does not matter much in practice since nothing else looks like opal; we still need to determine whether a particular specimen is natural or synthetic opal or an imitation (there are some very effective ones around) – opal can also be colour enhanced.

The types of opal

The play of colour is a diffraction effect and the colours produced are exceptionally bright and clear since they do not arise from selective absorption. The play of colour is usually seen against a dark or light background, the names *black opal* and *white opal* being in regular use and undisputed. Black opal is by far the more highly prized and thus most synthetic and imitation opal manufacturers aim at the reproduction of this variety. Quite often the play of colour is seen against a milkiness or cloudiness which is correctly known as opalescence, a noun which is often (incorrectly) used to signify the play of colour.

The varieties *fire opal* and *water opal* are also highly valued: fire opal's body colour ranges from red through orange to yellow, with or without a play of colour: in water opal the drops of colour are seen in a transparent colourless background as if they were suspended in water.

Many other names are used for colour and patterned varieties of opal but as no two people agree about any of them and because they are of no value in testing they will be ignored here.

Opal with no play of colour is *common opal*: where opal occurs as a silica replacement of organic or other material (shell, plant material) it is said to be *pseudomorphous* after that material. Many opal pseudomorphs after fossils (belemnites in particular) are very beautiful.

Identification of opal

While in principle there is no reason why some of the standard gemmological tests should not be used on a suspected opal, the porosity of the stone means that it will take up some of any liquid used, so that refractometer tests are ruled out. As opal is not crystalline some of the other tests will be irrelevant. We need to fall back, as always, on magnification, though if you really want or have to use it, the hydrostatic weighing method of obtaining specific gravity can be used (described, with other tests, in Read, 1992).

Opal is silica with a variable amount of water, with a specific gravity generally near 2.10 (as we shall see, some imitations give a much lower figure). The refractive index is near 1.45. These figures vary to a small extent between colour varieties and locations but the differences do not affect testing. Some opals have been found to give a green fluorescence from included uranium-bearing minerals but again this is interesting but not significant.

The variable amount of water present in opal (1–21 per cent chemically combined) can affect its mechanical strength. Many opals develop a fine surface crazing which is not easy to prevent and impossible to reverse.

Some of the finest examples of opal replacing wood come from the Virgin Valley opal deposit in Nevada, USA. The form of the original twig is easy to see and the play of colour is exceptional: however, some examples at least are not very stable and tend to craze – keeping specimens in water or glycerine is sometimes recommended.

The secret of synthesizing opal was discovered only after the nature of the natural structure was fully understood in the mid-1950s. The play of colour is caused by regularly stacked silica spheroids forming a three-dimensional array which diffracts white light passing through it into its component spectrum colours. This array is silica in natural opal but such an array can be formed from a variety of other materials – one will be encountered below. Synthesizers are able to control to some extent the colour which will predominate in their product – this is why many synthetic opals show a good deal of red which is the colour most in demand. Sphere size determines the colour seen.

Opal is imitated by many different materials but most people who have seen a number of opals would not be easily taken in. By far the commonest is glass which can be made to look like anything you want it to.

Composites: opal doublets and triplets

The opals *doublet* and *triplet* are familiar to all jewellers and are an excellent way of using material which is often of very high quality. It often seems as if the finest opal occurs as the very thinnest coatings on sandstone rock and it is preferable to cut such material together with the sandstone. These are called doublets.

The opal in many cases needs some form of protection (the hardness of opal is a little over 6) and the opal on the rock can be covered by a transparent convex cap, made in the best qualities of rock crystal, or by glass or plastic as the quality abates. In addition to protecting the opal the cap forms a convex lens which will magnify the coloured patches. The doublet can most easily be recognized by its characteristic flat upper surface: opal is hardly ever faceted (we have seen a brilliant cut example) but the cabochon form suits it perfectly, water opal in particular having a steeper dome than other varieties. Fire opal is most often faceted, however.

Examination of the individual patches of colour in natural opal will show parallel lines: the similar patches in a synthetic opal show a highly characteristic hexagonal pattern which will be discussed further below.

As it plays an important part, synthetic opal will be examined before the various imitations. It can be used in a doublet or triplet. The first synthetic opal appears to have been made by Pierre Gilson in the late 1960s and I (MO'D) remember seeing a most spectacular specimen in which the spectrum colours (all were present) did not form the usual country or state-map pattern but rather appeared in long sub-parallel bands. The Gilson opal which is now familiar came onto the market in 1974. White opal appeared first, then black. The quality is excellent.

As previously mentioned, the colour patches in the Gilson and other synthetic opals show an underlying honeycomb-like pattern which can easily be seen by the microscope, which will also show whether or not the colour patches are the upper surfaces of columns which extend to the base of the stone. If they do not the stone will be one of the two forms of composite – this test applies to natural opal too.

As black opal is considered the most desirable it is the most frequently imitated, often quite ingeniously. Some composites are given a dark base to show off the play of colour as in natural black opal, the base being black glass or even black or dark onyx. The refractometer can of course be used on a flat base if you are sure that it does not form part of the upper section if it seems likely to be natural opal. If the opal is covered by a plastic cap the investigator should be able to see scratches on the surface.

Again with the intention of approaching black opal from less promising beginnings, white opal with a reasonable play of colour can be effectively darkened by the incorporation of carbon which enters the porous interior and shows under magnification as dark spots. How the carbon spots get there can arise from a number of different methods but probably the commonest, as the materials are easily available, is for used car sump oil to be heated with the stones immersed: the hydro portion of the hydrocarbon volatilizes leaving the carbon behind. Brown paper can also be used as a source of carbon (it is burnt).

Whatever the method used to darken the opal, two interesting problems remain, one easily solved: the insoluble problem is why no name has been given to the product which has for years been known as *treated opal matrix*. Identification is easy since the spots of carbon can be seen with the hand lens – if the investigator is on the look-out for them.

Opal with no or little play of colour has been immersed in colourless oils or waxes: glycerine has been used but tends to ooze out of the stone and turn the surface sticky. None of these processes gives a convincing result. Various impregnations have been attempted over the years but are not really worth while.

We found above that opal is often found with the appearance of wood or fossil, including shells, belemnites and bones: silica has taken over the original organic material. The opal is to be *pseudomorphous* after the original. Specimens are widely collected but not imitated as far as we know.

Imitations of opal

Opal imitations made from glass are not entirely convincing but must have succeeded because in the main fine opal is not often seen and because they look quite attractive in themselves. Glass may be foiled by mother-of-pearl cemented to the base: a mother-of-pearl triplet can be quite convincing. Natural opal can merely be darkened on the back by any convenient substance. The 'Schnapperskin' triplet uses a dyed fish skin – examples are now hard to find. Small glass or plastic tubes containing a liquid (glycerine or water) make a convenient way of displaying small fragments of opal. Fire opal is easily imitated by glass.

If an opal imitation cannot be identified by magnification the thermal reaction tester can be used (with care). Plastics will ooze: they can also be easily scratched by a needle. A report by Nassau in 1994 describes a Brazilian plastic-impregnated opal which was found to contain crystals of a nickel–iron sulphide. This will not be an everyday occurrence nor will the relevant tests be available to the gemmologist but such items do show that work on the development of imitations is constant.

While the best imitations of opal are made from silica, other substances have been used from time to time. One such was made from latex (artificial rubber) which was able to provide an array of spheres: it had the very low SG of 1.0 and felt very light. Another imitation made from a copolymer of styrene and methyl methacrylate had an RI of 1.465. It was identified by IR spectroscopy. Yet another had RI 1.48 with SG 1.17.

Examination of two other imitations of opal, both made from plastic, showed that one did not have the hexagonal patterning in the colour patches while in the other the colour domains were scattered in such a way as to suggest natural opal. The main constituent of the latter appeared to be colourless polystyrene but both specimens showed an absorption region extending from 590 to 565 nm: the SG was 1.18 and the RI 1.485. Under LWUV a bluish-white fluorescence could be seen.

When you have examined a great many opals any peculiarity of colour is more likely to attract attention. Many synthetic opals appear to have a blush of a single colour which sweeps across the stone as it is tilted. We have already noted the very early Gilson opal whose striped colour pattern was apparently not often repeated – this specimen did not show a blush. Gilson also produced some synthetic fire opal with an orange body colour, some specimens showing a play of colour. Water opal has also been produced by Gilson – this has been very successful.

Slocum stone

One of the best imitations of opal around today is Slocum stone, a glass containing very thin laminated pieces of metal foil which at times do quite closely resemble opal: in most specimens the resemblance is not very close but all Slocum stones look very attractive and a parcel full of mixed stones seen under a single source of light so catches the eye that they appear to be something important! This is one of the best imitations of any gemstone that we know and the stones are full of personality!

Slocum stone, which has been marketed under the name Opal-essence, has an RI of 1.49–1.50, SG 2.4–2.5. Under magnification gas bubbles and other features of a glass can easily be seen. The foil shows a purplish colour by transmitted light.

In general synthetic opals do not exceed and in some cases do not match the properties of the natural material (RI, SG and hardness). While some natural opal will show a yellow fluorescence under LWUV light this does not always happen, nor can it be used as a test to distinguish natural opal from its imitations or from the synthetic material. The hexagonal patterning should always be looked for (some texts call this 'chicken wire' or 'lizard skin' as mentioned earlier). The material 'Opalite', a common glass imitation of opal, is very often used as the backing in composites.

Reports of interesting and unusual examples from the literature

Items in this section have been chosen to illustrate points made in the chapter and to bring one-off items to your notice.

A milky white opal on a dark blue sodalite base formed a doublet to give a black opal appearance – both materials were of Brazilian origin. Another interesting example showed a pinfire effect (very small specks of different colours) characteristic of Brazilian opal from the state of Piauí: this natural material is reported to be robust compared to opal generally and also notably resistant to heat. In the imitation a green colour was predominant and the body colour was uniformly black. The stone, tested by GIA and reported in *Gems & Gemology* (Fall, 1990), gave an RI of 1.45 for the dome and 1.56–1.57 for the base.

Under LWUV light the dome fluoresced a very strong yellow–green while the base fluoresced a very strong chalky bluish-white. Reactions to SWUV light were similar but weaker and there was no phosphorescence. The surface yielded to the point of a pin and the SG was 1.91. The whole stone was coated with a transparent substance which was thicker on the base and contained gas bubbles. With a hotpoint an acrid smell characteristic of plastic was produced.

The rather uncommon chatoyant opal can be enhanced by fitting on a colourless chatoyant cap which acts as a condensing lens. Some Idaho opals are chatoyant and the effect is most attractive.

Opal pseudomorphous after shell is interesting enough to make a simulant likely. One such is made from cemented chips of white opal, boulder opal and matrix in a transparent colourless binder which was found to be a plastic.

GIA reported an assembled black opal which was found to be a triplet with a wavy separation plane and a fairly flat top: this was natural and had an uneven surface. An ironstone backing was joined to this surface with a cement tinted to match the colour and appearance of the ironstone. As the cement filled the uneven surface of the opal top a single-stone appearance was achieved. The joining material melted when lightly touched with the thermal reaction tester.

Gilson transparent brownish-orange opal described by GIA in 1985 resembled the finest Mexican fire opal. However, from the side the stone showed thin colourless top and bottom layers with no play of colour. The centre section was brownish-orange with different colours confined to distinct areas. When viewed through the colourless areas the characteristic structure of synthetic opal could be seen. The sections showed variable hardness and in their response to

UV light the colourless surface area was easily marked with a pin and flowed when the needle of the thermal reaction tester was held directly above it.

Opalite often produces its colour effect from a mosaic with a clear top and wax-like base. The mosaic in the case of a specimen reported by the *Journal of Gemmology* in 1993 consisted of pieces of natural opal. The adhesive used to hold them together was found to phosphoresce.

The manufacture of an encapsulated opal using Mexican material from Jalisco was discovered at the 1991 Tucson Gem & Mineral Show. A slice of colourless, white or orange opal was contained in an oval single cabochon (with a flat back) made from acrylic resin. It appeared that some of the liquid resin was poured into a dome-shaped mould, followed by the insertion of a slice of opal with its base coated black to give a contrast to the play of colour. A second thinner resin layer is used to seal the assembly and to provide a base. The material was said to be made in Jalisco.

A plastic imitation of opal was found on sale in Thailand. It contained clearly spherical bubbles and showed an RI near 1.57. There was a strong chalky bluish-white fluorescence under LWUV. The hand spectroscope showed fine absorption lines across the entire spectrum. The thermal reaction tester produced an acrid smell revealing that the material was a plastic.

A blue enhanced opal was on sale at the 1992 Tucson Show, the starting material rumoured to be Brazilian. The stone was very porous and was a chalky white hydrophane opal with a weak play of colour. Hydrophane opals need to be immersed in water to produce the full play of colour. The Spring 1992 issue of *Gems & Gemology* describes a specimen in which the rough material was said to have been soaked in a mixture of potassium ferrocyanide and ferric sulphate to give a dark blue colour. When dry the material is placed in a slightly warmed plasticizing liquid of methyl methacrylate with some benzyl peroxide. This closes the pores and clarifies the opals to near-transparency. The rough is removed and cleaned for fashioning into cabochons before the mixture solidifies. The blue body is sufficiently dark to appear black. Specimens feel like plastic and will show the blue colour under intense light.

An effective translucent simulant of natural opal showed an unusual purplish-brown colour with no inclusions of any apparently natural material (this would be the pink sandstone characteristic of Lightning Ridge, New South Wales) and the typical structure of synthetic opal similar to that of the Gilson material. It seemed likely that the opal simulant was made in South Australia.

Dyed opalized sandstone and conglomerate is being offered in Australia, the material being a naturally occurring slightly porous silicified sandstone or sandy conglomerate in which the clay matrix and some quartz grains have been partially replaced by precious opal cement. The whole piece has been treated to darken the colour and enhance the play of colour. Opal-cemented sandstone often shows columns not unlike those seen in some Gilson synthetic opal. The columns are easiest to see when the specimen is viewed at right angles to the bedding. The material is readily distinguished from other forms of opal by the presence of abundant sand-sized quartz grains or clasts. The material does not take a polish so will most likely form part of a composite or be coated with a clear polymer.

Gems & Gemology, 45(3), 1996, gives three methods in which opals can be enhanced. Stones may have their lost water replaced under vacuum conditions or they may be artificially coloured or impregnated. Porosity, especially in opal from Andamooka, South Australia, allows dyeing: this can also be carried out on Brazilian and Mexican hydrophane opal. Cracked opals can be treated with oils and resin to improve translucency and play of colour. CIBJO rules that dyed or resin-treated opals must be specifically declared.

A synthetic opal from China is green or pink and shows regular and sharply defined colour patches with the characteristic hexagonal synthetic opal patterns and columnar structure.

The opals are made of closely packed silica spheres about 320 nm in diameter: the RI is 1.450–1.468 and SG 1.80–1.90. A synthetic opal from Russia closely resembles natural opal and is composed of strongly distorted silica spheres with an approximate diameter of 240 nm. The RI is 1.440–1.450 and SG 1.74–1.86.

Fire opal from the area of Campos Grande, Rio Grande do Sul, Brazil, has an orange–red to fire-red colour. The RI is 1.440 and the SG 2.01. Colour zoning and swirls can be seen.

Fire opal from the Shapane Mountains of Turkey is reddish-orange and yellowish-red, similar to Mexican fire opal. The RI is 1.442–1.446 and the SG 2.00–2.02.

Blue opal with no play of colour is from the Acari copper mine, Peru. The colour is thought to derive from finely dispersed inclusions of Cu^{2+} and the properties are in the same area as for other natural opals.

Pink opal from Peru is reported to show RI 1.45–1.49 and the SG 2.16–2.22 with a hardness of 6. It is translucent to opaque and shows a pinkish-white luminescence under LWUV light. Adjoining veinlets of colourless opal fluoresce bright green in SWUV light.

Opal of a greenish-blue colour, also reported from Peru, contains chrysocolla inclusions and the colour is probably due to copper. Specimens are collected in a copper-mining area. The blue opal has RI between 1.430 and 1.452 with some specimens showing an anomalous birefringence up to 0.005. The SG is between 2.01 and 2.08; specimens are collected in a copper mining area. Analysis (probably X-ray diffraction) has shown that the Peruvian pink opal is a mixture of palygorskite, opal and chalcedony but the cause of the colour was not known at the time the report was published.

Opal from the Banten area of West Java, Indonesia, shows a 'reverse hydrophane' effect in that the play of colour disappears when the stones are immersed in water, reappearing after about 1 hour after drying out.

Opal with dendritic inclusions has been reported from south-eastern Zambia. It is transparent to translucent, yellow to brownish-yellow with dendrites of manganese oxide.

A notably porous green opal from Serbia showed green through the Chelsea filter and gave an RI of 1.46 with SG 2.20. Inert to SWUV light, the material showed an even, faint chalky green under LWUV light. The absorption spectrum showed cut-off edges at 620 and 450 nm. Some healed fractures and veining could be seen under magnification. Nickel was found to be present and probably influenced the colour.

An opal imitation marketed as Opalus in Australia is made from a plastic foil in a colourless plastic material cemented to a black background.

A synthetic fire opal with a play of colour has been made by Kyocera: it shows the usual columnar structure and hexagonal patterning within the colour patches. RI is 1.459–1.460 and the SG 1.87–1.91. The IR spectrum showed a peak ascribed to an artificial resin. The colour of the opal is mainly orange with some specimens showing a yellow–green colour flash and some others a definite red play of colour.

A fire opal from Opal Butte, Oregon, USA, has an arrangement of silica spheres of different diameter up to 250 nm with a high content of calcium. The RI is 1.431 and SG 2.05. Some organic material may be included.

Opal from a number of locations was briefly reported in *Revue de gemmologie* in 1999. They included a mauve opal from Androy in the south of Madagascar, blue–green opal from Acari, Arequipa, Peru, pink opal from Montana, USA, Quincy Cher, France and from Mapimi, Mexico. A yellow opal is reported from Saint-Nectaire, Puy-de-Dôme, France.

A new opal nomenclature has been formulated by the Opal Nomenclature Sub-Committee of the Gemmological Association of Australia. In the category *Types of natural opal* three forms of natural opal are listed which are treated only in so far as they have been cut and polished. These types include opal having a substantially homogeneous chemical composition, boulder opal and matrix opal. The next category, *Varieties of natural opal*, describes the opal's body tone (the face-up relative darkness or lightness, ignoring the play of colour and its transparency).

Body tones N1 to N4 are used for black opal, N5 and N6 for dark opal and N7 to N9 for light opal. An opal with a distinctly coloured body has a hue notation appended to its body tone classification. Transparency codes A, B and C are used to cover all forms of diaphaneity ranging from transparent to opaque.

The nomenclature, though not associated with the descriptive classifications for natural opals, also describes opal treatments, composite natural opal, synthetic opal and imitation opal.

A new version of Opalite described in the *Journal of Gemmology* for October 1993 shows a clear plastic or acrylic layer covering the whole surface of the cabochon and has a refractive index of 1.50 compared to the 1.44 of natural opal. It does not display the phosphorescence shown by the majority of white opals. When the flat back can be seen and examined a green foil-like flash shows that the specimen cannot be natural opal.

Impregnated black opal rough with a play of colour gave two separate refractive indices of 1.44 and 1.50, the latter value being attributed to the impregnating material. The SG was 1.82, slightly lower than the lowest values obtained from Kyocera treated synthetic opal. Specimens gave a faint orange fluorescence under SWUV light – this was consistent with Kyocera test samples. Under magnification the characteristic lizard-skin effect and columnar structure could be seen and FTIR analysis showed several absorptions between 6000 cm^{-1} and 4000 cm^{-1} which are absent from natural opal. The specimen was presumed to be an impregnated synthetic opal.

A Hyperfine Analytical Network Instrument has been used to plot diffraction spectra; distinguishing natural from synthetic opal and characterizing rare and unique specimens may be possible by taking spectrometric data from diffraction patterns.

Imitation opal in which the play of colour emanated from a holographic projection from an aluminized grating sandwiched between plastic covers gave an RI of 1.48 and SG 1.2. These values were taken from the plastic cover. The grating can be seen under the microscope.

The uncommon opal varieties cacholong and prasopal have been reported from the Oman mountains near Muscat/Oman. The stones have a notably high porosity and are sometimes used for holding scents.

Opal's clarity and durability are said to be favourably altered by impregnation with a resin of similar RI.

Chapter 6

Quartz

The quartz family of gemstones has a simple chemical composition (SiO_2) and can be divided into single crystalline and crypto-crystalline varieties. The single crystalline quartzes include such old friends as rock crystal, amethyst and citrine with rose quartz and a green transparent variety that has gone under several names over the years.

The crypto-crystalline quartz varieties include chalcedony and agate with many other varieties all distinguished from one another either by colour, patterning or both. We shall make no attempt, even the smallest, to list the names that have been used for them, preferring to keep the oldest ones and those about which there seems to be the least ambiguity. For the student, crypto-crystalline means that the material is made up of extremely small individual crystals, completely resolvable only with high magnification. This structure affects some of the physical and optical properties of the specimens but not in so great a way as to make testing difficult.

With so large a range of varieties both transparent, translucent and opaque, the testing of the quartz gemstones might be thought to pose a number of problems. However, as material is so plentiful there is little impetus towards imitating any but the most desirable varieties such as amethyst and citrine, both of which have been synthesized. There is little need to imitate such materials as jasper or agate but where the odd one exists it will almost invariably be glass.

Rock crystal

Rock crystal can occur in very large crystals, this facilitating the creation of large carvings, crystal balls and even skulls, as well as faceted stones, caps for the best quality opal triplets and so on. Crystals are common enough to be collectable for their own sake without big business getting to the deposits first, and the finest ones, clear and showing a multiplicity of crystal faces, have been imitated by glass. A very good example is the 'Herkimer diamond' found in New York state, USA – this has a small book devoted to it, the book having a small crystal fixed to the back cover so that the reader can view it from the front as if looking down a mine shaft to see the crystal below! (Claude H. Smith, *Let's Hunt for Herkimer Diamonds*, 1950). Rock crystal has an easy to read RI of 1.544–1.553 with a DR of 0.009 and SG 2.651. We have cited these figures to the third place of decimals for no pedantic reason but to show the reader that quartz, with its simple composition, is straightforward internally. The hardness is 7 on Mohs' scale and the dispersion low. These constants apply to amethyst and other varieties of single-crystal quartz.

In fact quartz crystals play some tricks on the investigator and some of the synthetic varieties, especially of amethyst, are not very easy to distinguish from their natural counterparts. We shall describe synthetic quartz later but while we are still not too far away from discussion of the crystals we ought to remember that they play an important role in micromounting and general crystal collecting and the reader should try to look at as many as possible.

Rock crystal shows no absorption spectrum but may contain inclusions so large and attractive in themselves that large, clear pieces with included golden to yellow–brown rutile (Venus hair stone) or green tourmaline or actinolite crystals are used in large-scale creations for commercial premises and other places where they will catch the eye. These pieces cannot be said to be easily imitated and scarcely need testing. Rock crystal on a smaller scale and with completely clear material easily available makes excellent spheres often used as beads in necklaces. Testing for rock crystal spheres is one of the most satisfying of all gemmological tests (providing the material turns out to be rock crystal rather than glass!). The spheres act as their own condensing lens so that when they are placed between crossed polars on the polariscope and correctly positioned in the right orientation they show the almost unique uniaxial interference figure of quartz – a black cross with slightly splayed arms reaching the edge of the field and with a hollow centre filled with coloured concentric circles radiating outwards.

No other transparent gem material shows the hollow centre – readers are unlikely to encounter fashioned berlinite which has the same crystallographic structure as quartz. This test would show very well between crossed polars if a crystal ball was the subject but most are too large to go on the polariscope. The same effect can be seen with faceted stones and with coloured quartz single crystals but may be harder to see if the body colour is strong.

A simpler test for rock crystal spheres is to see whether the subject is rock crystal or glass: merely introducing a card so that the edge can be seen doubled through the ball – make sure to move the card in several different directions – or you may be able to see lines on a page doubled, again in the appropriate orientation.

If the comparison with rock crystal is made with glass, as it is most likely to be, glass, as a poor conductor of heat, will feel warmer when touched lightly with the tip of the tongue compared to rock crystal or any other crystalline material, since crystals conduct heat more effectively than amorphous substances. Glass does not actually feel warm but is less cold than rock crystal. In the days when heavy liquids (i.e. denser than distilled water) were useful adjuncts to gem testing it was useful to keep a small amount of bromoform diluted to 2.65 with a piece of rock crystal suspended but unattached as an indicator: specimens of unknowns could be introduced into the liquid and their behaviour compared with that of the rock crystal: bromoform may now be used only under strict laboratory conditions.

Rock crystal will show a perceptible birefringence on the refractometer while

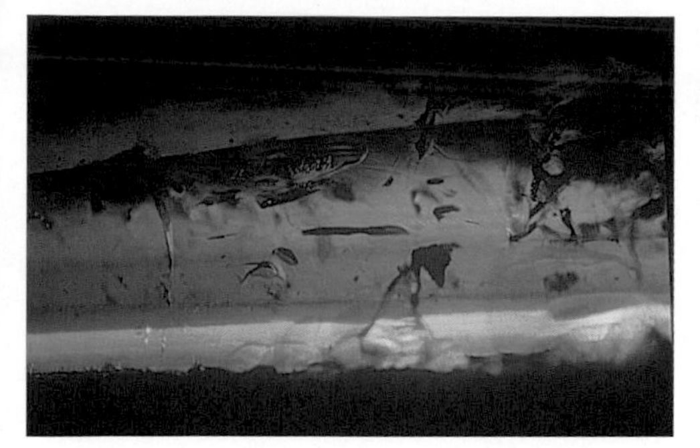

Plate 1 Swarowski hydrothermal emarald-two-phase inclusions

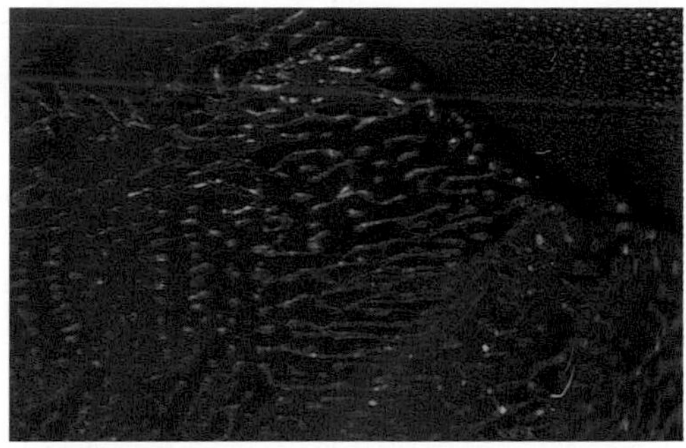

Plate 2 Flux feather close-up in flux-grown ruby

Plate 3 Synthetic moissanite-doubling of opposite facets

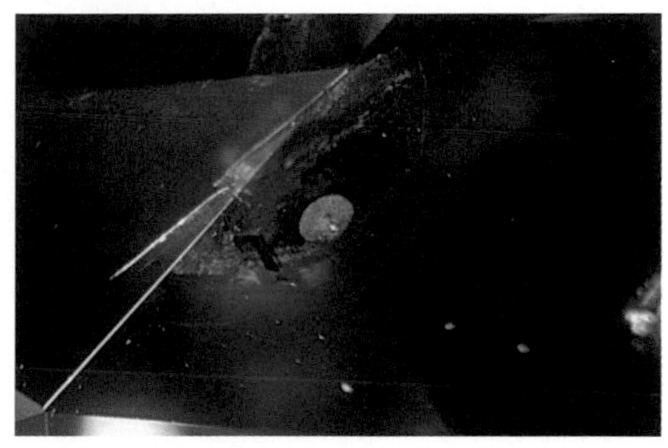

Plate 4 Laser-drilled diamond

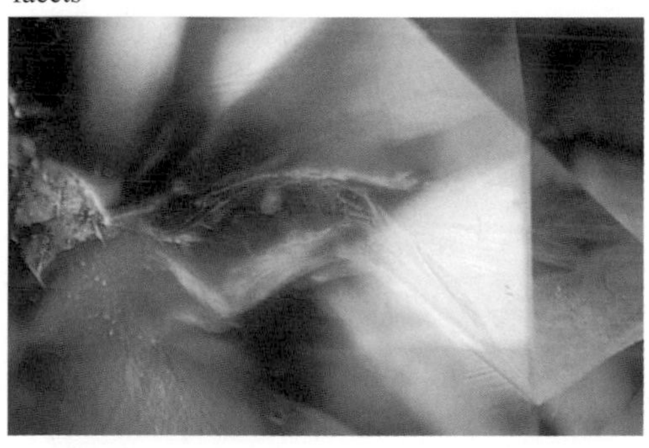

Plate 5 Infill in diamond

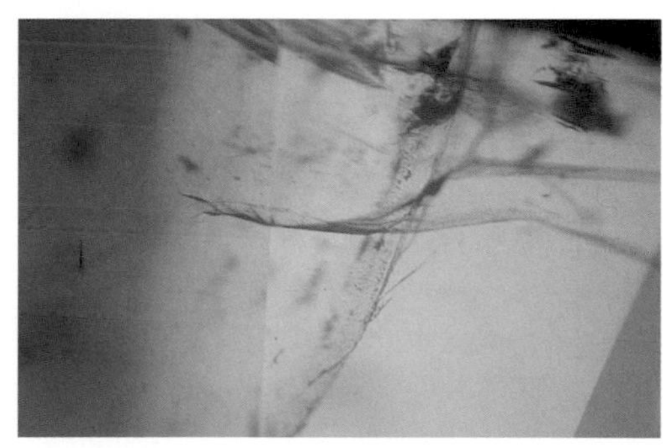

Plate 6 Glass infill in diamond

Plate 7 Monochromatic flash in cubic zirconia

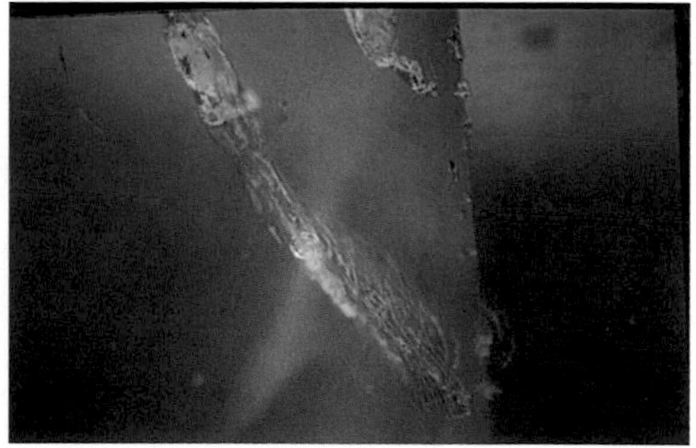

Plate 8 Cavity in emerald filled by Opticon

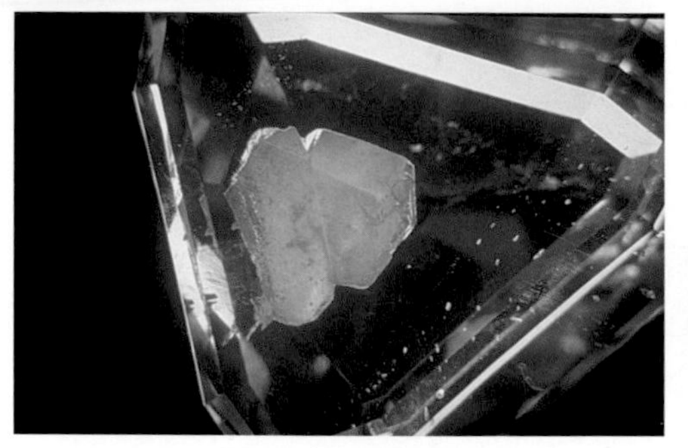

Plate 9 Albite twin crystal in topaz

Plate 10 Aqua aura and untreated rock crystal

Plate 11 'Mandarin' (spessartine) garnet from Nigeria

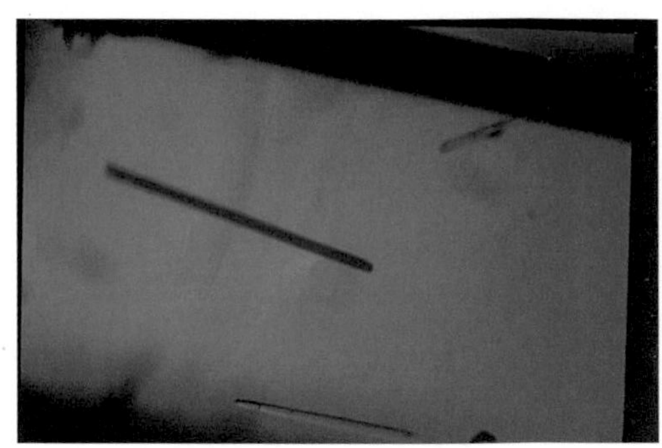

Plate 12 Acicular inclusion in a natural emerald

Plate 13 Tremolite crystals in natural emerald

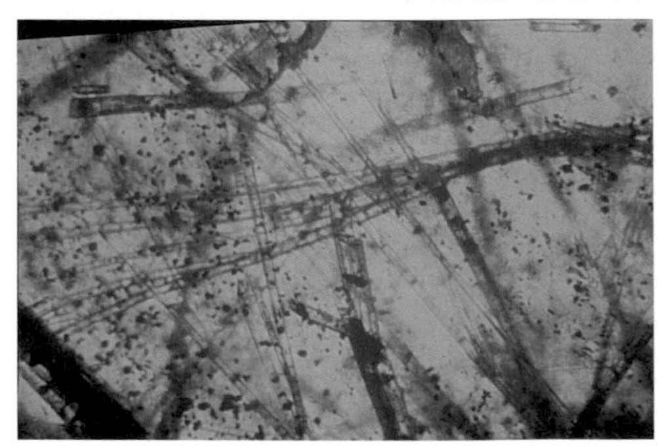

Plate 14 Inclusion scene in natural emerald

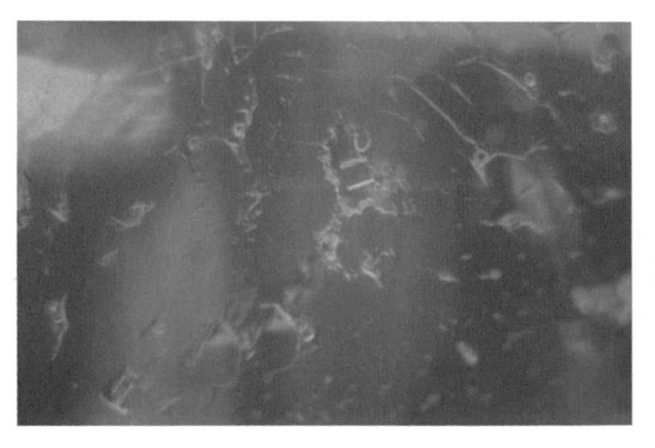

Plate 15 Three-phase inclusions in natural emerald

Plate 16 Signs of devitrification in glass

Plate 17 ‘Rain’ in natural aquamarine

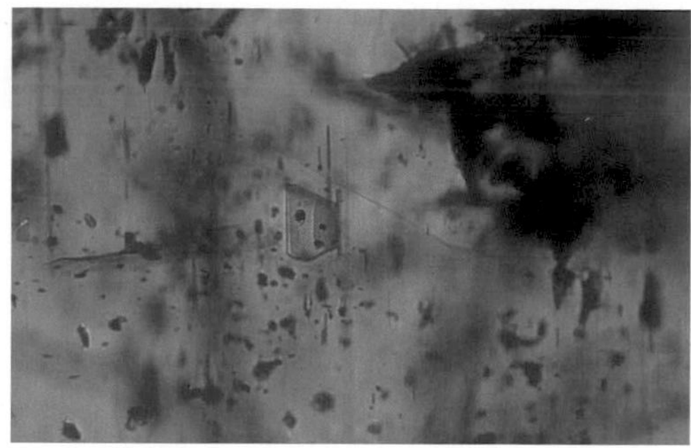

Plate 18 Phlogopite crystal in Madagascar aquamarine

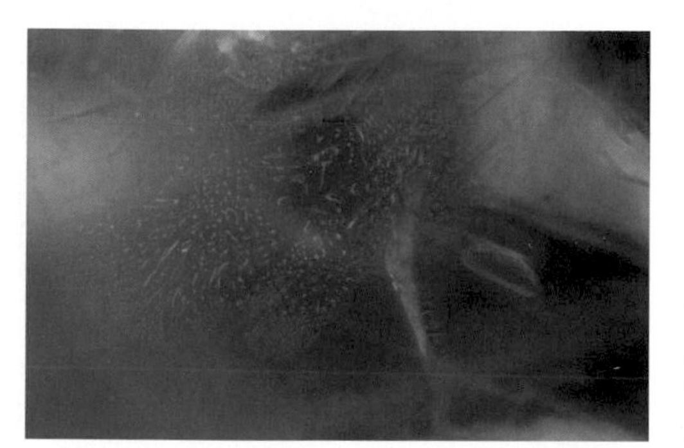

Plate 19 Synthetic aquamarine showing 2-phase inclusions and feathers

Plate 20 Inclusions in Sandawana emerald

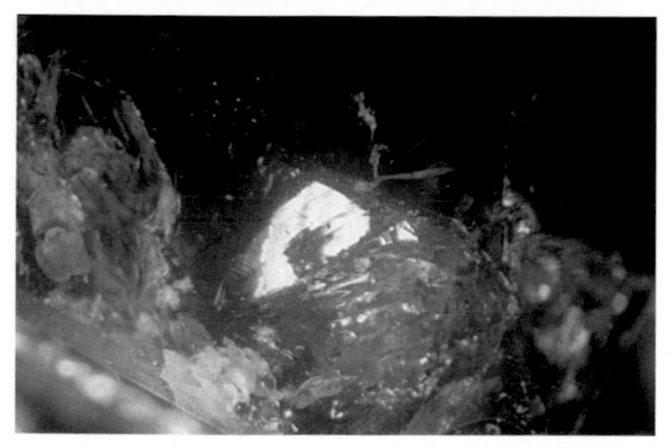

Plate 21 Oil in emerald

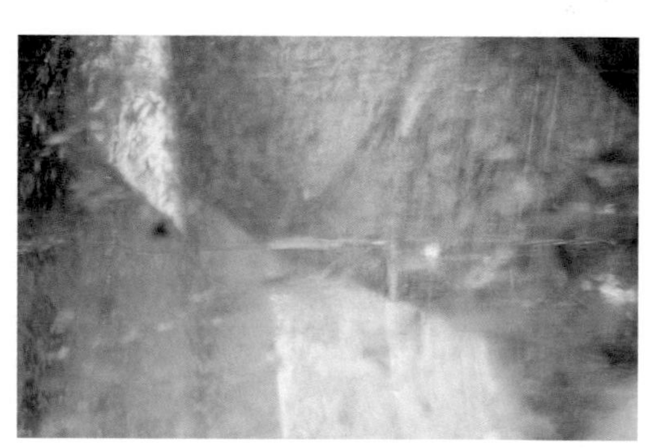

Plate 22 Flash effect in filled emerald

Plate 23 Russian hydrothermal emerald showing seed plate

Plate 24 Seiko synthetic emerald

Plate 25 Bubble trails in synthetic flame-fusion sapphire

Plate 26 Diffusion treated sapphires: untreated: treated, unpolished: treated, polished

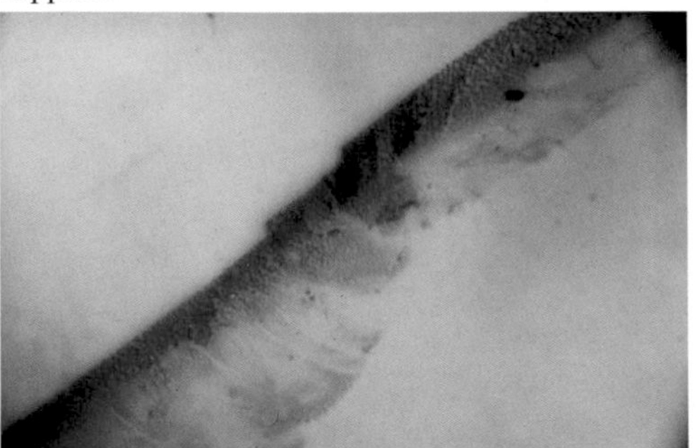

Plate 27 Facet edge of diffusion-treated sapphire

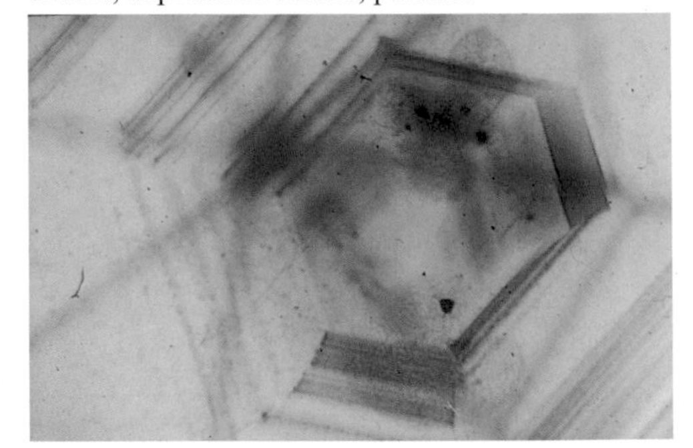

Plate 28 Zoning in natural yellow sapphire

Plate 29 Bohmite in natural sapphire

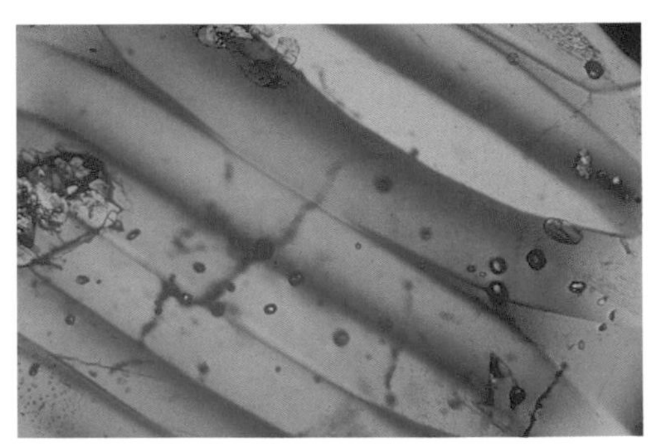

Plate 30 Twin planes in natural sapphire

Plate 31 Rutile crystals in natural sapphire

Plate 32 Zircon crystal and zoning in Sri Lanka sapphire

glass will show none and if a faceted specimen is being examined under the microscope doubling of back facet edges and inclusions will not be difficult to see.

Transparent to translucent quartz may contain inclusions other than rutile and tourmaline. Some specimens contain blood-red crystals of hematite, others may be chatoyant from included fibres and rose quartz sometimes gives a six-rayed star often best seen by transmitted light (star corundum appears best by reflected light) but I have seen a good deal of star rose quartz looking good with reflected light.

All these varieties can quite easily be distinguished from imitations but most of them are not imitated, though we have recently seen a convincing rose quartz imitation made from translucent pale pink GGG!

Smoky quartz

The variety smoky quartz could be taken as rock crystal with an interesting 'smoky' colour (if you are used to brown smoke!). It would be interesting to know where the name originated. More to the point, the material is still sometimes offered as smoky topaz, mostly at gem and mineral shows I imagine. The colour of smoky quartz is caused by an aluminium impurity together with either natural or artificial irradiation. There is no way of distinguishing between the two for the gemmologist and it is not worth devoting research time to the problem, if there is one. Heating removes the colour. Though we shall meet the comparison again, the smoky quartz RI and SG differ considerably from the 1.62–1.64 and 3.53 of topaz.

Amethyst

The quartz variety which can give the most trouble to the gemmologist is amethyst, which is also the most valuable form of quartz. Reaching large sizes on occasion, amethyst occurs in a variety of shades, many of which are attractive and appeal to different markets. Though sometimes announced as a new discovery (gemmologists are constantly discovering changes of colour when, in fact, most coloured minerals vary quite a lot according to the nature of the viewing light) amethyst is usually a reddish-purple in tungsten light and bluish-violet in daylight or fluorescent light. The name Siberian is still sometimes used for the best colour (presumably not in Siberia) and both Uruguay and Rose de France are occasionally encountered – the latter sometimes given to a synthetic Verneuil-type corundum of near amethyst colour. For the record, this would have RI 1.76–1.77, SG 3.99–4.00.

Natural amethyst is almost always colour banded and many crystals call for considerable skill on the part of the lapidary to make the best of it. An amethyst-coloured specimen without colour banding should arouse suspicion. The name *prasiolite* has been given to a rather dull transparent green material which, though it may occur naturally, is usually obtained by heating amethyst to 400–500°C – not high quality amethyst, of course. It should not be confused with any other gemstone or any other confused with it – in any case it is easily tested and if a really awkward example turned up a mistake need not be too expensive!

Citrine

Citrine is the best-known coloured transparent quartz variety after amethyst and may in fact have started life as that colour though this is of no importance in the commerce of gemstones. Citrine can be very beautiful and though some specimens may closely resemble topaz the greater hardness and refractive index of topaz makes some difference as perceived by the eye. Topaz, admittedly, may contain some rainbow-like interference colours showing that cleavage has been initiated inside the stone: quartz has no pronounced cleavage. The RI and SG of topaz, at 1.62–1.64 and 3.53, effectively separate the two species. The colour of citrine comes from Fe^{3+} as an impurity: Fe^{2+} gives the green colour in prasiolite.

The best citrine should be strongly coloured either a rich yellow or orange–red (the faintest suggestion of orange or even red always gives a desirable colour). At one time dealers used the name Madeira or Rio Grande for reddish-brown citrine, these denoting citrine derived from the heating of amethyst (dealers must have been hanging around at the mine to know that!). Today such names hardly occur (Palmyra was used for the darker orange–yellow citrines obtained by heating amethyst).

In both citrine and amethyst, fashioned stone sizes can be quite large. While citrine can be imitated by synthetic corundum, though not too much of this comes up since there would not be much profit in growing and marketing such stones, especially now that it is possible to synthesize quartz varieties by the hydrothermal method.

Quite a number of natural yellow stones, as well as *topaz*, can be confused with citrine, though in commercial terms no great harm would be done by a misattribution. The colours of citrine vary so greatly that almost any transparent yellow stone, apart from diamond, could be confused with it. Here are some of them.

Yellow or golden beryl has quite a citrine look about it and the RIs of both are low compared to that of topaz at about 1.62 (beryl is 1.57) with SG around 2.7 for beryl.

Both natural and synthetic yellow *corundum* could be taken for citrine but the higher RI and SG (1.76, 3.99) and, hopefully, a visible iron absorption spectrum or even, with some Sri Lanka yellow sapphires, an apricot-coloured fluorescence under LWUV light should make the distinction easy.

Yellow scapolite does give the investigator quite a problem; not only do the RI and SG overlap but the marked yellow fluorescence shown by Burmese yellow scapolite and some of the material from Tanzania may not be noticed. These is no distinctive absorption spectrum for either species so the tester must determine which of the two stones is positive or negative on the refractometer – see Peter Read's *Gemmology* (second edition, 1999).

It is possible that some yellow varieties of *fire opal* might be mistaken for citrine if seen from a distance and not handled (when the lower specific gravity of 2.10 would make the difference apparent). Opal's lower hardness ensures that facet edges are rarely as sharp as they would be in citrine. If there is any suggestion that a yellow unknown may be fire opal a refractometer test is not recommended.

Some of the African bright yellow to orange *grossulars* can resemble citrine but show a higher RI at around 1.74 with no birefringence: hessonite, also a grossular variety will show a similar reading and citrine will not show the almost invariable hessonite inclusions of apatite and calcite crystals in profusion. Among other yellow–orange natural possibilities are *zircon* (much stronger birefringence plus high dispersion); possibly *chrysoberyl* when very dark orange (such attractive specimens do exist but have a higher RI and SG); *sphene* (titanite) is too birefringent and dispersive but a near-citrine-coloured example is in the Pain Collection of gemstones in the Natural History Museum in London – sphene can sometimes show a rare earth absorption spectrum.

Aventurine quartz

The name aventurine or aventurine quartz is given to a bluish-green translucent material marked by minute light-reflecting crystals of a chromium-bearing mica group mineral (the name fuchsite has been dropped as a species name at the recent revision (1998) by the Commission on New Minerals and Mineral Names of the International Mineralogical Association).

No other material seriously imitates aventurine quartz, the best of which comes from India. Some aventurine may show a chromium absorption spectrum though this will not be very strong. The rock quartzite of which aventurine is in fact a variety may, even when colourless, be fashioned to make use of the bright spangly appearance.

So far we have not mentioned the propensity of silica to replace other minerals and sometimes organic substances. The replacement of the asbestiform mineral riebeckite preserves the fibrous structure of the original mineral but in appearance only – the soft fibres themselves are completely replaced by silica. The silica is said to be *pseudomorphous after riebeckite*. This would give the material an interesting appearance by itself but when it is coloured, presumably by hydrous iron oxides, a variety of yellow to brown and black stripes gives the celebrated tiger's-eye. A blue form has been called hawk's-eye but is less arresting and less often seen. Heat treatment may give a wider variety of colours. Tiger's-eye can be bleached with chlorine to lighten the colour.

Rock crystal containing parallel inclusions of strong yellow fibres of goethite makes an attractive ornamental stone: it has sometimes been called cacoxenite quartz from the mineral of that name. The fibres can often occur as inclusions in amethyst.

Quartz can be chatoyant though the eye is normally perceptibly less sharp than that shown by chrysoberyl (the only cats's-eye that needs no qualifying adjective – thus quartz cat's-eye, tourmaline cat's-eye, etc.). With a flat back the quartz cabochons should give the quartz refractive index: at least a shadow-edge at 1.55 compared to chrysoberyl's mean RI of 1.75. Chrysoberyl has a specific gravity of about 3.71 compared to that of quartz at 2.65. If no refractometer is available, the spectroscope will show a strong iron band in chrysoberyl, at 444 nm. Quartz cat's-eyes are usually pale yellow or straw colour rather than the dark honey green–gold–brown shown by the finest chrysoberyl. The eye in quartz may be coarser as it is reflected from much less fine channels than the slender ones giving the same effect in chrysoberyl. Many chrysoberyls have a deeper base than quartz which is usually polished flat – the lumpy base is left to conserve as much weight as possible and thus increase the selling price. Quartz cat's-eyes, which may be a grey–green or brown colour, are not imitated in themselves nor have there been any reports of their synthesis.

Synthetic amethyst and citrine

Both amethyst and citrine are grown by the hydrothermal method, which leaves very few clues behind. Over the years there has been a good deal of discussion about how to identify synthetic amethyst – the absence of natural inclusions is the best clue.

To recognize a negative quality we have to learn to identify those inclusions which the natural material often shows. Natural amethyst may contain colour zoning and characteristic structures long known as 'tiger stripes' (gemstones have their own menagerie) which are formed by alternating stripes of liquid inclusions with negative (hollow) crystallites. Crystals of the iron oxide goethite form 'beetle legs'. None of these inclusions are found in synthetic quartz.

Hydrothermally grown amethyst can be identified by a technique in which the specimen's main axis is first placed parallel to a ray of light passing through the optical train of a microscope or to the main axis of the polariscope (i.e. between crossed polars). In most if not all natural amethyst interference colours reveal the crystalline phenomenon known as Brazil twinning whereas the synthetic material will normally show a succession of colour bands. Some Japanese-grown amethyst has been found to show flame-like or arrowhead structures. In the mid-1980s some gemstone dealers were reporting that up to 25 per cent of their stock of amethyst had been shown to be synthetic by the use of this test. The unusual material 'ametrine' (amethyst–citrine) which has been found naturally and is grown hydrothermally may also be tested by this method. GIA at one time reported that acicular crystals and spicules with quartz crystal caps had been noticed in synthetic amethyst.

The same problem that might be faced by anyone wishing to distinguish citrine from yellow scapolite will arise to a more significant extent when the transparent purple form of scapolite is confused with amethyst. The overlapping refractive indices and the necessity for obtaining the optic sign may involve the tester in some careful observations. Even though most scapolite is not really like most amethyst there will be the odd occasion when confusion may arise – however, such specimens are unlikely to be of the finest colour (scapolite is a quiet colour on the whole) so commercial considerations will not be too pressing. While the amethyst-like scapolite from East Africa may present one problem, the pink chatoyant scapolite from Myanmar may be confused with rose quartz which, however, prefers to display the phenomenon of asterism instead. Burmese pink cat's-eye scapolite is a rare and desirable gemstone.

Amethyst is well imitated by the many shades of purple or mauve sapphire whose RI and SG will be well within the corundum range. Some examples contain chromium and may show the Cr emission line in the red section of the absorption spectrum. Many sapphires of this colour display a striking colour change of a kind rarely seen in amethyst.

Gemmology textbooks seem to have missed this one but the synthetic corundum grown by the flame-fusion process and intended to imitate alexandrite (presumably by those who have never seen one!) looks much more like amethyst – if viewed in the appropriate direction and in suitable lighting conditions. This very interesting and attractive material shows a sharp absorption line at 475 nm at the blue end of the absorption spectrum but the colour change will reveal its true identity.

An interesting use of rock crystal is in the manufacture of simulants of famous diamonds so your diamond collection ought perhaps to be tested! As the specimens will be of some size if they reproduce the originals the polariscope could be used to find the unique interference figure.

The synthesis of quartz does not arise from a desire to create fine gemstones but to produce inclusion-, twinning- and strain-free crystals for electronic purposes, making use of the piezoelectricity of quartz (the development of an electrical charge on subjection to mechanical stress or the reverse, a measurable and consistent change in mass when an electric current is passed through a crystal). Colourless synthetic quartz is not fashioned except for crystal balls, figures, skulls and other elaborate designs but amethyst and citrine are grown by the hydrothermal method with their colour induced by the addition of appropriate dopants which will give colour centres as described below.

In 1973 one of the authors (MO'D) was able to examine a superb synthetic citrine of 49.28 ct, grown by Sawyer Research Products of Ohio, USA. This stone was a beautiful golden colour with RI and SG characteristic for quartz. Tiny groups of crystals could be seen inside and may have been the mineral aegirine, a sodium–iron silicate pyroxene which may have entered the growing crystal during the growth process. The colour of citrine and amethyst are caused not by the addition of colouring elements like chromium in ruby and emerald and iron in peridot and blue spinel but by the development of colour centres in which electrons may be absent from their normal sites or be present in sites where they would not be expected. This leads to electron movement which absorbs energies corresponding to those involved in the production of visible light. The formation of colour centres needs a trigger: in quartz this may be a trace of aluminium which initiates development of a smoky colour when the material is irradiated by, for example, gamma rays from a ^{60}cobalt source. A deep smoky colour will develop in around 30 minutes. This is considered to echo the process by which smoky quartz is found in nature. A brownish-green colour is sometimes seen in synthetic smoky quartz with the smokiness often occurring in well-defined bands. The colour is stable under normal conditions. A greenish-yellow colour may form when the irradiated smoky quartz is gently heated though, if heating is continued, a colourless state results in the end. This material contains no iron as natural citrine does though it has sometimes been given the name citrine. The smoky colour can be restored on further irradiation. Irradiating rock crystal may produce a greenish-yellow colour but some smokiness, even sometimes a blue colour, may occur during the heating of smoky quartz – the blue colour is stable to light.

If rock crystal contains traces of iron, irradiation may cause an amethyst colour to develop. Both types of iron, ferric (Fe^{3+}) and ferrous (Fe^{2+}) which give yellow or green colours respectively, can be irradiated to produce amethyst colour, the colour, as in the natural material, preferring to develop in certain areas of the crystal rather than others. The crystal grower wishing to produce gem-quality amethyst must take this into account when orienting the seed on which growth is to take place. Heating will dispel any lingering smokiness.

Amethyst fades somewhat when subjected to prolonged exposure to bright light and the process may be accelerated by simultaneous heating. Sometimes the colour needs to be lightened by heating for sales purposes, the citrine colour being reached when heating is continued. Further heating will develop the green colour of prasiolite (400–500°C). Amethyst and citrine can be darkened by heating to the range 500–575°C.

None of the above processes can be detected by the gemmologist but it is useful to know something of what may be going on in the world of crystal growth. As quartz is plentiful and not too expensive, mistakes do not have to be disastrous for the investigator. Rose quartz may be deepened in colour by irradiation or lightened by heating. Some rose quartz may fade in strong light while others seem unaffected. It has been suggested that two distinct types of rose quartz may exist. More significant is the recent appearance of pink translucent GGG (SG over 7, RI cannot be tested on the refractometer) in a rose quartz colour!

Some rock crystal can be made into very effective-looking simulants of both ruby and emerald by heating and quenching in an appropriately coloured dye – the material is known as 'crackled quartz' and the colours achieved are surprisingly attractive. They show, of course, the RI and SG of quartz rather than corundum or beryl. The dyestuff makes highly characteristic swirly patterns inside the stones which are sufficient for them to be recognized with a little familiarity – which is hard to achieve since these interesting products are rare. Some examples of polymer filling of citrine, smoky quartz amethyst and rock crystal have been reported.

The crypto-crystalline quartzes

The crypto-crystalline quartz varieties have long been treated to give sometimes (often) garish colours. The banded agates lend themselves particularly well to this type of treatment which has traditionally been carried out in Idar-Oberstein where the material occurs.

If any of the chalcedonies (convenient if not completely accepted generic name for the crypto-crystalline quartzes with a fibrous texture) is heated the usual result is for the colour to be darkened if pale and for pale colours to be milky white. Many readers will be able to cite exceptions – quartz is like that! Dyeing of banded agate has been known since classical times and is reported by Pliny in his *Natural History*, saying that boiling in honey, particularly in Corsican honey, makes all gems appear more colourful. Many varieties of chalcedony contain compounds of iron and specimens of this kind, on immersion in hydrochloric acid, will have these compounds dissolved, leaving a reddish-brown or yellow staining deposited on the inner surfaces. An iron nail was said to darken the colour. Honey was used because its acid content was sufficient to enhance the colour of poor quality material and to produce deep yellow, brown or black sections. Corsican honey was exceptionally acidic so that only heating would be needed to produce the desired darker colours from material placed in it. The honey treatment was known in Idar-Oberstein at least by the beginning of the nineteenth century. The inorganic dyes used then give more stable colours than those obtainable by the use of aniline dyes today, a point made by Nassau (1994).

From the testing point of view it is sufficient to say that the brighter banded agates and other chalcedonies may very well have been dyed and the paler ones not. Nonetheless some chalcedonies are dyed to resemble other ornamental materials and these should be watched for. A classic and now very rare study of agate dyeing is Dreher's *Das Farben des Achats* published in 1913 at Idar-Oberstein, right in the centre of the agate-staining business.

Such materials as black onyx which is not too common in nature may begin life as pale or colourless unbanded chalcedony subsequently darkened by carbon from the sugar-based dyeing process. Some dyed jasper has long been offered as 'Swiss lapis' in the hope that buyers will assume that it is true lapis-lazuli which has the higher specific gravity of 2.83 instead of the 2.65 of jasper. Some imitation cameos (true cameo is shell with a specific gravity around 1.29) are made from dyed chalcedony which is heavier.

Dyeing and other forms of treatment are sometimes used for special effects such as the creation of dendrites but testing is not a serious problem. Some green chalcedony can be dyed to imitate jadeite but will not show the absorption band at 437 nm present in many jadeite specimens. Some green dyed chalcedony may show a pink colour through the Chelsea filter if the dye contains chromium and in such cases faint chromium absorption lines may be seen in the red. A natural green chrome chalcedony called, locally at least, mtorolite, was discovered in Zimbabwe in the past few years and this shows a sharper chromium spectrum than that of green dyed chalcedony. Mtorolite has a quite characteristic quiet green, not at all like emerald, and was used in Roman jewellery. It shows red through the Chelsea colour filter.

Chrysoprase is a much brighter green colour which is derived from the presence of nickel. Specimens may show a faint absorption line at 632 nm in the orange. A variety of opal, also coloured by nickel, looks quite like chrysoprase but will have a lower RI of 1.45 compared to 1.55.

The remaining chalcedonies include:

- jasper, usually red to brown
- sard, red, brown
- sardonyx, red and white, brown
- bloodstone, dark green with bright red spots
- cornelian (carnelian), reddish-orange–yellow translucent and with a banded structure seen at its clearest when light is passed through it.

Moss agate and mocha stone both show very intriguing dendritic structures formed from a manganese compound – these are usually seen against a light brown or milky-white background.

There are many more varieties but as testing is not usually a problem details should be sought in more general books on gemstones.

Before leaving the chalcedonies it is worth mentioning an interesting phenomenon sometimes encountered on the refractometer when a specimen of chalcedony is tested. An unexpected birefringence of around 0.006 may be seen, unvarying in position as the stone is rotated. This has long been known as 'form birefringence' and is believed to be caused by particles of one refractive index being set within a body with a different RI.

Reports of interesting and unusual examples from the literature

Items in this section have been chosen to illustrate points made in the chapter and to bring one-off items to your notice.

As single-crystal quartz varieties are relatively large and plentiful it is not surprising that collectors have long sought well-developed crystals, particularly of rock crystal and amethyst. Since the European Alps provide a very wide range of specimens crystal collecting has long been a profession and the *strahlers* (*cristalliers*) continue to bring down crystals and crystal groups from Alpine clefts. The visitor to a gem and mineral show cannot fail to see many examples and it is worth being on your guard, not so much against other minerals being offered for quartz, but for repaired and/or assembled crystals and crystal groups. Over the years many papers in *The Mineralogical Record* have shown the reader how to identify fake mineral specimens and these papers should be consulted if possible: the *Record* is a very high class and respected journal.

In recent years a bright yellow to orange danburite of Russian origin has been reported. It is possible that some specimens may have been irradiated but at the time of writing this is not certain. Some of the stones I have seen could be taken for citrine but danburite has a higher refractive index and specific gravity, at around 1.63 and 3.00. Madagascar has recently produced some fine yellow danburite.

Also becoming prominent in recent years is a fine yellow apatite from Madagascar: specimens may show the usual rare earth absorption spectrum with two groups of fine lines, in the yellow and the green. The RI and SG are higher than those of quartz, at 1.63–1.64 and 3.2. East Africa has been producing fine yellow to orange tourmalines which can look quite like citrine but their strong birefringence of about 0.018 is sufficient to distinguish them from quartz.

The fluorites with their great range of colours (and also found in Alpine clefts) can look like any variety of coloured transparent quartz, especially if faceted. Their lower RI of 1.43 and higher SG of 3.18 as well as possible signs of cleavage seen as rainbow-like interference colours should be easy enough to test. Fluorite crystals look nothing like quartz, however, forming groups of cubes, the individual crystals easily cleaving to octahedra.

In an issue of *Gems & Gemology* a presumed dyed black chalcedony is described. The stone appeared black by reflected light but when examined by strong transmitted light from a fibre-optic source under dark-field illumination conditions the piece showed blue with semi-translucent edges. Inside the stone dendritic patterns showed that it was devitrified glass with a refractive index of 1.50. The stone also showed a cobalt absorption spectrum. This is an example of glass imitating chalcedony and is by no means unusual as a later report in the same journal showed.

This other example consisted of dyed black chalcedony beads, some with high lustre and some with a dull lustre. RI readings of 1.54 showed that all the beads were quartz. In strong transmitted light the beads with the lower lustre appeared to be semi-translucent while the others remained an opaque black. The duller beads showed parallel agate-type banding by transmitted light and a thin black layer incompletely covering the surface. Acetone removed this layer. The beads were surface-coated to resemble dyed chalcedony.

While much dyed black onyx is made by darkening lighter-coloured chalcedony using the sugar–sulphuric acid method, some pale material can be darkened by staining and electrolysis, a method also producing dendritic patterns attractive to the customer. In some examples copper salts are dissolved in water to make a saturated solution. Preformed chalcedony is then immersed to take on a bluish-green colour. When an electric current is passed through it a slowly spreading copper dendrite is formed. This does not resemble any natural product and should be immediately identifiable by its colour.

A very interesting rock crystal with an artificial three-phase inclusion was seen at the 1989 Tucson Gem & Mineral Show. The three-phase inclusion contained water, a gas bubble and a small faceted red or blue stone. Access for these exotic strangers to the interior of the crystal was achieved by drilling thin tubular columns into the stone from the crystal base. After the columns were filled with liquid the faceted stones were inserted.

A flat translucent blue–green chalcedony cabochon was noticed at the same show, the specimen containing a large dendritic inclusion, probably caused by soaking the porous chalcedony in a copper solution and then applying an electric current causing the copper to precipitate out. A small area on the surface showed where the current was applied.

A visit to a crystal healing shop will probably locate the rather attractive 'Aqua Aura'. It will usually be found as a small crystal group though we have seen faceted stones. The colour is an aquamarine-blue and arises from a thin gold covering of the crystals. The underlying rock crystal is colourless and the properties are consistent with quartz; the coating is said to be very difficult to remove. Distinction between the quartz crystal and aquamarine is not difficult and the reader is referred to Read, *Gemmology*, second edition, 1999, for diagrams. Some topaz has been treated in the same way.

Though reminders are often given in journals, purple sapphires still turn up in parcels of amethyst. These will no doubt be Verneuil synthetics as natural purple sapphire would be far too valuable to be used in this way. A chromium spectrum, even sometimes the chromium emission line, will give away some such sapphires – if you think to look out for them. When you take into account the quantities involved (a report in 1989 by the International Colored Gemstone Association stated that 20 000 ct of synthetic amethyst was arriving every month in New York from Korea) it is not surprising that dealers cannot look at every specimen. The same report stated that amethyst classed as 'Uruguay', a highly valued colour, was in many cases synthetic amethyst. Dealers at the time were warned to beware of material priced 10–20 times lower than normal.

'Rainbow quartz' is colourless rock crystal with what seem to be prominent interference colours on the surface, best seen by reflected light.

An assembled product has been made to imitate white plume agate (transparent with dark dendritic patterns). *Gems & Gemology*, Summer 1990, noted a colourless, near-transparent specimen of 61.39 ct which contained dark reddish to greenish-brown dendrites. The top was formed by a translucent convex cap glued to a flat light grey semi-transparent base. A fairly thick colourless layer containing profuse gas bubbles separated the two; this layer was easily scratched and was probably some form of epoxy resin and the specimen would have to be called a glass and dendritic agate doublet.

Dyed green quartz can make quite a good imitation of emerald. In the Fall 1992 issue of *Gems & Gemology* some of the examples current at that time were reviewed. Some stones owed their colour to fractures filled with dyestuff and some resembled oiled emeralds in showing a soft appearance. When specimens were placed table down about 35 cm above a white background and viewed through the base the green dye-filled fractures were easy to see. Standard gemmological tests showed the stones to be quartz and the absorption spectrum with a band extending from about 690 to 660 nm and characteristic of green dyed material showed that dyed quartz imitating emerald was the answer. Infra-red spectroscopy showed a series of sharp absorptions at approximately 2965, 2930 and 2870 cm^{-1} which are consistent with bands showed by the epoxy resin Opticon, widely used in the fracture filling of emerald. GIA believed that in this case a dye had been mixed with Opticon or some similar resin before being introduced into the fractures.

Some glass imitations of amethyst have been found to show a chalky blue fluorescence under SWUV light – natural amethyst is inert. This is not a complete test in itself but it serves as a warning to the tester that colour is not the only guide.

GIA examined and reviewed a synthetic quartz with a medium to dark-blue colour in the Summer 1993 issue of *Gems & Gemology*. The material, doped with cobalt, showed a cobalt absorption spectrum. Examined between crossed polars, the stone gave the usual quartz interference figure but this could be seen along the length of the main crystal axis, an effect not often observed. Using diffused transmitted light wedge-shaped zones of darker colour alternating with very light blue zones were visible (they were easier to see with the unaided eye).

A white opaque synthetic quartz of 63.69 ct seen at a gem show was found to display a milky-white colour in reflected light and yellowish-orange in direct transmitted light. It had an RI of 1.55, was inert to both types of UV light and gave an SG of 2.37 some distance from the quartz SG of 2.651. Under magnification a faint columnar growth could be seen. Porosity may be the cause of the lowered SG. EDXRF spectrometry showed that traces of chlorine, potassium, calcium and iron were present.

A blue opaque quartz has also been reported, the material (Russian made) being an aggregate of minute quartz crystals in parallel orientation.

Synthetic green quartz has a distinctive colour and no other synthetic or natural stone looks like it. Some examples have been said to resemble tourmaline. A specimen described by GIA showed parallel green banding and with some angular brown colours at right angles to the green banding. A similar specimen showed white pinpoint inclusions, with EDXRF showing the presence of potassium, iron and silicon (one specimen showed some chromium). GIA classified the material as synthetic green quartz.

Jade is well imitated by dyed green quartz. The metamorphic rock quartzite, consisting of about 95 per cent quartz grains, can be dyed effectively but will show the quartz RI of 1.55 against jadeite's 1.66. Dye can be seen concentrated among the grains and some specimens at least may show the dyestuff absorption band centred at about 650 nm. GIA found in one example that a substance similar to the synthetic resin Opticon had been used.

Quartz coated by glass was described by GIA in the Summer 1994 issue of *Gems & Gemology*. A strand of beads with a dark violet colour showed the normal properties of quartz and the drill holes showed that the coloured layer was thinner close to them. Some of the beads showed surface depressions like dimples in the vicinity of the drill holes, the depressions showing the colourless and rough-ground surface of the underlying material. On sawing a bead with its surface at right angles to the drill hole the surface layer was found to be about 0.07–0.10 mm thick. Undercutting showed that the surface layer was softer than the bead and had a hardness of about 5–6. Some of the surface layer was found to be brittle, producing conchoidal chips. The beads were quartz coated with a glass-like, perhaps enamelled substance. Cobalt was the cause of the colour.

An evenly coloured light blue crystal seen at a gem show turned out to be a piece of colourless quartz grown on a medium dark blue seed crystal wafer. The crystal faces were oriented to reflect the colour from the seed. The crystal showed the cobalt absorption spectrum.

A colour-zoned synthetic amethyst showed a very uneven distribution of colour, resembling 'leopard spots'. It is probable that growth took place on a seed plane cut perpendicularly to the main crystal axis. Amethyst coloration prefers some areas of the crystal as can be seen in natural specimens where the deepest colour can be found under the faces of the positive rhombohedron. Magnification showed that the darker-coloured regions grew as expanding rhombic pyramids from a point inside the crystal. Such a colour placement can be taken as proof of artificial origin.

Synthetic transparent pink quartz has been grown in Russia at least since 1992 and the name 'flamingo quartz' has been used. Comparing this with natural Brazilian pink quartz it was noted that colour banding in the synthetic material was always parallel to the basal faces of the crystal and often appeared uneven. In the natural pink quartz the banding was oblique to the axis and showed feathery edges. No twinning was found in the synthetic material and natural pink transparent single-crystal quartz is relatively uncommon.

A chrome green chalcedony known as 'Chuquitanica' by local Bolivian dealers has an RI of 1.530–1.550 with DR up to 0.005 and SG 2.56–2.57. There is no fluorescence under either type of UV light but specimens show bright red through the Chelsea filter. The absorption spectrum shows a small peak at 684 nm with broad bands at 605 and 416 nm – a distinct minimum can be seen at 510 nm. This is similar to the absorption spectrum shown by the chrome chalcedony (mtorolite) found at Mtoroshanga in Zimbabwe. Dyed green chalcedony intended to imitate chrysoprase shows brownish-red through the Chelsea filter.

A glass imitation of blue chalcedony turned up in the Taiwan gem market. It was made from light blue translucent glass which showed horizontal banding and a perpendicular fibrous pattern. The RI was 1.54 and SG 2.57. The specimens showed green through the Chelsea filter. Any gas bubbles or swirl marks were to some extent hidden by dense aggregates of very fine needles which were identified as wollastonite, a mineral appearing as a devitrification product.

The piece of rough glass appeared inhomogeneous and consisted of massive glass with a slaggy top zone. Most examples were dark blue or sky blue but next to the slaggy zone a green colour appears. The degree of translucency is not uniform. Most specimens have been cut as cabochons or bangles.

A synthetic ametrine has been produced in Russia by hydrothermal growth in alkaline solutions. It resembles the natural material found in Bolivia, which is that country's best-known gemstone, showing characteristic colour zoning and twinning. X-ray fluorescence tests showed higher contents of potassium, manganese, iron and zinc than natural ametrine while IR spectroscopy of the citrine portions showed more intense absorption in the 3700–2500 cm^{-1} range. The synthetic amethyst zones gave a weak but diagnostic peak at 3543 cm^{-1}.

A chatoyant quartz from Luodian County, Guizhou, China, has been found in brownish-green, light green and bluish-green colours. The very sharp eye is caused by parallel fibres of tremolite.

The 'power beads' on sale in some quarters may be dyed quartzite offered as cornelian, 'onyx' may be marble beads with a plastic covering and 'white howlite' may be magnesite.

Hydrothermally grown synthetic quartz of a brown colour and a shallow green layer has been offered as a simulant of andalusite.

Blue quartz crystals from a deposit near Malaga, Spain, appear to have gained their colour from inclusions of the mineral aerinite. Some crystals are reported clear enough to be faceted. A blue quartz from Madagascar owes its colour to inclusions of lazulite.

Colourless quartz specimens from Minas Gerais, Brazil, on irradiation with gamma rays, has appeared in a variety of colours, including pale yellow, yellow, greenish-yellow, olive-green, greenish-orange, orange and reddish-brown.

Chapter 7

The jade minerals

Jadeite and nephrite

To claim the name of jade a specimen must be either nephrite or jadeite. Any other mineral, however much its appearance suggests one of the jade minerals, cannot be called jade. Even the presence of a qualifier ('new' jade, jade from country x) does not alter the force of this rule. In the Far East, as visitors to Hong Kong (and especially to the jade market) will know, the commerce of jade is very considerable and simulants abound.

In general nephrite has the quieter and jadeite the brighter colours though anyone attempting to distinguish one type from another on colour alone would be liable to serious and possibly expensive error. Almost all or even all *archaic* Chinese jade is nephrite. Consultation of Hong Kong auction house jade jewellery sales catalogues will show that virtually all the lots are jadeite. Nephrite artefacts have to be sought in catalogues of oriental and other antiquities.

Another impression given by a visit to the jade markets will be that all jade is green. Most jadeite used in jewellery is green but there are other colours though these are not imitated to the same extent. Jade, and in particular nephrite, is venerated by the Chinese partly for its appearance, more for its texture but most for its toughness which enables pieces to accept the engraving of characters. The written word is held in great reverence in China so that any material likely to preserve writing indefinitely will be greatly valued.

Before looking at the jade minerals in more detail it is worth mentioning that the term 'Imperial jade' is not used by the auction houses or major dealers as it can so easily be debased. The catalogue entries will use such phrases as vivid emerald-green colour, exceptional translucency, vivid brilliant green. The criteria for the finest green jadeite are a combination of the finest even emerald-green colour with no flecks of white or any other colour, combined with high translucency. It is not surprising that the most ingenious attempts have been made to simulate material for which such large sums might be paid.

Jadeite

Natural jadeite is a sodium aluminium silicate, a member of the pyroxene group of silicates, with the green colour caused by chromium. It has an RI near 1.66 and SG around 3.33. Interestingly some specimens with a flat back which can be tested on the refractometer may show birefringence as well as readings of 1.654 and 1.667.

Members of the pyroxene group usually show a significant strong absorption band in the violet at 437 nm. Jadeite has a hardness of 6.5–7 but is very tough and breaks with a characteristically splintery fracture. Colours range from the predominating green and lavender to white, pink, black, yellow, orange, brown and other colours. Green material often shows patches of white, the combination being cleverly used by carvers. Pale jadeite may fluoresce a faint white in LWUV light, while some pale yellow and mauve specimens may show an intense blue–violet glow under X-rays.

A coarser-textured dark green variety of jadeite has been called Yunnan jade. Both this and the finest green jadeite come from Upper Myanmar whose government holds regular auctions in Yangon. The deposits of jadeite which have produced material of archaeological interest do not produce jadeite for contemporary ornamental use.

While it should be fairly easy to test a flat-backed cabochon of jadeite for refractive index the testing of carvings is much more difficult. With some good fortune a chromium absorption spectrum may be obtained from fine green examples – this will add to the 437 nm pyroxene band the characteristic absorption features of lines in the red which are more diffuse than those that might be shown by the type of emerald that might be confused with jadeite. The centre point of the doublet (seen as absorption) is close to 691.5 nm and there are weaker lines at 655 and 630 nm. While the violet is strongly absorbed as in most chromium-coloured stones the iron (pyroxene) band can be seen within the general absorption.

If the specimen is comparatively pale reflected light is the best source of illumination. Dark jadeite other than green may not allow the observer to distinguish the pyroxene band. Dyed green jadeite shows a rather woolly band in the red. Green jadeite does not show red but green through the Chelsea filter.

The skill of the carver can be appreciated once it is realized that although jadeite has a relatively moderate hardness of 6.5–7 it is very tough. This is because the material is composed of interlocking grains rather than existing as a large single crystal. This structure is the cause of the dimpled surface so venerated by connoisseurs of jade. When diamond powder is used as the polishing agent this type of surface is less seen and the piece may be less prized.

The collector is very unlikely to obtain good quality jade before it has been fashioned. Boulders which may or may not turn out to carry fine or acceptable colour come from the sources in Upper Myanmar to reach the State Gem Auctions in Yangon when the boulders, which are usually covered by a potato peel-like skin, are slit by buyers to ascertain the nature of the colour inside. Windows may also be cut for the same reason. The boulders, which are reduced to pebble size, can be obtained only through the auction process.

Jadeite was synthesized by the Inorganic Materials Laboratory of General Electric and reported in 1984. Examples did not reach the markets. The starting material was said to be glass obtained by melting sodium carbonate, alumina and silica to obtain the jadeite composition of sodium aluminium silicate. Chromium would be used to give acceptable colours but these have not yet been reported. Growth was said to have taken place in platinum crucibles at about 1500°C.

While synthetic jadeite has not been troublesome, imitations abound. The jadeite structure allows dyeing to be carried out with little difficulty and though the colours produced are sometimes a little too strong to be natural, the specimen needs to be carefully tested. Absorption bands from the dye are at 650 nm (a wide band) with a narrower band at 602 nm.

Nephrite

Nephrite is even tougher than jadeite though it is not high on the scale of hardness at 6.5. The refractive index is usually in the range 1.600–1.641 and the specific gravity in the range 2.96–3.02. Colours are just as varied as those of jadeite but are quieter. While jadeite is a member of the pyroxene group of minerals, nephrite has been found harder to classify through no fault of its own. Long considered a member of the amphibole group it is now placed as a variety of the mineral actinolite. This has led to nomenclature problems familiar in other species as over the years there have been nephrite cat's-eyes, actinolite cat's-eyes, tremolite cat's-eyes (tremolite is still a valid species in the amphibole group, forming an isomorphous series with actinolite) and so on. Sorting out such specimens in a mineralogical rather than a gemmological context needs equipment and a depth of knowledge which are not normally found in purely gemmological laboratories.

The best-known colour shown by nephrite is a dark green often compared to the green of spinach. A lighter green can quite closely resemble jadeite and is often associated with the New Zealand greenstone variety of nephrite though this can be a dark green too. Nephrite from Siberia is also a fine dark green, often showing black or dark spots.

Some nephrite shows an absorption spectrum with a doublet at 689 nm, a sharp line at 509 nm and other bands at 460 and 498 nm. The spectrum is less clear than that of jadeite and will only be shown by darker specimens. Nephrite shows no luminescence under UV light. Under the Chelsea filter nephrite shows green.

The colours of nephrite range from a creamy-beige (mutton-fat jade) through green to brown, the two last attributed to iron. A dark brown skin or rind is sometimes found. Yellow and black colours are also known. Nephrite may develop brownish stains after long burial and since archaic Chinese artefacts are nephrite brown staining is seen more often in nephrite than in jadeite. Interestingly there have still been no reports of either jadeite or nephrite being found in China proper (however this area might be defined), both minerals having always been imported. The large scale of this trade has meant that substitutes have always turned up somewhere along the line. Today nephrite is found in a number of countries, including the United States (Wyoming, Californian and Alaskan jade) and Canada where some beautiful green nephrite is found in British Columbia. Here the nephrite occurs with serpentine and it would not be surprising if there was not the occasional confusion. Many of the so-called nephrite cat's-eyes have come from Taiwan. These have shown an RI of 1.615–1.631 with an SG of 3.01.

Enhancement of colour

Heating can be used to lighten specimens which are considered too dark for easy sale. Jadeite containing iron compounds which may give yellow to brown colours can be heated to turn them brownish or reddish. Lavender jadeite is especially sensitive to heat, Nassau (1994) reporting that temperatures as low as 220–400°C can cause it to lose its colour though it can be stable up to 1000°C on occasion.

Neither of the jade minerals is routinely irradiated. Though some reports do turn up there seems to be little or nothing to be gained commercially by the practice. Dyeing is the commonest treatment. As jadeite is more porous than nephrite it is easier to dye and is more translucent. Pale green specimens are dyed to darken them and in recent years lavender and other colours have also been strengthened. Brown staining is used to enhance the colour of some carvings.

Some but not all of the dyes are stable in ordinary light, aniline dyes being quite often unstable. Fade tests though often advocated are not really practicable. The best way to test for possible dyeing is to examine the specimen for signs of dye concentration. These will be found, if present, in out-of-the-way places such as drill holes and will appear darker than the surrounding areas. If the piece is intricately carved look in the less prominent parts of the carving.

Jadeite and nephrite artefacts are often recovered from archaeological sites in which they have long been buried. Iron in the soil enters the piece through cracks to give a brown stain which can be removed by bleaching. Soaking in acid to achieve this should not affect the other colours in the stone. The acid treatment may be followed by soaking in water and neutralization with an alkaline substance, leaving the specimen pale. Increased porosity then allows waxes or polymers to be introduced. These may be coloured (green is the commonest colour) or colourless, the latter being used after the specimen itself has been dyed. This practice, with others, is not recommended as the integrity of the artefact may be spoiled and the signs of authenticity lost.

Bleached jade is now known as B-jade while the term C-jade is used for polymer-impregnated lower quality green jadeite. A-jade is untreated. Despite some guides, the presence of brown stains should not be taken as assurance that treatment has not taken place as some brown stains are sometimes deliberately left where they are.

Treatment is carried out sometimes at the expense of the specimen since some may be weakened and in others a yellow exudation from the acid used may be seen.

Wax is often used to conceal small fractures close to the surface. Paraffin wax is often used as it is easy to obtain and can be introduced almost anywhere on a specimen. Nassau (1994) warns that such treatment may escape the notice of the setter whose subsequent actions can then damage the specimen. Impregnation and coating by polymers helps to improve transparency and some pieces, usually cabochons, can be foiled.

Many processes are carried out in Hong Kong where concave mirrors with small openings have been used to act as the backing for jade pieces. Imperial jade has been simulated by placing a thin dark green polymer layer over translucent white jadeite. A better imitation of Imperial jade is the hollow cabochon, long famous among gemmologists and completely unknown in the Hong Kong jade market – I (MO'D) have never been able to find one! But then, I did enquire about them!

The hollow cabochon is intended to be taken for Imperial jade and the colour is very fine. A pale piece of jadeite is hollowed out to give the shape of a cabochon and another piece fashioned to fit the hollow base. When the two are cemented together a green dye is first introduced into the hollow coating the interior. When complete the specimen appears to be solid though many cutters are said to leave a ridge where the base is joined. The stone looks solid and the colour is very like that of good quality jadeite; the translucency of fine jadeite is also reproduced. With the spectroscope the green specimens show both elements of the chromium spectrum as well as the woolly band in the red which seems always to be associated with dyestuffs. Some dark green natural jadeite may show a broad absorption band in the red.

Bleached jadeite may give a slightly lower specific gravity at 3.32 from 3.33. Impregnating material often shows gas bubbles and the piece may be fractured. Bleaching is not easily proved by standard gemmological testing. Imitations of jadeite made from glass will feel lighter in comparison with a range of specific gravities lower than that of jadeite.

Identification of the jade minerals

Testing for the two jade minerals may involve investigation of likely natural materials or of elaborate composites like the hollow cabochons already encountered above. Green is obviously the most desired colour and we should remember that dyed jadeite may show pink or even red through the Chelsea filter – this does not happen with natural jadeite or nephrite. While the spectroscope helps with the identification of some dyes at least, other treatments are harder to detect. It may be said at this point, although this is not intended as a gemmology teaching manual, that the spectroscope gives acceptable results only after some time has been taken and a lot of experience gained. Anyone relying on the spectroscope for a good deal of gem testing knows that constant adjustment of the lighting and specimen is needed for a reliable identification. The spectroscope is especially useful in the testing of the jade minerals and patience is needed since great sums of money are often involved.

Natural minerals likely to be confused with the jade minerals

Natural materials likely to be confused with jade (either type) include *emerald* as a very possible substitute for that emerald-green translucent jadeite once known as Imperial jade. As we have seen this designation is no longer used but cabochons of this type of jadeite could be mistaken for emerald. Emerald has a lower specific gravity, at around 2.70 at most, than the 3.3 of jadeite. If a suitable surface is available the refractive indices of the two materials are appreciably different at 1.57 (emerald) and 1.66 (jadeite). The non-emerald types of beryl can also be mistaken for jadeite. No nephrite shows precisely an emerald-green colour.

The cryptocrystalline nickel-coloured green *chrysoprase* is sometimes suggested as a possible jade(ite) simulant but to my mind there is too much blue in the green for the colour to be a serious contender. Nonetheless, not everyone buying jade knows about all the possible substitutes and customers tend to believe what they are told. Chrysoprase is not the only quartz family variety to provide a serious imitation of jade(ite) so a comparison of the properties is worth remembering: chrysoprase has an RI 1.55 and an SG 2.65.

A very convincing substitute for either of the green jades is the garnet family mineral *grossular* in its variety *hydrogrossular*. The RI is 1.73 and the SG 3.48. Very characteristic of hydrogrossular coloured green by chromium is a strong absorption band at 630 nm with another at 461 nm. Most if not all massive grossular fluoresces a strong orange under X-rays. While hydrogrossular varies in its properties with origin, none of it should be confused with either of the jades when tested. The name Transvaal jade turns up in the literature on occasion.

The mineral *vesuvianite* (sometimes known as *idocrase*) is chemically and structurally similar to grossular and thus may play a part in jade mineral identification. As in the case of grossular, vesuvianite's properties do not overlap with those of either of the jades. The name californite has been used for massive green vesuvianite mixed with grossular; its refractive index is in the region of 1.71–1.72 and the specific gravity, nearly overlapping with that of jadeite, is in the range 3.25–3.32. The absorption spectrum with the strong band at 461 nm and the weaker one at 528.5 nm distinguishes vesuvianite. It is just possible that someone may offer a brown massive vesuvianite as jadeite – some brown vesuvianite from Canada may show a rare earth absorption spectrum. Vesuvianite has a hardness of 6.5 and shows no luminescence.

Serpentine as a mineral group is now known as kaolinite-serpentine in current mineralogical lists (see *Fleischer's Glossary of Mineral Species,* 1999). The name arises from the characteristic markings shown by many specimens and several varieties can give quite convincing imitations of one or other of the jade minerals. The variety *bowenite* may be yellow, yellow–green, blue–green or dark green and is translucent, the best greens quite easily confused on colour grounds with the finest dark green nephrite. One of the varieties of New Zealand bowenite has a specific gravity of 2.617. In general bowenite has a refractive index near 1.56 with specific gravity about 2.59. Bowenite is quite hard (5) and translucent for a serpentine. It may include small groups of dark green chlorite group minerals. The name Tangiwai (tears) may be encountered in New Zealand. If you are on a nephrite-buying trip there, note that 'Tangiwai' is not nephrite but bowenite! Bowenite is found in China and is very commonly used there for all kinds of ornament.

The mineral *antigorite* (kaolinite-serpentine group) includes a fine ornamental variety *williamsite*: this has a refractive index of about 1.57 with specific gravity 2.61. Some examples have been reported to fluoresce a weak green in LWUV light. The mineral *clinochlore* may also resemble nephrite but is very soft at a hardness of 2–2.5. It has an RI of 1.58–1.59 and SG 2.7.

The rather attractive and somewhat oily green mineral *prehnite* may also be a pale yellow but so are some of the jades. With familiarity the appearance of prehnite becomes distinctive as it also shows a radiating fibrous structure. As well as the pale yellow material prehnite can also be pale green or blue–green: the RI is in the range 1.61–1.66, most usually near 1.63 while the SG is usually 2.87–2.94. There may just be a faint brownish-yellow fluorescence under UV light and X-rays but this is unlikely to trouble the gem investigator. There is no useful absorption spectrum.

The material known as *saussurite* is formed by the alteration of plagioclase of which it is partly formed with some epidote group species: it may be white or mottled green or brown. The RI is around 1.55–1.56 for lighter stones and 1.68–1.70 for darker areas. The whitish material has a specific gravity of about

2.75, the darker specimens ranging up to 3.25. One specimen of saussurite is reported in the literature as showing the characteristic clinozoisite absorption band at 455 nm in the blue–violet. The hardness is 6–7.

It is possible that a customer new to jade buying and faced with such a scene typical of the markets of Hong Kong may be deceived by a piece of *aventurine quartz* but the unmistakable spangly appearance ought to give it away. If the piece is cut as a flat-backed cabochon the RI of near 1.55 will be enough to separate it from the jade minerals which in any case it will not closely resemble. If as sometimes happens the spangles (of the green chrome mica) are not apparent and the piece does not have a flat back it may be possible to use the Chelsea filter, through which the piece will appear red, or to find some elements of the chromium absorption spectrum, especially the doublet in the deep red accompanied by a line in the orange.

The material *verd-antique* is a serpentine marble (a rock) and is green with veins of whitish calcite. It will have lower properties than either of the jade minerals and may effervesce with dilute acids – not a recommended test but useful to test for calcite. It may show an absorption band at 465 nm. Also called *Connemara marble* is a serpentine showing greenish-white banding with a mean RI of 1.56 and SG 2.48–2.77.

A rock made up of fine-grained chrome mica with brown fine-grained rutile is known as *verdite*. Sometimes small traces of ruby are found in it. Verdite has a refractive index of near 1.58 and specific gravity about 2.7–2.9. Some exercise of the imagination would be needed to mistake it for either of the jades, but, as always, it depends on your experience and what others tell you. There is a plethora of old names which are a great nuisance though many of them have their own charm. We have tried to omit them since no two people agree on their meanings.

The very soft substance *steatite* (soapstone) is not a valid mineral species but an impure talc rock. It has a hardness of 1 on Mohs' scale and can thus be scratched by a finger-nail. It is also sectile. The reader may very well think that there could be no confusion between this material (best called *soapstone* from its greasy feel) and either of the jades but when viewed through the glass of a showcase its softness is less apparent. The colour most likely to confuse is greenish to brownish-yellow. Again, it depends on the mental picture the buyer has of 'jade'. The RI is near 1.55 and the SG 2.7.

A most attractive material with the established name *maw-sit-sit* comes from Myanmar and consists of white albite with a chromium mineral which gives a fine green. Various shots have been made at establishing the true composition but this need not bother the customer too much. It has a lower RI (at about 1.53) and SG (mean 2.77) than either of the jade minerals. The hardness is about 6 on Mohs' scale, coming from the feldspar, the major mineral constituent. Quite high prices are asked for this very beautiful material and it frequently accompanies jadeite in the specialist markets in Hong Kong and elsewhere.

Fluorite has one of the greatest colour ranges of all minerals but is almost invariably transparent and unlikely to be mistaken for either of the jades. It shows a very easy cleavage and is far from tough. It is only 4 on Mohs' scale. Nonetheless it has been used for carvings (great skill is needed to avoid the cleavage). It may show reddish or pinkish through the Chelsea filter and most examples if not all will fluoresce a pronounced violet under LWUV light. Glass imitations turn up but can be recognized by the usual swirliness and gas bubbles.

While those examples already described provide the better imitations of the jade minerals not all of them would be serious contenders. Some cited in textbooks would be even less likely but you can never be sure and there is always someone who 'knows about jade'. For the benefit of these experts we can include the zinc carbonate *smithsonite* and the green potassium feldspar *amazonite* (microcline). The latter has a quite unmistakable colour never shown by either of the jades, whether or not the whitish banding is present. Smithsonite as a carbonate will effervesce with dilute hydrochloric acid (not that you would immediately turn to such a test). Smithsonite has an RI around 1.62 (admittedly in the jade area) but with SG well above that of both the jades at 4.35. Microcline has an RI of 1.53 and SG 2.56, both well below either of the jade minerals.

Reports of interesting and unusual examples from the literature

Items in this section have been chosen to illustrate points made in the chapter and to bring one-off items to your notice.

GIA reported a composite imitating jadeite in the Summer 1996 issue of *Gems & Gemology*. The piece was a statuette weighing 239.37 ct and was made from plastic and calcite dyed to look like jadeite. When tested on the refractometer strong birefringence could be seen which suggested that a carbonate was present: the SG was measured at 1.98 and under magnification the piece appeared to have white grains in a groundmass of transparent colourless or green material. The general appearance was quite convincing, with a resinous lustre.

Another composite was described in the Fall 1996 issue of *Gems & Gemology*, this specimen at first sight appearing an entirely fine green but on closer examination showing a very thin green layer on the top and another, thicker white layer beneath, the thicknesses being 0.1 mm and 2.2–2.3 mm respectively. The layers were joined by a cement containing gas bubbles and with a slightly yellowish colour. The darker upper layer showed mottling and contained numerous colourless veins. The white lower layer showed a distinct crystalline structure. RI readings were 1.64 and 1.74 from the two layers which also gave a characteristic chromium absorption spectrum with the hand spectroscope. X-ray diffraction analysis showed that both layers were jadeite though the differences in the RI could not be accounted for.

A drop of concentrated hydrochloric acid placed on a cleaned surface of jade will be drawn by capillary action beneath the surface of Type A (untreated) material and the point of application will be marked by an aureole for some minutes afterwards. If the piece is Type B the acid will remain on the surface until it evaporates because the polymer impregnation has sealed the surface. In Type C jadeite the markings are shown up by the magnification of the acid droplet.

A quite common jadeite imitation seen by GIA took the form of a 'moss-in-snow' colour necklace of 8 mm beads which turned out to be quartzite selectively dyed. The RI was 1.55. The beads did not show red through the Chelsea filter nor did they show the usual absorption band at about 650 nm which most dyestuffs give. With the hand lens dye could be seen concentrated in cracks: it could be easily removed by an acetone-soaked cotton swab.

Dyed quartzite has also turned up as an imitation of nephrite. GIA describes a necklace of dark green 10 mm beads which were offered as 'imitation nephrite'. The RI of 1.55, the lack of a red colour when the beads were examined through the Chelsea filter and the presence of the absorption band near 650 nm showed their true identity. Interestingly the dyestuff might be expected to give a red colour through the filter.

The thermal reaction tester will cause the surface of plastic-impregnated jadeite to melt or at least soften. The fine depressions characteristic of untreated jadeite are said to be less apparent in impregnated specimens.

A 'jade' sphere with a mottled dark reddish-brown colour grading into medium yellowish-brown and with a good deal of the surface a strongly mottled medium yellow–green to very dark yellowish-green showed birefringence on the refractometer. Through the Chelsea filter the greenish areas appeared greyish while the brown ones showed reddish-brown. The hardness was close to 3 and there was no apparent luminescence. The specimen was predominantly calcite with some serpentine and had been selectively dyed and coated with paraffin or wax.

Infra-red spectroscopy usually gives away the presence of a synthetic resin used for impregnation as it shows a band at 2900 cm^{-1}. GIA reported a lavender cabochon of 15.86 ct which standard gemmological tests showed to be jadeite. Under magnification small cavities each containing a transparent colourless filling material could be seen. IR spectroscopy proved this material to be synthetic resin.

While lavender and the other colours of jadeite are often the target of those engaged in colour enhancement, by far the greatest efforts are put into green jadeite as this is considered the most desirable colour. This means that all green jadeite pieces and certainly those of any appreciable size or other significance need to be looked at carefully. The eye is unable to detect treatment in many cases but it is unreasonable to expect every piece to be sent for testing so that some examples progress quite far into the jewellery chain. There seems no answer to this!

Varnish used to improve the surface is reported in the Fall 1994 issue of *Gems & Gemology*. The piece was a large bead of greyish-purple colour coated with a layer of mottled green varnish. Some of the coating had spalled off, showing the underlying colour. The refractive index of the coating was measured at 1.52 against the 1.66 of jadeite and the specific gravity was 3.29 as against 3.33. The thermal reaction tester caused the coating to melt and the piece gave a strong chalky-blue fluorescence under LWUV light with a weaker reaction to SWUV light. Jadeite cannot be realistically tested by its fluorescence or lack of it but some specimens at least will show a spotty yellowish-white.

A coated jadeite pipe on which some of the coating had been chipped proved to have been coated with two layers. The presence of the pyroxene absorption band at 437 nm showed that the specimen was jadeite. The lower of the two coatings was mottled green and the upper one either colourless or a uniform very light yellow. UV light showed up the double layering, both layers giving a yellow–green with different intensities. The jadeite showed no fluorescence. IR spectrometry gave the organic polymer spectrum. If the IR spectrum had been the first and only test used the specimen could well have passed for B-jade.

A treatment leaving what has been called a 'beehive' structure behind is reported in the Winter 1994 issue of *Gems & Gemology*. The resulting structure is stated to be one of the chief ways of identifying jade following treatment by bleaching and impregnation (B- to C-jade). The bleaching, in this case of a high quality green and white patterned bangle, was carried out with acid to remove iron staining and left behind a honeycomb pattern formed by grain boundaries. The voids left by the removal of the impurities were filled with a neutral-coloured polymer or wax, making the specimen more translucent and uniformly coloured. Grain boundaries could still be made out under the microscope but the impregnation process made them less sharply defined.

GIA has suggested that different lighting techniques should be used routinely in the testing of jade-like specimens. This is particularly important in the identification of the honeycomb-like structures described above. Reflected light gave better results than transmitted light from a fibre-optic source. Filled cavities flush with the polished surface were also seen better with reflected light. When they were approached by the thermal reaction tester the contents burnt slightly, a characteristic polymer reaction. The bangle just described floated in di-iodomethane, showing that its specific gravity fell in the range 3.20–3.25. A figure in this range would be low for untreated jadeite but could indicate B-jade.

A test not often used with gem material is the sound they make when gently tapped: a pair of tongs is useful for this test and the specimen should not be in contact with any surface. Untreated jade gives a ringing sound whereas treated specimens sound dull by comparison. It is also useful to remember as we said above that a drop of hydrochloric acid will sweat on untreated but remain unaltered on the surface of treated jadeite.

A double-strand necklace with both treated and untreated jadeite beads is reported in the Spring 1995 issue of *Gems & Gemology*. All the beads were found to be jadeite using standard gemmological tests but some beads in each strand fluoresced a weak to moderate yellow under LWUV light while the remainder did not fluoresce. Most polymer-treated jadeite will fluoresce but some untreated material may also show a weak yellow fluorescence under LWUV light. To find both types in a single piece of jewellery is unusual.

An example of dyed nephrite was given in the Spring 1995 issue of *Gems & Gemology*. Five specimens of oval mottled green cabochons ranged in weight from 6.38 to 7.12 ct and closely resembled good quality jadeite. Gemmological testing gave a refractive index of 1.61 and specific gravity of 2.95–2.96, these figures being in the nephrite range. Dye concentration in cracks could be detected and with the hand spectroscope an absorption band in the red was seen (the dyestuff indicator).

A glass imitation of jadeite was reported from Vietnam in 1995 by GIA staff. The material was apple-green with some inclination towards yellow, showing a mottled appearance with slightly circular whitish areas with deeper colour and enhanced transparency around them. The piece was full of spherical gas bubbles, some of which appeared as surface cavities, indicating glass. Under magnification the whitish areas were seen to be associated with bundles of fibrous inclusions – this effect is probably due to devitrification, often confused with natural inclusions.

The material known as Victoria stone or meta-jade, which first appeared during the late 1960s in Japan, bears no resemblance to either of the jades. This has angular fibrous patches of lower transparency in a mainly transparent groundmass. The green material reported from Vietnam had an RI of 1.51 compared to the 1.48 of meta-jade and showed an absorption spectrum with general absorption above 590 nm and below 510 nm (meta-jade shows general absorption above 560 nm and below 480 nm). No fluorescence could be detected. Neither the green material nor meta-jade showed any change of colour through the Chelsea filter. The Vietnamese material was found to contain rubidium, yttrium and zirconium, elements which have not been found in meta-jade.

Serpentine can easily be confused with the jade minerals. It is both fibrous and fine-grained where the jade minerals have interlocking or felted crystals which make fashioning more difficult than with serpentine. GIA in 1995 reported a green and black serpentine resembling nephrite. Standard gemmological tests gave an RI near 1.57 (nephrite would be about 1.60) and a specific gravity of 2.63 as against the 3.0 of nephrite. The serpentine's lower hardness and toughness was shown by many fine scratches and rounded edges on small fractures. Some serpentine will give an indistinct mottled chalky-blue fluorescence. Black inclusions of magnetite were found in the serpentine, sufficient in number for the specimen to be attracted to a hand-held magnet. However, there are many varieties of serpentine and not all will behave in the same way.

A translucent mottled green carving in a closed setting which concealed its depth gave a spot RI reading of 1.66 and showed chromium lines in the absorption spectrum. These features all pointed to jadeite. However, magnification showed that it was a thin hollow shell of jadeite filled with a transparent colourless material containing gas bubbles. The filler was exposed in one place on the surface and was shown by Fourier transform infra-red spectroscopy (FTIR) to be a polymeric substance of a kind which had previously been reported as a jade simulant. The thickness of the shell was about 0.05–0.10 mm.

A boulder of jadeite with the traditional windows cut in it while retaining its crust, on examination, showed that the windows were in fact thin slices of better quality jadeite placed in a fabricated crust.

The visible spectrum of Xinjiang Hetian jade showed that the green colour arose from the total iron content with an absorption peak at 450 nm. The colour indices of this jade showed that it has a low saturation and low brightness.

The interesting name black-skin-chicken jadeite is used to denote jadeite with small amounts of omphacite and accessory minerals. The black colour is considered to arise from minute inclusions of amorphous carbon with some metallic oxides and sulphides. The colour is sometimes patchy.

Fourier transform infra-red spectroscopy methods can be used to test B-jade non-destructively.

Cathodoluminescence can be used to distinguish between natural jade, maw-sit-sit and chloromelanite, treated and untreated materials.

A mottled lavender, green and orange bangle assumed to be jadeite gave a spot refractive index reading of 1.54 and showed dye concentrated both in fractures and between individual grains. A dark absorption area in the deep red could be seen with the hand spectroscope but there were no other visible absorption features. The lavender areas fluoresced a medium pink under LWUV light and the whole piece fluoresced a medium to strong bluish-white under SWUV light. FTIR testing disproved that the bangle was jadeite and suggested quartz. It also showed the presence of polymer impregnation. The piece was a dyed quartzite – it is unusual for this material to have been impregnated.

A relatively new variety of jadeite from northern Myanmar has been called

Hte Long Sein (full green) jadeite. The colour is bright green but the transparency is low due to profuse internal fractures which are normally filled with an epoxy resin whose presence can be detected by IR spectroscopy. The material is coarse- to medium-grained granular with an RI 1.66 and SG 3.30–3.31. It shows green through the Chelsea filter and is inert to both forms of UV light. It has been found to contain between 1.3 and 2.6 per cent Cr_2O_3 and in some specimens a loose texture allows easy impregnation with epoxy resin. The colour and clarity ranges are large.

Violet jade from Turkey has RI 1.65–1.66, SG 3.10 and is inert to both forms of UV light. Samples are composed of jadeite, quartz, feldspar, muscovite, pyrite, Mn and Fe oxides. The rock should be defined as an impure jadeite.

An interesting variegated banded brown, white and light green bead necklace was made from naturally coloured jadeite. It had probably been cut from the outside layer of a naturally coloured jadeite boulder.

A large white translucent stone could have been mistaken for nephrite as its SG of 2.84 was consistent with that mineral. It was in fact aragonite.

Nephrite from Chuncheon, Korea (the trade name is Chuncheon Ok or Baek Ok (Chuncheon jade or white jade)) is monomineralic in the higher grades, consisting almost entirely of tremolite, while poorer qualities are polymineralic, containing small amounts of diopside, calcite and chlorite group species. The colour varies from greenish-white and pale yellowish-green to pale green, rarely deep green, with specimens being uniformly coloured and translucent. This is not the material sometimes known as 'Korean jade', which is serpentine. The RI is 1.62, SG 2.96–3.01 with hardness 6–6.5. There is no response to either type of UV light.

X-ray photoelectron spectroscopy (XPS) is a non-destructive technique for distinguishing unambiguously bleached impregnated jadeite (B-jade) from natural and untreated material. The impregnating substance can be identified by chemical analysis at the surface. A study of 16 jadeite samples showed that some had been impregnated with wax, polymer or both. XPS results were supported by routine gemmological tests and complemented by IR spectroscopy.

A necklace of highly polished black beads sold as black jadeite showed a spot RI midway between 1.60 and 1.70 with SG 3.36. X-ray diffraction analysis proved it to be a ferrohornblende, an iron-rich member of the amphibole group of minerals.

It is unusual to find nephrite dyed green but some examples have been reported. They showed the characteristic dyestuff absorption band in the red centred at 660 nm.

Material offered as 'Siberian blue nephrite' at the 2000 Tucson Gem & Mineral Show was a submicroscopic mixture of quartz, tremolite and the blue amphibole potassian magnesio-arfvedsonite. The blue colour is attributed to Fe^{2+}–Fe^{3+} charge transfer. The Russian name for this material is Dianite.

Chapter 8

Chrysoberyl: alexandrite, cat's-eye and other varieties

Introduction

The mineral chrysoberyl has many good things going for it: no mineral with such varieties as alexandrite, cat's-eye and a brilliant hard transparent yellow–green can be ignored.

Chrysoberyl is an oxide of beryllium and aluminium and the substitution of some of the aluminium by chromium is the cause of the colour of alexandrite (the reason for the colour change will be mentioned below). The mineral is both hard, about 8.5 on Mohs' scale, and very tough, giving a high lustre and sharp facet edges. There is no easy cleavage. Crystals forming apparently hexagonal shapes are in fact multiple crystals (twins) joined according to twinning laws: when the individual crystals are particularly distinguishable they are often too thin to cut but high prices can be asked for well-shaped examples for which the name 'trillings' is used.

Yellow–green chrysoberyl

The transparent yellow–green stones (as well as alexandrite and cat's-eye) have a refractive index in the range 1.744–1.770 with a birefringence of 0.008–0.010 and a specific gravity of 3.71–3.72. The colour is caused by iron and most specimens show a strong band at 444 nm where the violet takes over from the blue. As the colour of the specimen darkens from a bright yellowish-green to golden-yellow to brown this band increases in strength: where the stone has a strong colour two additional bands can be seen at 505 and 485 nm in the green–blue.

The yellow–green chrysoberyls show pleochroic colours of yellow to brown; the commonest inclusions are liquid-filled cavities containing three-phase inclusions though stepped twin planes are also seen. The chatoyancy in cat's-eye is caused by light reflected from short needles and tubes parallel to the main crystal axis.

Alexandrite

Alexandrite at its finest is one of the classic gemstones together with the finest translucent green jadeite, black opal and Burma ruby. The colour change (caused by a shift in the central G absorption band) from red to green when the light by which it is viewed alters from a raspberry-red in incandescent light (tungsten light) to an emerald-green in daylight or fluorescent tubes. Candlelight is often quoted as a good medium in which to look at alexandrite but in fact many alexandrites are small and dark and would appear black, perhaps with the occasional red flash in these conditions. While alexandrites from Russia do show quite a good colour change, Zimbabwe specimens are at least as good in this respect (but are more included) and stones from a fairly recent find in Brazil have been acclaimed worldwide for a really spectacular colour change. Chromium is the cause of the colour.

While the colour change of alexandrite is from red to green, the mineral also shows strong pleochroism in which red, orange and green can be seen in different directions, though two colours only at any one time. The pleochroic colours though similar to those of the colour change are not connected with it.

Cat's-eye

Dealers and connoisseurs prefer the cat's-eye variety of chrysoberyl to have a honey-coloured background (there are many colours of honey!) against which the sharpest possible eye is placed by the lapidary as a blue–white line of light crossing the cabochon parallel to its long direction. The stones are usually dark enough to show the absorption band at 444 nm which without any further testing will separate chrysoberyl cat's-eye (the name cat's-eye without a distinguishing adjective always implies chrysoberyl) from similar examples of quartz or tourmaline.

Synthetic and imitation chrysoberyl

Having introduced the most valuable varieties of chrysoberyl we can review those materials with which they may accidentally or deliberately be confused.

Simulants of cat's-eye

The closest competitors of cat's-eye, quartz and tourmaline cat's-eyes, show no useful absorption spectrum in the visible whereas chrysoberyl shows the absorption band at 444 nm. Many gem species show chatoyancy, though, even if it occurs only by accident and is not commonly repeatable in the same species. In times past a quick test between chrysoberyl and quartz cat's-eyes could be made, if a spectroscope was not handy and you knew one of the specimens was chrysoberyl. Both specimens could be tested for specific gravity by immersion in the heavy liquid di-iodomethane when this type of test was less hedged about with safety restrictions. In this solution which had a specific gravity of around 3.32 quartz would float and chrysoberyl sink.

Convincing imitations of cat's-eye are provided by thin glass rods which can form masses of hollow parallel tubes. When a cabochon is made from this material, light is reflected from the tubes to give a very reasonable eye. The material Cathay stone is very convincing, at least from a distance: it is made by arranging thin glass optical fibres in fused mosaics. The fibres are made from several distinct glasses and are stacked in cubic or hexagonal arrays giving a very sharp eye as up to 150 000 fibres per square centimetre can be accommodated. Neither of these two ingenious products will show the chrysoberyl absorption at 444 nm and the stones are distinctly light compared to chrysoberyl.

Synthetic alexandrite

Because alexandrite commands such high prices all kinds of materials have been used as imitations and it has also been synthesized, though not until comparatively recent years. Taking the synthetic material first, it is grown almost invariably by the flux-melt process and thus will contain particles of flux, forming twisted veils resembling liquid feathers, showing a bright metallic lustre: the refractive index and specific gravity will echo those of the natural material (1.74–1.75, DR 0.008–0.010, SG 3.71). Chromium is added to the starting materials to give the colour.

Creative Crystals Inc. of San Ramon, California, reported the growth of a synthetic alexandrite in 1973. Both crystal pulling and flux growth were said to be used though all commercially offered examples showed flux inclusions. The Japanese firms Seiko and Kyocera have also made alexandrite, which has also been grown for laser use: this would be by crystal pulling to give the necessary rod shape. However, some alexandrite grown for laser use is too pale to be used as a gemstone. It may be insufficiently chromium-rich.

The synthetic alexandrites show dust-like inclusions and the characteristic twisted veils of flux particles. One Czochralski-pulled stone is reported to have shown a strong red fluorescence under LWUV and SWUV light as well as under X-rays. Randomly oriented needles and lath-shaped crystal inclusions have been reported.

An early report on the Inamori synthetic alexandrite in 1987 discussed a cat's-eye variety which showed a distinct colour change. The eye was broad and of moderate intensity. Specimens were dark greyish-green with a slightly purple overtone under fluorescent lighting with the eye showing a slightly greenish-bluish-white. The stones were dull and somewhat oily. They appeared to be inclusion-free and under a strong incandescent light, and looking in the long direction, asterism could be seen with two rays weaker than the eye – such a phenomenon has not been reported from natural alexandrite.

Properties were the same as the natural chrysoberyl though when stones were placed over a strong light source a strong greenish-white transmission luminescence could be seen (it could also be seen in sunlight and in other forms of artificial light). This is the cause of the stone's oily appearance. Under SWUV light specimens gave a weak, chalky opaque yellow luminescence near the surface with a reddish-orange luminescence underlying it. Under LWUV light both natural and synthetic stones behave in the same way but the effect under SWUV has not been reported from natural material. Under magnification parallel striations could be seen along the length of the cabochon and with the diaphragm closed down it could be seen that the striations were undulating rather than straight – another feature not seen in natural alexandrite. Whitish particles oriented in parallel planes were associated with the striations and were the cause of the chatoyancy. No natural inclusions were present in the Inamori alexandrite.

Imitations of alexandrite

For many customers alexandrite appears to be quite a large stone of somewhat amethystine appearance with a definite colour change from a slaty purple to a raspberry red. This is a flame-fusion (Verneuil) synthetic corundum doped with vanadium which when examined with the spectroscope shows a strong vanadium absorption line at 475 nm in the blue. Though other absorption features are present this band is diagnostic. Most court cases featuring supposed (or genuine) alexandrites centre on this synthetic corundum. This says something about the general public awareness of gemstones (and about the level of attainment of many 'gemmologists' who have advised unfortunate clients).

If you want a Verneuil alexandrite imitation you would be better off with the spinel than with the corundum since here there is a real change from a dark green to a dull red. These are in fact quite rare stones and have the spinel properties rather than those of chrysoberyl: refractive index of synthetic spinel 1.728 and of natural alexandrite 1.74 at the lowest. The comparative specific gravities are 3.64 and 3.71. There will be no birefringence and the spinel will show between crossed polars (if the light source is sufficiently strong) the anomalous double refraction long called tabby extinction.

It is possible for alexandrite to be imitated by *glass* but no glass imitation is really convincing. As always, familiarity with the stone imitated will usually alert the careful connoisseur. Most synthetic products will contain traces of flux and the stones made by Seiko, who have used a flux-free method of growth, floating zone, show neither flux nor natural-looking inclusions and so are unlikely to be confused with natural alexandrite.

A glass imitation was reported in 1973 by GIA: it changed from an amethyst colour under incandescent light to a steely blue in daylight. It contained minute crystallites. A glass composite was made from pieces of red and green glass fused together, the two were separated by a plane in which gas bubbles could be seen. Instead of showing different colours under different lighting conditions this piece showed different colours in different directions.

Reports of interesting and unusual examples from the literature

Items in this section have been chosen to illustrate points made in the chapter and to bring one-off items to your notice.

A Russian synthetic alexandrite was described in 1995, the investigator having examined about 200 crystals which were grown in Novosibirsk and sold in Bangkok. Cut stones showed prominent growth zoning as closely packed parallel lines. Some stones had an intense red core with a lighter red rim; between the two was an even more intense red boundary. Alexandrites with any patches of intense red should be viewed with suspicion. Hexagonal outlines are reported from a large number of crystals of these Russian stones.

A flux-grown alexandrite with no diagnostic inclusions was reported in 1995. The stone, of 1.08 ct, was a brilliant-cut with properties identifying it as alexandrite but no routine test could establish whether it was natural or synthetic. In the end infra-red spectroscopy showed that no water-related absorptions were present – these are normally found in the natural material.

An alexandrite of 3.59 ct examined by GIA was found to contain numerous short, thin and highly reflective needles as well as abundant dust-like particles. The needles appeared mostly to be straight though some were slightly curved and curved colour banding was observed to be parallel to the girdle plane when the stone was immersed, though the curvature was slight. Through one end of the stone the banding was straight but curved at the other end. The colour change was from red–purple in incandescent light to both green and purple in fluorescent light. The identity was established as alexandrite by standard gemmological tests but chromium was detected by EDXRF and vanadium in very similar proportions to those found in reference samples grown by the Czochralski pulling method. IR spectroscopy showed that water was not present – this is a characteristic feature of melt-type synthetics.

The name Allexite has been used for a synthetic alexandrite manufactured by the House of Diamonair who state in their publicity that their product is Czochralski-grown. A stone examined in 1992 and reported by GIA showed a strong colour change from reddish-purple under incandescent light to bluish-green in daylight or fluorescent light. This type of change is shown by the finest Brazilian alexandrites. In visible light a strong red luminescence (red transmission) could clearly be seen. The refractive index was measured at 1.740–1.749 with a birefringence of 0.009. A strong red fluorescence was observed under LWUV light and a more moderate but similar reaction under SWUV light. The specimen showed red through the Chelsea filter and curved growth striae could be seen under magnification.

Alexandrites are not often fracture-filled but one example was described in 1992 by GIA. The filled cavity was in the base of a natural alexandrite and was in a large negative crystal. It could easily be seen when the stone was examined though the table and its presence was proved by the presence of a gas bubble. When the thermal reaction tester was brought near the filling the material softened and some smoke was produced, suggesting that the filler was a polymer rather than glass.

A flux-grown alexandrite described by GIA in 1988 contained unidentified crystals as well as minute, well-spaced whitish pinpoint inclusions. There were no traces of flux and the stone had a good colour change with a strong red fluorescence under LWUV light as well as an oily appearance found in other synthetic alexandrites. Crystal inclusions are rare in flux-grown material.

A sample of 200 Russian flux-grown synthetic alexandrites (which were mostly crystals) showed that most were cyclic twins (described earlier). X-ray fluorescence analysis showed traces of germanium, molybdenum and/or bismuth in many of the specimens: these elements have not been found in natural alexandrite. Absorption bands due to water and/or hydroxyls are absent from flux-grown synthetics.

Cathodoluminescence spectroscopy was found to be useful in establishing the origin of two alexandrites tested in the London laboratory of GAGTL. Most natural alexandrites show a fairly low CL response while flux-grown specimens show a much higher intensity. Some natural alexandrites, however, perhaps originating from the 'Malisheva' mine in the Sverdlovsk area of Russia show a CL intensity more characteristic of the synthetic material. One stone was thought to be of this type while the other showed a CL response more characteristic of most natural alexandrite.

Blue alexandrite from Malacacheta, Brazil, has been called locally 'peacock-blue alexandrite'. Strong pleochroism gives blue to green or greenish-yellow and the colour change gives blue or greenish-blue in daylight and a reddish-purple in incandescent light. The effect arises from a critical balance between Cr^{3+} and Fe^{3+} substituting for Al^{3+} in the sites with mirror symmetry: the concentration of Cr^{3+} in the sites with inversion symmetry is negligible.

Synthetic hydrothermally grown alexandrite made at Novosibirsk, Russia, shows distinct swirl-like growth inhomogeneities, some of which can be seen with the 10× lens.

Alexandrite from Andhra Pradesh, India, was pale to dark green with a pronounced colour change, while chatoyant chrysoberyl was a pale yellow with a whitish sheen. Alexandrite from Madhya Pradesh was a dark green with a strong colour change while yellow chrysoberyl from the same state had a slightly greenish tinge. Sillimanite, quartz, mica and rutile have been generally observed as inclusions, while crystals of apatite have been found in Andhra Pradesh material only.

Russian synthetic alexandrite and phenakite have been grown at the same time by flux-melting. Very dark to black alexandrite crystals supported yellowish phenakite crystals with the alexandrite crystal faces showing a marked roundness characteristic of synthetic origin.

The crystals grew as irregularly developed trillings and the alexandrite-type colour change could be seen only at the edges. Those inclusions which could be seen appeared as trigonal or distorted blackish material with a submetallic lustre, perhaps hematite, while characteristic flux veiling was abundant. Growth features and colour zoning seemed to be confined mostly to the outer rim and were not very common. RI measured on flat faces gave values between 1.750 and 1.758 with a birefringence of 0.008: there was no response to either form of UV.

Synthetic yellow–green, bluish-green and pink chrysoberyl have been grown using vanadium as the colouring element for the greenish stones and titanium for pink material.

A 4.58 ct cat's-eye reported by GIA had the very high RI of 1.78 (spot method) with SG 3.81. It was said to come from Orissa, India. When a flat face was polished on the specimen the RI was found to be in the range 1.780–1.793 with DR 0.013. Chrysoberyl was confirmed by X-ray powder diffraction.

Chapter 9

Topaz

Topaz is the archetypal yellow stone but not all topaz is yellow. 'Yellow' ranges from golden through a paler yellow to a brown or smoky colour: some topaz is pink and a good deal of topaz in the market today are shades of blue. Topaz relies completely for effect upon its body colour rather than for phenomena; fortunately the body colour is strong and is supported by a reasonably high refractive index. Topaz is the standard 8 on Mohs' scale of hardness and takes an excellent polish.

The properties of topaz do not vary greatly: certainly they do not really cause confusion with other major species: we shall see similar properties as we proceed. The refractive index and specific gravity are roughly correlated with colour and with the ratio of (OH) (hydroxyl) to (OH + F) in the composition.

The specific gravity is in the range 3.50–3.53 for pink stones, 3.51–3.54 for yellow stones, 3.56–3.57 for colourless stones, 3.56–3.57 for blue ones.

The refractive index also varies with brown stones from Ouro Preto, Brazil, in the range 1.629–1.637 with a birefringence of 0.008 and SG 3.53. These stones are rich in (OH) and may contain some chromium. Fluorine-rich bluish stones from Russia have RI in the range 1.609–1.619 with DR 0.010 and SG 3.53. Pink specimens from Mardan, Pakistan, have RI 1.632–1.641 with DR 0.009 and SG 3.53.

Most commonly when pleochroism is observed in gemstones it will usually be two shades of the same colour but in general this is not a very helpful test for topaz. Some of the chromium-bearing material from Ouro Preto may give an orange fluorescence between crossed filters or under LWUV. When these are heated to bring out the pink colour this response is stronger: a weak emission line may be seen when the stone is fluorescing (fluorescence spectrum), placed at 682 nm. In pink sapphire the emission line is in a different position.

Nonetheless, interesting though this phenomenon is, fluorescence would not be the first test of a brown to yellow unknown. We have to remember that there are not very many small faceted topazes around as large crystals are usually available so that taking the refractive index is usually possible. Topaz is not usually carved (the easy, perfect basal cleavage militates against this possibility) nor cut into cabochons. The cleavage in topaz is at right angles to the vertical axis of the crystal and is easy to start – a knock on a glass dressing-table will often cleave a stone: if cleavage has begun rainbow-like bands of interference colours may be seen inside the stone.

Topaz, an aluminium fluosilicate, is hard to synthesize and I have seen no reports of this being successfully accomplished for crystals of commercial size. Imitations provide the greatest threat.

The closest match to the usual yellow to brown topaz is provided by citrine which has lower hardness, refractive index and specific gravity at 7, 2.65 and 1.544–1.553 with a birefringence of 0.009. Citrine has no easy cleavage and does not show interference colours inside the stone. It shows no luminescence and between crossed polars the interference figure will show a black cross with arms which do not meet at the centre, being replaced by coloured concentric circles. This is quite unlike the figure shown by topaz, in which there is no cross but harder to distinguish, black boomerang shapes sometimes placed back to back.

Star stones are very rare in topaz but occasionally one turns up: cat's-eyes seem rarer. When the liquid bromoform was considered safe for use outside strict laboratory conditions it was possible to keep a piece of quartz as an indicator in the jar when the liquid was diluted to 2.65. Tested against the suspended quartz, topaz would quickly sink. However, the use of bromoform is now against health and safety regulations in many countries, though it may be used in laboratory conditions so that it has not completely disappeared as a testing medium.

Natural stones that may be confused with topaz

Other than *citrine* or *rock crystal* (as an imitation of colourless topaz) a number of gem species may be taken for topaz. Yellow *sapphire* has a higher refractive index and specific gravity at 1.76–1.77 and 3.99–4.00. Yellow sapphire may fluoresce (iron-poor Sri Lanka stones usually give an apricot fluorescence under LWUV) or, if richer in iron, give an absorption spectrum of three bands in the blue with the strongest at 450 nm. Yellow topaz shows no absorption in the visible.

Yellow *orthoclase* has a less bright appearance with a lower mean RI of 1.53 and SG 2.56. It shows two absorption bands in the blue and in the violet and its lower hardness makes it an altogether quieter stone. Yellow *chrysoberyl* is not often quite the same colour as topaz but occasionally there could be confusion: chrysoberyl will usually give a strong absorption band where the blue and violet meet, at about 444 nm, whereas yellow topaz will show no absorption in the visible. Yellow *apatite*, like orthoclase, is not the strong yellow of topaz and will usually show the rare earth absorption spectrum with two groups of fine lines, in the yellow and in the green. Orthoclase does not show an RI/SG anywhere near those of topaz: its mean RI is near 1.53 and SG near 2.6. Apatite has a mean RI of 1.63 and mean SG about 3.32.

Yellow *zircon*, like topaz, occurs in a variety of shades but most specimens will give the highly characteristic absorption spectrum of uranium, with fine lines and bands extending over the whole extent of the visible spectrum. The strongest absorption band will be in the red at 653.5 nm. If the yellow zircon has been produced by heat treating brown crystals – as most will have been – the absorption spectrum will show much less strongly, though the 653.5 nm band will still be the most prominent. Reflected light will be the best way of viewing the absorption spectrum: the RI of yellow zircon will not be obtainable on the refractometer and the SG will be higher, at about 4.69, than that of topaz.

The rather rare gemstone *sinhalite* may sometimes appear in a topaz-like yellow. Like zircon, it too can easily be distinguished by its absorption spectrum which will show, when the colour is sufficiently deep, quite strong absorption bands in the blue area, at 493, 475, 463 and 450 nm. If the RI is required it should be easy to obtain, at a mean 1.69: the SG will be near 3.48. *Spessartine* does not look quite like topaz but a mistake could be made: here again the absorption spectrum should be looked at first. It will show a strong band at 432 nm with several other bands – nothing like topaz. Topaz colour would be quite pale for the majority of spessartines and the absorption spectrum correspondingly harder to see. The RI and SG of spessartine could not be confused with those of topaz: they are in the region of 1.80 and 4.16.

Where the garnets are concerned, the yellow *grossular* found in East Africa makes a better 'topaz' than spessartine. The grossular can be a golden-yellow to brown; this time there is no distinctive absorption spectrum but the RI and SG should make the distinction clear; mean figures have been reported as 1.734 and 3.604. Some of these yellow stones fluoresce orange under X-rays.

Yellow *spodumene* is a rather greenish-yellow in many cases but some stones could be taken for topaz. Spodumene will show the 'pyroxene' absorption band near 437 nm with another band at 432.5 nm.

The golden and yellow varieties of *beryl* have RI and SG below those of topaz, at about 1.57 and 2.68. They show no distinctive absorption in the visible. The rare gemstone *brazilianite* has RI 1.613, just below that of topaz, and SG 2.98, well below. The DR is greater, at 0.021.

Yellow *danburite* will almost always be much too pale to be confused with topaz but irradiation and heating of some Russian material to a much stronger, sometimes near-orange colour has been reported and some presumably untreated stones from Madagascar can just reach a pale golden colour. Danburite has a mean RI of 1.63 (in the topaz range) and SG 3.00, well below it. The DR is slightly less, 0.006 compared to 0.008.

Tourmaline can be found in topaz-like shades of golden to yellow or brown but its DR is much higher, at a mean of 0.018, than that of topaz at 0.008. *Glass* will inevitably turn up but will not usually reach the relatively high RI/SG shown by topaz. Glass may also form the major part of a doublet but again will usually show lower RI/SG figures.

The blue colours of topaz have been much more sought after in recent years since the irradiation of pale blue material was found to produce a much more arresting darker blue. The treatment cannot be detected by the standard gemmological tests but treated stones can almost be spotted by the eye alone, their appearance being highly distinctive and somewhat metallic. The colour so far has been found to be stable and the stones are not radioactive.

Blue topaz is rarely like blue *sapphire* though it would be difficult to explain exactly why the colours are sufficiently different to make separation of one from the other by the eye alone possible – though a surmise would not be good enough for a customer nor for certification! Blue topaz when pale could easily be mistaken for aquamarine, however, and it can easily be separated from that stone by careful measurement of its refractive index, where possible.

Aquamarine will show a mean RI of 1.58, DR of 0.006 and SG 2.70, compared to the mean 1.62, 0.010 and 3.56 of blue topaz. *Aquamarine* will not show the signs of cleavage that blue topaz may very well show and will give absorption bands at 456 nm in the blue and at 427 nm in the violet, the latter being quite strong in large stones.

Maxixe and Maxixe-type beryl (see Chapter 4) can look much more like a dark blue treated topaz but will show an absorption spectrum where blue topaz will not. The Maxixe beryl spectrum is not easy to miss, nor is the pronounced pleochroism. In 'normal' (not Maxixe) aquamarine the deeper blue colour can be seen only with the dichroscope (which gives the darker and lighter blue side by side when the stone is viewed in the correct orientation for this to be seen) while in the Maxixe stones this colour is normally seen through the table facet, making the similarity with dark blue topaz even more dangerous. The absorption spectrum will give it away but in case the spectroscope is not available blue beryl's RI and DR will confirm its identity.

Again, blue topaz is not very like blue sapphire, the untreated material being on the whole too pale and the darker treated stones 'not quite right'. But it all depends what you are told! The gemstone buying public has to believe the sellers as they have no obvious means of proving them wrong. For the gemmologist the absence of an absorption spectrum in the visible is highly suggestive of blue topaz, especially when the specimen is dark, metallic and more or less inclusion-free. Otherwise the higher RI and SG of sapphire (mean figures 1.77 and 3.99) will have to serve as a basis for testing. Blue sapphire will show a stronger pleochroism.

Still with natural materials, other causes of confusion with blue topaz may arise with less common species which may include *iolite* (pleochroism much too strong, RI/SG lower at 1.54/2.59); blue *tourmaline* (DR much higher than that of topaz, at 0.018); *benitoite* (high DR of 0.047); *kyanite* (mean RI 1.72).

Blue *spinel*, often the cause of confusion, has a higher RI and SG, at 1.718 and 3.60: it shows no birefringence and no pleochroism. If testing these properties does not appeal the spectroscope will soon tell you that the unknown is not blue topaz and blue spinel's rather complex-looking (iron) absorption spectrum is not easily confused with any other. A variety of blue spinel found in Sri Lanka contains cobalt and shows red through the Chelsea filter – this is a very rare stone and does not closely resemble blue topaz. The much commoner iron-bearing blue spinel does not show red through the Chelsea filter.

The spinel scene looks complicated! Blue cobalt-doped synthetic spinel grown by flame-fusion is very common but could only just be confused with topaz: its RI and SG are 1.728 and 3.64. This one contains cobalt so through the Chelsea filter it will show a bright red.

Synthetic flame-fusion *corundum* can be a fine blue but this of course is meant to be mistaken for natural blue sapphire and will show RI 1.76 and SG 3.99 as well as a faint iron absorption in the blue.

Glass with RI between 1.50 and 1.70 and SG almost anywhere could be made to imitate blue topaz but should be not too difficult to identify.

Pink topaz may be obtained by heating some of the brownish material found in Brazil for the most part (or worth mining there). It is, however, despite some textbooks, found in the most magnificent dark pink to orange colour (Campari?) at Katlang in Pakistan, which I (MO'D) can vouch for, having mined there. This material is quite unlike the pink topaz usually found on the market with vivid and stable colour properties in the topaz range and no other stone remotely resembles it.

This material is exceptional, though, and most pink topaz is quite easy to confuse with some of the other gem species, more so than blue topaz. Pink sapphire shows a chromium emission doublet in the red which is much stronger than the weaker fluorescence line, which may also be part of a doublet, in pink topaz. Nonetheless, if the investigator is not used to the spectroscope and some of its results, confusion may arise when any chromium emission lines are seen as corundum is bound to be thought of first. Fortunately the refractometer will take care of this particular identification problem.

There are several other natural gem species which show pink and although the pink is sometimes rather pale this is the case with pink topaz too. They can be reviewed in no particular order, as follows.

Pink *scapolite* has mean RI and SG 1.54 and 2.59, well below those of topaz. Some stones incline to violet but this is not close to the Pakistan deep pink to orange topaz. The pink variety of spodumene is *kunzite* with a mean RI of 1.67, rather high for topaz, and SG 3.18, quite a lot lower. Like topaz, kunzite has a very easy, perfect cleavage but normally one would hope to avoid it. Kunzite has no characteristic absorption spectrum but shows strong pleochroism with purple to violet in one direction and near colourless in another. Topaz in general may show a variety of fluid inclusions but is generally free from solid ones: cavities containing two immiscible liquids are characteristic of topaz and some of the Pakistan pink crystals contain script-like shapes made up of liquid inclusions. Topaz may also show angular growth zoning with angles echoing the orthorhombic shape of the crystal. As Gübelin (1992) states, topaz is poor in guest minerals but may contain, on occasion, crystals of monazite, muscovite, spessartine, albite, quartz and brookite.

The pink *beryl*, morganite, can look quite like pink topaz although its refractive index is lower at around 1.59 with SG about 2.80. It has a higher birefringence than the other varieties of beryl, at 0.008, the same value as topaz. It is possible that amethyst of a certain colour could be mistaken for topaz but the RI will quickly show its true nature. Pink spinel may resemble pink topaz but will show the spinel refractive index of 1.718, or 1.728 if the specimen is synthetic.

Irradiated blue topaz

The nature of the treatment which turns pale blue topaz to a darker blue is normally by placing the material in a linear accelerator (Linac) because it can provide higher energies than those available from gamma-ray irradiation. Early experiments with Linac caused some stones to become radioactive for a short period but the problem was soon overcome and the market was never flooded with radioactive stones. Blue topaz thus irradiated is usually then heated but material turned blue in a nuclear reactor may not need subsequent heating. The name London blue is used for stones with a somewhat inky dark blue colour. Reactor-treated stones may retain radioactivity for 1–2 years and release onto the market is legislated for in many countries. Reactor-treated stones with such names as Super blue, Swiss blue and American blue have been on the market and thought to have been reactor-treated, Linac-treated and then heated to lessen inkiness.

As with quartz crystals, topaz with a thin gold coating has been marketed under the name Aqua topaz. A surface iridescence can be seen.

Reports of interesting and unusual examples from the literature

Items in this section have been chosen to illustrate points made in the chapter and to bring one-off items to your notice.

Topaz has been used in composites and one such example was reported in the Winter 1995 issue of *Gems & Gemology*. The stone was a 1.48 ct faceted oval imitating the fine blue Paraíba tourmaline which sells for extremely high prices. Topaz forms both crown and pavilion, the blue colour coming from the cement joining the two parts together. Viewed from the side the deception can be seen: when set the low DR of 0.01 is far too low for tourmaline.

An irradiated green topaz has been on sale under the name Ocean Green Topaz with colour ranging from light to medium tones and from yellowish- and brownish-green to a more saturated green to slightly bluish-green. Specimens are said to have come from Sri Lanka.

Gamma-ray spectroscopy has been used to identify radionuclides in irradiated gemstones, particularly in blue topaz. The GAGTL laboratory in London encountered a steel-blue topaz with 20 counts per second on a Geiger counter – this would need care in handling. Sets of perfectly matched blue topaz should be carefully examined although any radioactivity will diminish fairly quickly.

The name 'Imperial topaz' is given to all topaz mined in the Ouro Preto area of Minas Gerais, Brazil, even though colours range from yellow and orange to pink and red, the latter being the most keenly sought. Heating brownish-yellow and orange material changes the colour from peach to pink. Fluid inclusions hinder the heating process as they may cause cracking. Heat-treated stones show a stronger fluorescence under SWUV than natural ones but so far a statistical sample has not been achieved.

Thermogravimetric studies of 'Imperial' topaz from Ouro Preto show that heating up to 1000°C appears to bring about no change in composition but the colour changes from reddish-brown or orange–yellow to colourless. At 1200° the first loss of mass is reported to occur, this being ascribed to the loss of (OH) groups, and a second loss at around 1320°C when mullite (aluminium silicate) is formed.

Sherry-coloured topaz from the Thomas Range in the west of Utah is rich in fluorine and crystals range up to 3 cm in length. The colour is reported to fade in strong sunlight.

Topaz found in the pegmatites at Klein Spitzkopje, Namibia, has RI 1.610–1.620, these figures being more often associated with topaz found in rhyolite. A slightly higher SG may be due to an enhanced fluorine content.

A bluish-green cobalt-diffused topaz gave a negative reading on the refractometer: a natural bluish-green topaz showed marked colour zoning and no radioactivity (some green topaz has been found to be radioactive).

Topaz of pink, orange and red colours was found to have gained them by a form of sputtering which could easily be scratched. A bluish topaz had been coated with a cobalt-rich material which could only be scratched by a point of hardness 8 or above. In the Winter 2001 issue of *Gems & Gemology* synthetic topaz crystals are described. They were made experimentally by the Institute of Experimental Mineralogy, Chernogolovka, Russia, by the hydrothermal method. A semi-transparent light-grey crystal weighed 23.66 ct and a semi-translucent brown specimen 27.48 ct. Two-phase inclusions were observed. Colour stability has not been reported on so far. RI was 1.610–1.620 and SG near 3.57.

Chapter 10

Tourmaline

Colour varieties and properties

Like the garnets, the tourmaline minerals form a group. Although there is not the same close intermingling between intermediate species, some exists. All species are borosilicates. Unlike some of the garnet gemstones, their tourmaline counterparts cannot be distinguished by their colour and to the gemmologist or jeweller it does not matter to which species an individual tourmaline belongs and the species' names are unfamiliar; they are elbaite, dravite, uvite and liddicoatite, all gem tourmalines belonging to one or other of them. The black tourmaline, schorl, which is sometimes cut but more often collected for its crystals, is also a separate species.

Tourmaline has the greatest colour range of all the gem minerals and fine qualities are found in all of them – until the last few years blue was perhaps the poor relation but with the coming of the superb Paraiba blue tourmaline this has changed. This is coloured by copper.

Tourmaline, like the silicate garnets and topaz, is very difficult to synthesize in sizes above a few microns and at the time of writing there are no crystals of gem size. Tourmaline species are imitated, however, as we shall find later. In many ways tourmaline is the ideal gemstone: adequately hard (close to 7.5) and with no easy cleavage, not too rare but rare enough to make it desirable, and with excellent, sometimes superb, colours, it can always command a sale. Watermelon stones have pink core and green rind.

No tourmaline species gives a really convincing absorption spectrum and most do not luminesce: fortunately most specimens behave really well on the refractometer, where they show a marked mean birefringence of 0.018. The usual refractive index range is 1.610–1.675 and the specific gravity range is 3.01–3.12. To some extent these two constants vary with colour though it is not really a vital testing criterion. For the record red to pink stones usually fall in the SG range 3.01–3.06, pale green stones 3.05, brown stones 3.06, dark green 3.08–3.11, yellow–orange 3.10, black 3.11–3.12.

Some (but very few) dark green faceted tourmalines have been found to show up to six shadow edges: they can be removed by repolishing the stone but as they do not affect the stone's appearance they might well be left as they are for general interest and display. Examples I have seen over the years are all rather too dark a green to make the finest gemstones.

Colourless tourmaline is very rarely found but all the coloured varieties show notable pleochroism, light and dark shades of the same colour being the most usual effect. Observers should note that in tourmaline absorption of the ordinary ray (see gemmological textbooks) may be sufficiently complete for only one shadow edge to be seen on the refractometer so that the specimen could be taken as singly refractive: rotation of the specimen in any optical test is vital. One specimen of tourmaline showed a fine deep green and a deep garnet red through a dichroscope, showing that pleochroism is not always light and dark of the same colour in tourmaline.

Though in general tourmaline does not luminesce a few have been found to respond to UV. Under SW some brown, yellow and golden specimens from Tanzania give a strong yellow and some Brazilian pink stones a blue or lavender colour. Tourmaline is quite heavily included with many stones showing flat, liquid-filled films which can appear black in reflected light. Also found are elongated cavities, sometimes containing two-phase inclusions, lying parallel to the length of the crystal. When the cavities are sufficiently dense they may give rise to chatoyancy (the cat's-eye effect) though the eye is rarely if ever as fine as that found in the best quality cat's-eye chrysoberyl. Examination appears to have shown that green stones are the most likely to contain growth tubes while needles occurred most often in blue and red stones: while the growth tubes are filled with tourmaline with other minerals, the needles were found to be either tourmaline or hornblende.

Of the gem-quality minerals of the tourmaline group, elbaite shows the greatest range of colours while dravite is usually dark. Uvite provides some brown and green stones (a chrome-uvite from Burma is a very fine green): liddicoatite shows a variety of colours including some fine multi-coloured crystals which when sectioned across their length show triangles of different colours among which red is especially prominent. These come mainly from Madagascar and show an RI of 1.621 and 1.637.

The causes of the different colours in tourmaline can be established but for the gem tester they do not always help with identification. As stated earlier, the absorption spectra are not diagnostic but here are a few examples. Some red tourmalines obtain their colour from iron and others from manganese: some of the red stones will show two narrow absorption bands in the blue at 458 and 450 nm with a fine line at 537 nm within the general absorption of the green. If the stone is a dark red and the lines in the blue are observed first the observer should not jump to the conclusion that the stone is a dark ruby of the Thai (Siam) type.

In green tourmalines most if not all of the red is absorbed down to about 640 nm and an iron band can be seen at 497 nm. It can also be found in blue tourmaline. Some green tourmaline of a near emerald-green colour may contain chromium or vanadium or both and the chromium-bearing specimens will gave elements of the chromium absorption spectrum.

While most of the commercial tourmaline of different colours comes from Brazil and shows properties which fit into what we have already said, it is worth mentioning some isolated examples of production from other countries where individual or unusual properties may assist in testing. In 1984 a report on some intense yellow gem-quality tourmaline from Zambia showed that the material belonged to a series with elbaite, the sodium (lithium, aluminium) aluminium borosilicate and a variety of similar composition but with some manganese (about 6.3 and 6.9 per cent manganese oxide) present. The stones gave an RI in the range 1.622–1.648 with birefringence 0.023–0.025 with an SG of 3.13. A red tourmaline with a very high RI of 1.623–1.655 appeared to be an iron-rich dravite (the SG was 3.07 and the stones came from Osara, Narok, Kenya). Another high-reading red tourmaline was reported from the Chipata area of Zambia: this had an RI of 1.624–1.654 with a birefringence of 0.030 and a mean specific gravity of 3.05: the pleochroic colours were bright red and a dull brownish-red.

A dark green variety of elbaite found in the Klein Spitzkopje area of Namibia turns to an emerald green on heating: elbaites of a chrome green from Tanzania will give a strong red through the Chelsea filter though other tourmalines remain inert. These are dravites, which species is also represented by a reported blue–green stone with RI 1.616–1.637 and pleochroic colours blue–green to light yellow–green from East Africa. The so called 'cranberry tourmalines' seem first to have been reported from the pegmatites at the Itataia mine in Minas Gerais, Brazil. Brown tourmaline from Sri Lanka may be uvite or dravite, the uvite with the RI readings of 1.619–1.648, the dravites 1.623–1.640. The respective birefringences are 0.016–0.021 and 0.017. SGs are for the uvite 2.85–3.24 and the dravite 2.92.

GIA examined a yellowish-green tourmaline with a DR of 0.028 and showing an absorption band at about 423.5 nm, with a stronger one extending from about 420 to 418 nm with a cut-off at about 410 nm. This absorption pattern was probably due to manganese. The stone contained many gas- and liquid-filled inclusions.

We have already mentioned that the inclusion picture of the tourmaline gemstones is one of some richness. Tourmalines are a classic gemstone of the late pegmatitic to hydrothermal growth periods: crystals can reach quite large sizes and contain growth channels parallel to the main crystal axis as well as the classic 'trichites', networks of hair-like capillaries formed by fluid residues. We have already noted that the trichites appear to be found in tourmaline of all colours while the growth tubes seem confined to blue, green and pink crystals. No solid inclusions characterize any of the tourmalines.

Identification

Tourmaline is not the hardest gemstone to identify and a specimen with a bi-refringence of about 0.018 and RI in the range 1.62–1.64 is certainly more likely to be tourmaline than any other major species. Two brownish-red stones are reported to have shown a birefringence of 0.015 and also a biaxial interference figure but this is exceptional. When observing the behaviour of an unknown gemstone on the refractometer the specimen should as a matter of course be rotated: if the unknown is a tourmaline only the shadow edges corres ponding to the lower RI reading will move and if the table facet (being the one tested) has been cut in a direction at right angles to the optic axis of the stone (in tourmaline this corresponds to the vertical crystal axis) neither of the shadow edges will move during a complete rotation, remaining fixed at the full birefringence position throughout. This is an additional aid to identification.

With the varied colouration of tourmaline in particular the lapidary may find that a better stone will be obtained by placing the table facet at an oblique direction to the vertical crystal axis and in such cases one of the shadow edges will move to approach the other but never completely correspond with it. Finally (this is a favourite examination topic so for the sake of students the full story should be told) the table facet being placed parallel to the vertical crystal axis will cause the shadow edges, again during a complete rotation, to move from a position of coincidence to one of maximum separation and back again.

All this may be hard to carry out with a very small specimen but if it is coloured (and if tourmaline, it will be) the dichroscope can be brought into service. Tourmaline normally shows light and dark of the same colour though in very dark stones the effect cannot be seen as one of the two refracted rays is absorbed strongly: this happens particularly with brown and green specimens. Very pale stones will be harder to test in this way and greens seem particularly difficult, emerald-green specimens resulting from heat treatment being perhaps the worst.

In this event the observer should first examine the stone under magnification when, after trying out a number of directions, doubling of opposite facet edges and of inclusions will be seen if the stone is a tourmaline. We may also remember that coloured stones looking dusty in a lighted jeweller's shop window are likely to be tourmaline since crystals of tourmaline are pyroelectric – they develop an electrical change on heating and thus attract dust.

Natural materials which may be confused with tourmaline

Tourmaline with so many colours can be imitated by almost any other gemstone or artificial material though really serious attempts at imitation centre around the most valuable tourmaline varieties, in particular the magnificent blue Paraiba tourmaline from Brazil. Dispensing with glass first, tourmaline's birefringence at once makes the distinction and no more needs to be said.

We have already compared red tourmaline with the darker rubies which will show the ruby absorption spectrum and have a higher RI and SG in the region of 1.75–1.76 and 3.99–4.00. Ruby shows a much lower birefringence at 0.009 compared to 0.018 of tourmaline. Rubies will contain mineral inclusions which will not be seen in tourmaline and will fluoresce red under LWUV – though this effect might not be perceived with very iron-rich rubies. Red spinel can also be distinguished from red tourmaline by its chromium absorption spectrum, lack of birefringence and fairly constant RI of 1.718 and SG of 3.60. Some of the red garnets may also be confused with red tourmaline: they will show no birefringence and have higher RI and SG in a wide range from, say, 1.71 to 1.83 and 3.58 to 4.31.

Green tourmaline may be confused with *peridot*: the appearance can be quite similar and both species will show a marked birefringence, tourmaline around 0.018. The DR of peridot, however, is much greater, in the region of 0.036 accompanying an RI of 1.654–1.690 and an SG of around 3.34. The pleochroism of peridot is much less marked than that of tourmaline.

More seriously, some of the chrome green tourmalines could be mistaken for *emerald*: here again tourmaline has a much higher birefringence than emerald's 0.006, a higher refractive index and specific gravity than emerald's 1.56–1.57 and 2.71 (average figures). Some of the chrome green tourmalines show red through the Chelsea filter while other green specimens remain green.

Standard gemmological tests can also be used to distinguish chrome green tourmaline from *tsavolite*, the rich chrome-green variety of grossular garnet. Tsavolite has a much higher RI and SG, at 11.74 and 3.68, and of course there is no birefringence. Inclusions too will distinguish both emerald and tsavolite from green tourmaline, which includes very few mineral but profuse liquid inclusions.

Some green *fluorite* can resemble tourmaline when cut but fluorite is softer, cleaves easily, usually fluoresces purple under LWUV and shows no birefringence. Should matters reach the refractometer, fluorite has the quite low RI of 1.43. Taking the SG may knock the fluorite specimen and start a cleavage but the mean value will be quite close to that of some tourmalines, at about 3.18.

The other colours of tourmaline can be distinguished from other stones by standard gemmological tests and in particular by tourmaline's birefringence. Yellow tourmalines are more birefringent than yellow *sapphires* which show only about 0.009, and those from Sri Lanka, at least, are likely to show an apricot-coloured fluorescence under LWUV. Other yellow sapphires, containing more iron, will show a distinctive absorption spectrum of three bands in the blue at 471, 460 and 450 nm, the latter much the strongest. Nothing like this can be seen in yellow tourmaline.

Yellow *apatite* will usually show the rare earth (RE) fine-lined absorption spectrum of didymium (this is a coined name to avoid having to cite neodymium and praseodymium each time you need to). Otherwise, apatite could overlap with some of the tourmalines in refractive index but not in birefringence which is no more than 0.013 at most. The pleochroism is not so notable as in tourmaline but yellow apatite may give a variety of fluorescent effects, not shown by tourmaline.

The rare yellow sodium aluminium phosphate, *brazilianite* has RI in the range 1.609–1.623 and quite a high birefringence 0.019–0.021. The SG is in the range 2.98–2.99. It would sometimes be necessary to check the specimen's behaviour on the refractometer if confusion with yellow tourmaline arises; when correctly oriented on the instrument, both shadow edges of brazilianite will be seen to move as the specimen is rotated: this cannot happen with tourmaline.

Yellow and pink *topaz* can be confused with tourmaline if only the refractive index is taken. The RI of topaz is in the same range as that of tourmaline but the SG is higher at 3.53: the birefringence of topaz is only 0.008–0.009 and the marked cleavage is commonly indicated by rainbow-like interference colours inside the stone. Pink topaz may show at least one of the emission lines of chromium when examined with the spectroscope.

The beautiful yellow *orthoclase*, the potassium variety of feldspar, has a refractive index in the range 1.518–1.539 with a DR of 0.008. The specific gravity will be in the range 2.55–2.63. With the spectroscope many specimens of yellow orthoclase will show two iron absorption bands, a strong one at 420 nm and weaker ones at 445 and 420 nm.

Golden, or more often yellow, *beryl* could look very like tourmaline were it not for the latter's much higher birefringence and lower refractive index. *Citrine* may also look like yellow tourmaline but again can be distinguished by its lower DR of 0.009. No tourmaline, unlike sapphire, has been found in amethystine colours yet.

On the grounds of colour alone, some natural blue *spinels* could be taken for blue tourmaline but that would be before spinel's lack of birefringence and pleochroism are revealed in the application of standard gemmological tests.

Spinel has an SG of 3.60 and RI 1.718: the blue variety, like the blue tourmaline, is unlike aquamarine or blue topaz but will show an interesting and quite complex-appearing absorption spectrum comprising a strong absorption band in the blue centred at 459 nm with a narrow, also strong band at 480 nm. Further bands can be found in the green at 555 nm, in the yellow at 592 nm and in the orange (632 nm).

Neither yellow *danburite* nor yellow *chrysoberyl* really look very like tourmaline as they lack its high birefringence, danburite's DR being about 0.006 and that of chrysoberyl about the same. The RI of danburite is in the tourmaline range. Chrysoberyl has a much higher RI at 1.74–1.75. It usually shows a strong absorption band in the deep blue at about 444 nm where the blue merges into the violet.

Synthetic materials which may be confused with tourmaline

Some of tourmaline's colours have been imitated by Verneuil-type flame-fusion synthetic crystals of corundum or even spinel but stones cut from the characteristic boules have the same properties as their natural counterparts though with a complete lack of those mineral or liquid inclusions which characterize tourmaline and other natural gem minerals. Tourmaline and the other silicate gemstones cannot be grown by the flame-fusion method (gem-sized tourmaline crystals have not yet been grown in other than extremely small sizes which would need a microscope's assistance to be visible) so all the corundum and spinel crystals of tourmaline-like colours are imitations. Interestingly a dark green Verneuil spinel grown, it was said, to imitate dark green tourmaline, shows flashes of red and, when faceted, makes a more than adequate alexandrite imitation: these dark green spinels are rare and collectable.

The water-melon tourmalines with their pink centres and green surrounds have not been successfully imitated, perhaps because the natural materials are not too hard to find, although they are not cheap.

In recent years tourmalines of an amazingly fine bright blue, quite unlike the normal rather quiet blue tourmaline, have come from the Brazilian state of Paraiba. The prices asked for these admittedly remarkable stones are very high indeed and it is not at all surprising that they have been quickly imitated, along with reports of the mine being closed and taken over by the army – always a sure sign of something exciting having been found. Inclusions of copper and of the copper oxide tenorite have been reported from some of the blue–green stones, the minerals appearing as yellow specks.

One such imitation took the form of a beryl on beryl triplet with a blue adhesive layer.

Reports of interesting and unusual examples from the literature

Items in this section have been chosen to illustrate points made in the chapter and to bring one-off items to your notice.

A colour-change tourmaline changed from green to red with an increase of path length of light through it. The specimen, described in the *Journal of Gemmology*, came from the Umba Valley in Tanzania and showed strong yellow to blue–green in transmitted light: when two specimens are superimposed upon one another the combined colour becomes red even when the sample orientation or type of illumination is altered. A change of path length of light through the specimen has been suggested as the cause of the phenomenon.

Fine bluish-green tourmaline from the farm Neu-Schwaber in Namibia had RI 1.621–1.640, DR 0.019, SG 3.07.

Tourmaline found in Nigeria shows a wide range of colours with red, green, pink and bicoloured examples. The RI is 1.622–1.640 with DR 0.018, SG 3.06. Round birefringent so far unidentified minerals and trigonal growth zones are reported as inclusions.

Yellow tourmalines from East Africa are most commonly dravite–uvite compounds with dravite dominant. The iron content of gem-quality specimens tends to be low, higher contents of iron and titanium making the stones cloudy. Yellow and green zoning is often indicative of tourmaline from East African deposits.

Yellow tourmalines recently mined in the Northern Areas of Pakistan gave a spot RI reading of 1.63 and SG 3.00. The crystals contained partially healed fractures, feathers and dust-like clouds.

Multi-coloured tourmalines with a high content of bismuth have been found in the Kalungabeda area of the Lundazi district of Zambia. Colours include pink, orange, green and yellowish green with some cut stones showing two colours. Since the rough appears to crack easily the green rind has to be retained to give physical stability. Growth bands, either angular, straight or wavy, are the most distinctive features. No bismuth minerals have been found as inclusions. EDXRF shows that the highest bismuth content is in the pink centres and the lowest in the green rims.

The colour of the copper-bearing blue and green–blue tourmalines from the state of Paraíba, Brazil, was discussed in *Canadian Gemmologist*, 17(4), 1996. Amethyst-violet or emerald-green colours are attributed to Mn^{3+} and Mn^{2+}–Ti^{4+} charge transfer combined with Cu^{2+} absorptions.

Crystals of a chrome tourmaline from the Umba Valley of Tanzania show a wine-red colour when viewed in certain directions but this is neither a pleochroic nor a colour-change effect. The change from green to red depends primarily on the path length of the light traversing the specimen and probably too on the Cr and V content. The red colour appeared when specimens were more than 15 mm thick. The name ‘Usambara effect’ has been proposed.

An imitation of water-melon tourmaline proved to be stained red quartz surrounded by other mineral fragments and cement enclosed in slices of blue–green tourmaline.

A bright blue copper-bearing tourmaline has been reported from the Brazilian state of Rio Grande do Norte and has been wrongly called ‘Paraíba tourmaline’ since there is some resemblance to this variety.

Hsa-Taw tourmaline comes from south-east Burma close to the border with Thailand. Properties are in the accepted range for elbaite: the material shows colour zoning, trichites, possible mica and undetermined solid black materials.

Pink, yellow and green tourmaline from the Jagoda mine area in Zambia contains characteristic capillaries and growth tubes.

A canary yellow tourmaline from Malawi reported in the Summer 2001 issue of *Gems & Gemology* is said to have been heat treated to obtain the most arresting yellow colour.

Chapter 11

The garnet group of gemstones

The garnet group of minerals comprises fifteen species at the time of writing but only six can claim to have any ornamental significance. The minerals concerned are *pyrope*, *almandine* and *spessartine*, *uvarovite*, *grossular* and *andradite*. The garnets have a common formula $A_3B_2Si_3O_{12}$ where A may be iron, calcium, magnesium or manganese and B may be aluminium, iron, titanium or chromium. There is complete series of solid solutions between some of the species but not between all of them. All have similar structures, being cubic. On the whole it does not matter which species of garnet is being tested or offered for sale – the main aim is to find out whether or not the specimen in question is one of the garnets or something else. Some of the garnets can be indistinguishable from one another on colour grounds alone; the eye cannot distinguish between a dark red garnet which may be almandine or an iron-rich spessartine.

The red garnets

Looking first at the red garnets, pyrope and almandine form a solid solution series from Mg-rich pyrope, $Mg_3Al_2(SiO_4)_3$ to Fe-rich almandine, $Fe_3Al_2(SiO_4)_3$.

A *pyrope* with no trace of iron would be colourless and natural specimens are virtually unknown. Iron in nature is so ubiquitous, however, that as pyrope is forming iron always enters the garnet structure so that a low-Fe garnet would be a pleasant red and a very iron-rich one so dark a red that in reflected light it might appear black. This is why very dark red almandines are often cut as hollow cabochons known as carbuncles.

On occasion a red garnet may contain some chromium and will then be a much brighter red: generally found in small sizes and used in suites of jewellery the stones were long known as Bohemian garnets. Another variety of red garnet is sometimes, rather confusingly, called rhodolite, a variety between pyrope and almandine, the colour being light to medium purplish-red, even approaching amethyst colour on occasion. Some authorities claim that rhodolite should have a magnesium to iron ratio of 2:1 but dealers will use the name indiscriminately. Some of the almandines, especially material from Idaho, USA, show well marked stars with usually four but sometimes six rays.

Most of the finer red garnets come from somewhere near the mid-point of the series between Mg and Fe aluminium silicates and as we have seen the red darkens as the composition approaches the iron-rich almandine. Some gemmologists have coined the name pyrope–almandine for specimens with a place in this series but this sounds suspiciously like pedantry and because the gemstone trade showed no interest the name has tacitly been dropped.

Pyrope has a hardness of 7–7.5, RI 1.730–1.766, SG 3.7–3.9. Brighter red stones, including the Bohemian garnets but also some of the specimens from Kimberley, South Africa (small ones are found with and in diamond), show a broad absorption band in the yellow–green near 575 nm which covers two of the three strong almandine bands which could otherwise be seen. These are of course iron bands and if they were not there pyrope would be colourless. Some of the brighter stones may show some chromium lines in the red.

Pyrope forms a series not only with almandine but also with *spessartine*, the manganese aluminium garnet $Mn_3Al_2(SiO_4)_3$ which we shall meet later. As we shall see, some pyrope–spessartines show a colour change from a darkish-yellow–green to a claret-red.

Almandine is a darker red than pyrope due to its iron content which also raises its properties: RI 1.76–1.81, SG 3.95–4.3. Most almandines show an RI of 1.78 or above which will give a confusing situation when round 1.79, the reading that the current contact liquid will give. In any case many almandines will be well above this reading and some other form of test will be needed. The hardness is 7.5.

Fortunately the spectroscope helps us considerably since almandine gives a classic absorption spectrum consisting of three broad, strong bands, in the yellow at 576 nm, in the green at 526 nm and in the blue–green at 505 nm: there are other, weaker bands. The band at 505 nm is particularly strong and persists not only in the more Mg-rich, Fe-poor red garnets of the pyrope–almandine series but can be seen also in many examples of spessartine. The rhodolite variety always shows the almandine bands and apatite is a common inclusion though the colour of rhodolite makes it easier to distinguish foreign crystals than in the very dark almandines, so that either may contain a fair number of apatites.

To summarize, the red garnets in general show a variety of inclusions, pyrope and almandine often containing crystals of apatite, rutile, spinel, zircon, mica and quartz: though interesting they do not give so good a distinction from other red stones as the absorption spectrum. Rutile may give a four- or six-rayed star as already mentioned.

The manganese silicate *spessartine* forms a series with both pyrope and almandine. With little iron content spessartine is a beautiful pale orange. As the iron content increases spessartine shows a darker orange, reddish-orange and as we have seen, finally a dark red indistinguishable from almandine. The RI of spessartine is usually in the range 1.79–1.81 though specimens from Amelia, Virginia, which are a very fine deep orange, give 1.795 and some Brazilian stones 1.803–1.805. The hardness is 7.25 and the SG 4.12–4.20.

While spessartine shows the almandine bands, especially the persistent band at 505 nm as soon as it begins to include iron, manganese too has a distinctive absorption spectrum. This gives two bands in the blue at 495 and 485 nm with a stronger band at 462 nm and a particularly strong one at 432 nm. There are bands also at 424 and 412 nm, the latter very intense.

Inside spessartine the commonest inclusions are liquid shreds or flags with little in the way of mineral crystals.

Natural stones resembling the red garnets

The red–orange garnets are not too easy to confuse with similarly coloured gemstones and any confusion is more likely to be accidental than deliberate. Spinel which shows a wide range within the colour red is the most likely to be confused with the lighter reds of the pyrope–almandine series: the RI of spinel is fortunately easy to obtain and will usually be at 1.718, well below that of the red garnets, though examples up to 1.730 have been recorded. Neither spinel nor garnet group minerals show birefringence. Should the spectroscope be the first choice of the investigator red spinel will show chromium emission (coloured) lines in the red not seen in any of the red–orange garnets, nor will red spinel show the ubiquitous 505 nm almandine band. Both pyrope and red spinel show a broad central band, centred at 540 nm in red spinel and at 575 nm in the yellow–green in pyrope.

Should an ultra-violet light source be available red spinel will be found to show a red fluorescence in LWUV light, an effect never shown by the red garnets as their iron content prevents luminescence. The garnets will not show any red in the crossed filter test, which is the illumination of a specimen in blue light, the effect being viewed though a red filter. Specimens whose colour is due to chromium will glow red through the filter, the red not having come from the blue light but from the specimen.

Thai ruby may be confused with some of the darker red garnets but many of these iron-bearing rubies are now heated to give them a more 'Burmese' look. In any case the refractive index, if it can be measured, will quickly show too low a reading for almandine; in most cases, though, some overlap is possible. More importantly, the refractometer will show the birefringence in the ruby, an effect not seen in the garnets.

Though we shall meet the product again, almandine is used as the 'garnet' in the garnet-topped doublet. While red flashes can easily be seen in green or blue GTDs, if the specimen under investigation is red problems can arise – the usual ones of 'not thinking doublets'. The presence of short rutile needles in the crown of the stone and nowhere else should be a warning and an unexpected absorption spectrum – the band at 505 nm will be there – in a stone where it would not be found if it was natural.

Some of the darker red *tourmalines* quite closely resemble the red garnets and while testing them for their refractive index the gemmologist should watch for their unmistakable birefringence around 0.018 (garnet shows none) and their much lower RI in the area 1.62–1.64. The colour can be quite close but tourmaline will show some pleochroism provided that the gemmologist takes care to view the specimen in a number of directions. Dark red stones are not too easy to test on the dichroscope.

Two very rare garnet-like transparent red stones are transparent red triclinic *feldspar* and *chondrodite*. These are very much specimens for the collector but the resemblance to garnet is considerable in both cases. The feldspar will give a much lower refractive index (none of the feldspars have an RI above 1.59) and the chondrodite's highest RI reading will not exceed 1.64. Both the feldspar and the chondrodite will show some birefringence.

Other colours of garnet

Spessartine can most easily be confused with another of the garnet group minerals, *hessonite*. This is a variety of *grossular garnet*, the calcium aluminium silicate, $Ca_3Al_2(SiO_4)_3$ which occurs in a range of colours. Hessonite, most commonly yellow–orange–brown, is less rare than spessartine but can look very attractive, profuse inclusions of tiny crystals of calcite giving it a 'treacly' appearance: crystals of apatite often fill specimens and zircon is also found. The inclusion picture in hessonite is nothing like that of spessartine in which liquid inclusions play so prominent a part.

Hessonite's refractive index and specific gravity will be near 1.743–1.748 and 3.65. The hardness is 7.25. It shows no absorption spectrum of value, unlike spessartine. Apart from spessartine, hessonite most closely resembles *citrine* from which it can be separated by its higher RI (citrine reads 1.544–1.553) and included interior (citrine shows prominent colour banding). While some of the fancy *corundums* can show a hessonite colour (or virtually almost any other) their higher RI of 1.76–1.77 and perceptible birefringence of 0.008 (hessonite shows none) should make separation quite easy.

Both spessartine and hessonite might casually be mistaken for *zircon* which can show similar colours. Zircon's high birefringence of at least 0.059 should give it away. *Sphalerite* is magnificent when cut in its finest orange colour but shows quite unmistakable high dispersion and a high SG of 4.09 which will make even the smaller stones feel heavy compared to hessonite though less so compared to spessartine which will be higher.

It is possible that a fancy *chrysoberyl* could be taken for spessartine – I have seen such stones, usually rather dark, from Myanmar and Sri Lanka but they are rare. Should one be encountered the RI will be just too low for spessartine but would fit into the hessonite range: however, chrysoberyl will show a birefringence of around 0.009 while the garnets will show none.

It is possible that a hessonite could be mistaken for yellow *topaz* which will not show the profuse apatite inclusions and will have a lower RI in the range 1.62–1.64 with an SG of 3.53. Yellow topaz shows no absorption spectrum, which is enough to distinguish it from any spessartine. The transparent yellow scapolite with RI overlapping with that of citrine does not closely resemble either of the orange–yellow garnets.

Hessonite is only one of the varieties of *grossular* which also include the very fine chrome green *tsavolite*, the lighter green vanadium garnet, usually called tsavolite too and the translucent green *hydrogrossular*, described in Chapter 7 on the jade minerals as a possible jade simulant and occasionally found in shades of pink.

Tsavolite

Tsavolite is by far the most commercially important variety of grossular and could naturally be confused with emerald and with *demantoid garnet*, a variety of *andradite garnet* which we shall meet below. We mentioned when discussing *emerald* that the inclusions in tsavolite are unlike those to be found in emerald, consisting mainly of lines of whitish dots and crystals of apatite and actinolite. In any case tsavolite has a much higher RI and SG than beryl, with a mean RI at 1.74 and SG of 3.61 compared to emerald's mean RI of 1.57 and SG of 2.71. Neither does tsavolite show anything like the emerald absorption spectrum, with only faint traces of the chromium lines and general absorption region. While the colour is not quite like that of most emeralds (rather dark) this is certainly not a means of distinguishing them and the RI comes into its own here. Emerald will show pleochroism and tsavolite will not but with paler stones this is a difficult test to rely upon.

Demantoid garnet: other green garnets

The distinction between tsavolite and demantoid, the chrome-green and highly important variety of andradite, can appropriately be examined here. Andradite has the composition $Ca_3Fe_2(SiO_4)_3$.

Demantoid very rarely occurs in large crystals and small-faceted stones are the rule. While the RI of tsavolite can easily be measured, demantoid's reading of 1.89 cannot be obtained on the standard refractometer and SG tests are likely to fail because of the small size of most of the specimens: the SG of demantoid is fairly constant at 3.85. On the other hand the absorption spectrum is more available and the inclusions are even better guides to identification. The hardness is 6.5.

The absorption spectrum of demantoid combines elements of Fe and Cr, the iron providing a very strong band in the deep blue at 443 nm. In paler specimens it can be seen standing alone but in darker ones, probably the majority tested, it will show as an 'early cut-off' to the spectrum. The chromium elements are a doublet in the red, deeper than in most Cr-coloured green stones, at about 701 nm (in fact this is usually the only line of the doublet to be visible) with other lines at 640 and 621 nm.

Demantoid provides one of the most familiar inclusion scenes in the whole of gemstone testing. The effect of hair-like, often curved crystals of one of the amphibole group asbestiform minerals is often referred to as 'horsetails'. The fibres, which are almost invariably seen, or at least parts of them are, often appear to originate from a crystal of chromite. This is a diagnostic test and no further work should be needed – this cannot be said for many other gem species.

The colour of demantoid is distinctive and its effect enhanced by a high dispersion value of 0.057 (higher than that of diamond, at 0.044). This cannot be tested by standard gemmological methods but it should not need to be.

Natural and synthetic materials which may be confused with green garnet

The calcium–chromium garnet *uvarovite* occurs in very fine emerald-green crystals. They are usually opaque, sometimes translucent, and usually so well formed that they are worth more in their natural state. The occasional very dark green cabochon or faceted stones have RI in the range 1.74–1.87 and SG 3.71–3.77. The hardness is near 6.5.

Among the *synthetic (non-silicate) garnets* the one material that actually is known to have (temporarily) been mistaken for demantoid is a very bright green chromium-doped synthetic garnet $Y_3Al_5O_{12}$ (YAG).

This material has a somewhat metallic, brilliant appearance where demantoid is a softer green but you cannot test impressions! Nor can you test the RI of YAG or any other of the synthetic garnets with the standard refractometer. However, when the spectroscope is brought into use, the investigator will be able to say, 'it's certainly not demantoid, but what is it?' This was in fact said (or something like it) when the first chrome-green YAG appeared on the market.

The specimen gave a very beautiful and unexpected spectrum in which both chromium and the rare earth neodymium combined to give a rich pattern of fine lines and bands quite unlike the spectrum of demantoid or of any other green stone previously encountered. For the record this material has an RI of more than 1.83 and an SG more than 4.55 so demantoid is quite ruled out. The analogous material gadolinium gallium garnet (GGG) ($Gd_3Ga_5O_{12}$) may also resemble demantoid when appropriately doped but again the RI and SG are much higher at about 1.97 and 7.02: the dispersion is about 0.045, close to that of diamond but below that of demantoid.

Apart from the synthetic garnets green *zircon* could be confused with demantoid but the low green zircons have much less life and though most will show very little birefringence there will still be some doubling of the back facets and of inclusions providing the stone is examined in the appropriate direction. Not all green zircons are metamict. Some may show a rather vague absorption band in the red, not easily confused with the chromium doublet in demantoid, and sometimes a few of the other zircon absorption lines.

Another possibility of confusion may be thought to arise with *sphene* (titanite) which is sometimes though quite rarely coloured a fine green by chromium – in my opinion this green is much darker than that of most demantoids (though undoubtedly fine) and in any case sphene shows a strong birefringence in the range 0.105–0.135. If this is not enough upon which to make a diagnosis, some sphene may show an absorption spectrum with two groups of fine lines, one group in the yellow, the other in the green, which is characteristic of yellow–green sphene but may be found in the lighter of the definitely green stones. However, the birefringence should be enough to distinguish sphene from demantoid.

Birefringence is also clear enough in *peridot* to make distinction from demantoid quite easy – only the small, bright, chromium-bearing peridot from Hawaii is likely to give the wrong impression at first though the dispersion is well below that of demantoid.

A more dangerous confusion with demantoid (if only it were less rare!) would arise with the chrome-green variety of *spodumene, hiddenite*. There is a definite resemblance here: stones seem always to be small and the colour is a bright yellowish-green very like that of demantoid. The green is due to chromium, too. Hiddenite has an RI in the range 1.65–1.68 with a mean birefringence of about 0.018. These figures are obtainable on the refractometer – those of demantoid cannot be. Like green zircon and all the other possible simulants of demantoid, hiddenite will not show the horsetail inclusions. If the spectroscope is used the chromium doublet in hiddenite is less far into the red with a mean wavelength of about 688 nm. Some hiddenite may fluoresce but this is an unreliable test on its own for the green material.

It would be surprising if *tourmaline* with its variety of colours did not provide one to cause confusion with the green garnets. A fine chrome-green tourmaline was reported from East Africa and from Namibia and while they do not show the dispersion of demantoid the colour can look quite close if the stone should be cut as a small brilliant (they are far more often cut as long rectangles). The high birefringence of 0.018 will be easy to see under magnification, tourmaline will not show horsetail inclusions nor so marked a chromium-iron absorption spectrum.

Other green stones come nowhere near demantoid but as always it depends on what you are told by the seller. Green *sapphire* is far too dark and lacks notable dispersion and *fluorite* may look convincing for the moment but will show an easily measurable RI at 1.43 and, of course, purple fluorescence under LWUV light. There is no close resemblance to demantoid here.

While *glass* will imitate almost anything and demantoid is no exception, the absence of the horsetails and lack of an absorption spectrum will rule it out. Green *diamond*, on the other hand, may seem an unlikely substitute but it does have a high dispersion (though lower than that of demantoid) and its RI cannot be read on the standard refractometer. But no horsetails will be visible in diamond (which will have to be a rather unusual green to be a successful imitation), its facet edges will be noticeably sharp (those of demantoid, with a hardness of only just over 6, will be much less sharp) and it will not show the recognizable absorption spectrum.

It is not common for the garnets to be treated or colour-enhanced in any way as they stand very well on their own. There are no gem-sized synthetic silicate garnets.

Reports of interesting and unusual examples from the literature

Items in this section have been chosen to illustrate points made in the chapter and to bring one-off items to your notice.

While the absence of recognizable inclusions in the synthetic garnets is a help towards identification, a specimen of YAG has been reported to have shown some features that may help the gemmologist. It may very well be that this was an isolated example, however: the stone was a dark green YAG of 15.45 ct which was coloured principally by chromium. Inside were elongated gas bubbles within fine blue-coloured layers and distinct slightly curved parallel graining. Small crystals were randomly distributed through the specimen, each crystal surrounded by stress fractures. It was believed by GIA, who investigated the stone, that it may have begun life as a reject from optical materials fabrication or to have been grown by horizontal crystallization, a variation of the floating-zone method.

A report in the Summer 1992 issue of *Gems & Gemology* described an emerald-green YAG which weighed 5.56 ct and gave a negative reading on the refractometer. The SG was 4.55: the stone contained natural-appearing inclusions which proved to be combinations of flux with gas bubbles.

A gem-quality green andradite recently found in Arizona showed no horsetail inclusions.

A new find of spessartine in Nigeria produced specimens with a spessartine component of 89–95 per cent with varying small amounts of pyrope and almandine. Variation of the almandine content accounts for a variation of colours between yellow, golden yellow and brownish-orange. Absorption spectra of the yellow and golden yellow specimens showed considerable similarity to those of spessartine from Kunene, Namibia, while the brown–orange stones looked more like those from Ramona, California. The Nigerian stones contained healed cracks representing thin liquid-filled cavities or fingerprint-type feathers. RI was between 1.801 and 1.803, SG between 4.15 and 4.22.

Colour-change garnets from Madagascar have been divided into five categories: all belong to the pyrope–spessartine series and contain some vanadium which is responsible for the colour change. Group 1 specimens remain blue under both types of lighting with only a slight change of hue: the colour saturation is strong. Type 2 stones change colour moderately to distinctly from bluish to purplish – these are the most commercially desirable of the colour-change garnets.

Group 3 stones change from yellowish-green to pink with medium to strong colour saturation. Type 4 stones change moderately from brownish-green to brownish-red with strong colour saturation. Group 5 stones change slightly from brown (daylight) to reddish-brown (tungsten).

A survey of inclusions in garnets was published in 1993 by *The Mineralogical Record.* In almandine, primarily found mainly in schists, the commonest mineral inclusions are apatite, biotite, ilmenite, monazite, rutile and zircon. Chromium-bearing transparent green andradites (demantoid) occur in chlorite schists and serpentinites, the commonest mineral inclusions being distinctive groups of white to yellow tremolite–actinolite series amphibole crystals (horsetails) radiating outwards from a central crystalline core of chromite.

Grossular is produced by contact and regional metamorphism of impure calcareous rocks: included minerals are actinolite, apatite, calcite, diopside, graphite, pyrite and scapolite. Pyrope is a mineral occurring in peridotites and kimberlites though most gem pyrope is recovered from sands and gravels formed by weathering of these rocks. Inclusions are diopside, forsterite and rutile. Spessartines are found in granitic pegmatites; these are the only gem garnets to contain liquid and gaseous two-phase and three-phase inclusions.

Demantoid reported from Namibia has SG in the range 3.61–3.85 and has been found to be pure andradite with only 0.02–0.13 per cent Cr_2O_3. Absorption spectra show bands of Fe^{3+} and Cr^{3+} and distinct colour zoning can be seen under magnification. No horsetail-type inclusions have been found.

Gem-quality grossular in yellow–green and orange–brown colours have been produced on a fairly large scale in Mali. Composition is intermediate between grossular and andradite and has been described by GIA Gem Trade Laboratory as grossular–andradite. The RI for the yellow–green and green specimens is in the range 1.752–1.769 and for the orange–brown specimens 1.773–1.779. An absorption band at 440 nm (attributed to iron) is found in both colour varieties and stacked parallel planes of growth zoning are always visible between crossed polars together with moderate to strong anomalous double refraction.

Demantoid found in Eritrea showed an RI above the range of the standard refractometer and gave an SG of 3.80–3.88. The usual horsetail fibres were present and the absorption spectrum showed characteristic bands of Cr^{3+} and Fe^{3+}.

Orange spessartine found in gneisses in Namibia gave RI 1.800–1.801, SG 4.08–4.12. Distinct growth zoning and hollow tubes were observed with an immersion microscope. The specimens are classed as spessartine-rich garnets of the pyrope–spessartine series.

Small, mostly transparent amethyst-coloured garnets with RI 1.792–1.798, SG 4.08–4.11 have been found at Fazenda Balisto, near São Valeria, Brazil. The name rhodolite was thought to be appropriate for them.

A transparent, near-colourless emerald-cut stone had properties not significantly different from those of grossular. EDXRF spectrometry and X-ray diffraction showed that it was the magnesium oxide periclase.

In a paper in *The Mineralogical Record*, 1993, the pink colour sometimes seen in hydrogrossular garnet was attributed to Mn^{3+} in the octahedral site.

Bright green inclusions seen in Arizona pyrope were found to be diopside, by Raman laser microspectroscopy.

Emerald-green grossular found near Bekily in southern Madagascar has RI 1.741–1.746, SG 3.58–3.62. Two broad absorption bands can be seen in the red and the violet. Inclusions of apatite have been observed. Some bright green grossular from Madagascar is known to be coloured by vanadium.

Some pyrope–spessartine garnets show birefringence which has been found to correlate weakly but positively with Ca^{2+} coexisting with Mg^{2+}.

Cabochon-grade uvarovite is reported from a valley in Bo Mi county of the Tibet Autonomous region of China, the crystals ranging in size from 7 to 21 mm.

Orange to cherry-red spessartine found in Azad Kashmir, Pakistan, has been marketed as 'Kashmirine'. It is composed mainly of spessartine with some almandine and grossular with RI 1.798–1.802 and SG 4.05–4.18. Crystals contain only a few healing cracks with fluids: some crystals of K-feldspar have also been included.

A paper in the Winter 2001 issue of *Gems & Gemology* describes pink to pinkish-orange malaya garnets from Bekily, Madagascar: the name Malaya has been used previously for garnets from the pyrope–spessartine series which have been predominantly orange. The stones show absorptions due mostly to Fe^{2+} and Mn^{2+}: elements in the absorption spectrum of synthetic iron-free spessartines and iron-free almandines have shown Mn absorptions at approximately the same positions as some Fe absorption maxima so that bands at 525 and 459 nm can be ascribed to both elements. There may be another overlap at 569 nm. Graphite and apatite crystals are noted as inclusions.

Chapter 12

Spinel

Spinel is an attractive gemstone with fine red and pink varieties and interesting blues. The spinel group of minerals includes more than 20 different species but we are concerned only with the magnesium aluminate which, perhaps confusingly, has the group name of spinel. Spinel is hard (8 on Mohs' scale) and has a single refractive index of 1.718 and specific gravity of 3.60. There is a number of famous named spinels such as the Black Prince's ruby in the Imperial State Crown in the English regalia.

Red spinel

Red spinel is coloured by chromium and is naturally confused with ruby which is harder, has a higher mean refractive index of 1.76 and specific gravity of 3.99 compared with spinel's 1.718 and 3.60. The more important mineral inclusions are parallel rows of spinel octahedra which may fill the whole crystal: hexagonal prisms of apatite are also found. Liquid inclusions are rare.

An unknown bright red stone ought to suggest the spectroscope as a first-line tool: some textbooks can be misleading in this area. Fine bright red spinel from Burma shows a chromium absorption spectrum: emission lines in the red may form a group with up to five individual lines: when this is the case the two central lines will be seen to be the brightest and the line at 686 nm the most prominent. If a blue filter is available it will assist observations in this region by diminishing the dazzling effect of the red light being passed. A strong source of light is needed and both reflected and transmitted lighting need to be tried. Verneuil-type flame-fusion red spinel (rare – see below) usually shows only a single emission line. In practice we have seen the group of emission lines (long known as organ pipes, though being the same length they would all have sounded the same note!) most effectively in some flux-grown red spinel rich in zinc. The magnesium aluminate (spinel) shares spinel group membership with a zinc aluminate, gahnite, with which it has a close relationship. It is also possible to illuminate an unknown red stone with a strong light passed through a blue filter (or a flask of copper sulphate if one is around). The fluorescence spectrum produced will show the emission lines quite well.

Away from the chromium emission lines the absorption spectrum of red spinel shows a broad absorption in the green centred at 540 nm but there are no lines in the blue. This absence is one of the best ways of separating red spinel from ruby if the refractometer is not available or, more likely, the stone's setting makes its use impossible.

The chromium content of red spinel also allows it to show a crimson colour under LWUV light with a diminished but similar effect under SW. The same colour is seen on irradiation with X-rays but such a test will rarely be necessary.

Confusion with other red stones can arise though the red of the finest spinels is not really like that of ruby (though equally attractive). The most likely cause of confusion, leaving aside synthetic red spinel for the moment, is with the brighter red garnets, especially those containing chromium.

The 'Bohemian garnets' which contain chromium usually show the persistent iron (almandine) band at 505 nm and a band in the yellow–green centred near 575 nm. Though present, chromium does not always show visible absorption bands in these garnets and if a close examination of the red end of the spectrum reveals nothing it is not always certain that chromium will be absent. The refractive index of the red garnets likely to be confused with red spinel are all higher, extending upwards from 1.73 at the lowest. Specific gravity will also be higher, from 3.65 upwards.

Bright red tourmalines will show strong double refraction and pleochroism where spinel shows none. If these tests cannot be carried out and there is no spectroscope handy, tourmaline's quite characteristic liquid inclusions should be a further aid to identification.

Red zircon's birefringence should be strong enough to distinguish it from a stone with none and the reddish variety of topaz has a lower mean RI/SG of 1.62 and 3.53.

Red glass is most likely to show an RI between 1.50 and 1.70 and, like spinel, will show no birefringence or any characteristic absorption spectrum. Spinel is not often imitated except perhaps by accident when someone is trying to manufacture ruby.

Synthetic spinel

Red spinel has been grown by both the flame-fusion and flux-melt methods though examples of the former are quite rare, even collectors' pieces. The reason is most likely to be that boules of spinel when doped by chromium tend to fracture and in a world where growers are used to a high throughput of material it is commercially unacceptable to spend time on problem pieces. Attractive and collectable crystals of both red and blue spinel have been grown in Russia by other means and gemstones have been cut from them.

The synthetic flame-fusion spinels show a celebrated effect when viewed between crossed filters on the polariscope – the name often used is 'tabby extinction' since the lighter and darker stripes seen are thought to resemble those on a tabby cat. Ruby will show a perceptible alternation between light and dark four times during a complete rotation though some of the red garnets may show an indistinct response which could confuse.

The gemmologist looking into a flame-fusion-grown spinel may at first be nonplussed since there may not be much to see. There will of course be no natural mineral inclusions and the well-rounded quite prominent gas bubbles familiar in flame-fusion corundum will not be profuse: nor will there be the curved growth lines associated with corundum. All is not lost, however, because there are some bubbles with shapes which with familiarity become easier to recognize. They may resemble hourglasses or furled umbrellas and can occur in parallel arrangements: they may appear to take up a hexagonal pattern when viewed in particular directions. Flat cavities containing a bubble of liquid or gas may be joined to neighbouring cavities by a tube. Inside the red boules or stones cut from them a very characteristic shuttering or Venetian blind effect can be seen.

Without a microscope or with doubts about interpreting what is seen, the gemmologist can turn to the refractometer where the usual synthetic reading of 1.728 is well above the 1.718 of natural spinel. If the identity of a boule or boule section is in question rather than that of a fashioned stone, complete spinel boules have a roughly square cross-section compared to their only rival, boules of corundum, which are rounder, faintly suggesting hexagonal symmetry. The surface of a spinel boule feels distinctly rough.

Boules of spinel (other than red ones) were grown to imitate other gem species, not the different colours of spinel itself. A wide range of colours is available, aquamarine, peridot and blue zircon seeming to be the natural species most often imitated. Colourless boules and their faceted stones are very likely to be used to replace other, even natural (even diamond) stones. The colourless synthetic spinels give a recognizable sky-blue glow under SWUV and an apparently diamond-set piece in which all the (usually small) stones glow this colour under irradiation should be regarded with considerable suspicion as a quantity of diamonds will by no means fluoresce uniformly.

Synthetic spinel's RI of 1.728 is far enough away from that of aquamarine (1.57–1.58) for the refractometer to act efficiently: spinel will show no birefringence so zircon, which cannot be tested on the standard refractometer, can easily be spotted in comparison by distinct doubling of opposite facet edges and inclusions. Peridot's high birefringence of 0.036 will show doubling quite well and in this case will give a refractive index reading between 1.65 and 1.69.

Blue to dark blue synthetic spinel is coloured by cobalt and will show red through the Chelsea filter. The red can be quite strong and is accompanied by a distinct absorption spectrum comprising three broad bands in the orange, yellow and green, with the central band the widest (it is the narrowest in cobalt glass). The bands in synthetic blue spinel are at 635, 580 and 540 nm: they may appear as a single absorption region in very dark blue specimens. The red seen through the Chelsea filter arises from the virtually complete absorption of the yellowish-green and transmission of the deep red.

Up to a few years ago a blue spinel containing cobalt would have automatically been labelled as synthetic as no natural cobalt-bearing magnesium spinel had been reported. In recent years, however, natural cobalt blue spinel has been found (though it is still very rare) in Sri Lanka and these stones also show red through the Chelsea filter while giving the normal spinel RI and SG.

Other colours of spinel

We have not yet looked at the common blue spinel which is coloured by iron rather than by cobalt. This is a rather quiet blue, due to iron, and the absorption spectrum often looks portentous to the beginner – and may be found in some of the fancy spinels (purple, mauve) found in Sri Lanka. Absorption bands can be seen in the blue, centred at 459 nm with a narrow band of about equal strength in the green–blue at 480 nm. Faint bands at 555 nm in the green, in the yellow (592 nm) and the orange (632 nm) complete the spectrum. The gemmologist should compare as many blue spinel and blue sapphire spectra as possible so that familiarity will avoid the confusion that may arise when the absorption is faint in a pale stone.

Blue spinel does not look very like sapphire but the RIs of the two, at 1.718 and 1.77 respectively, with spinel's lack of birefringence, should make distinction easy. It is hard, with the spectroscope at hand, to imagine that specific gravity will ever be needed for a distinction to be made but the comparative figures are 3.60 and 3.99–4.0. Blue spinel looks far more like the highly pleochroic cordierite (iolite) which has a much lower RI and SG, at about 1.55 and 2.63. Pale iolite will show less pleochroism and confusion is more likely to arise then.

We have already mentioned fancy spinels which seem usually to be found in Sri Lanka. Purple and mauve stones may give an iron absorption spectrum but occasionally a fine orange specimen turns up and such a stone may show signs of the chromium emission lines or at least one of them, as we saw in a very fine orange drop recently. The fancy stones can behave in unusual ways sometimes: purple to mauve stones may show a red fluorescence under LWUV (suggesting that chromium is present in them) and are inert under SW. Some may give a plum to lilac fluorescence under X-rays. Some other stones of the same colour may fluoresce orange to red under LWUV, not much under SW and green under X-rays. Some pale mauve and some pale blue or violet blue spinels may give a greenish fluorescence under LW.

These occurrences are interesting but we cannot imagine that a serious identification would have to rest on them alone when other standard tests are available. However, there are many collectors of fluorescent minerals of whatever kind.

In the same way spinel may occasionally show a colour change suggestive of alexandrite: a specimen changing from pinkish-purple under fluorescent light to purplish-pink under incandescent light was reported by GIA in 1984 and there are reported examples of changes from greyish-blue to amethyst colour (the GIA stone showed a moderate chalky yellowish-green under LWUV and was inert under SW – it was a Burmese stone).

Colour change in minerals and gemstones is usually held to be rare but this is not the case as many examples of different changes in different species are reported every year.

Spinel occasionally produces star stones and cat's-eyes but they are not a regular feature of the species and need no particular testing methods other than the standard ones.

In general testing of spinel presents few problems and it is not usually imitated. It may be worth mentioning the colourless variety once more in its role as a possible imitator of diamond: specimens of synthetic spinel and diamond immersed together in di-iodomethane, which has a refractive index of 1.745, will be distinguished from one another by the easy visibility of the diamond's facet edges compared with the near-invisible ones of the synthetic spinel. It remains to be seen how long some of these liquids survive health and safety regulations so it would be ideal if gemmologists learnt other ways of testing the major gemstones – fortunately this is usually possible so far.

Synthetic spinel may often be found forming the crown and pavilion (top and bottom) of a soudé stone which, if an emerald colour, will also have a green-coloured cement layer between the two.

Spinel forms more attractive crystals than some of the other major gem species, with perfect octahedra appearing on the market from time to time. When they are the finest red, as with specimens from Burma, they are probably worth imitating but we have not seen any so far. On the other hand a use for colourless synthetic spinel octahedra is easy to guess – as an imitation of diamond. Russian flux-grown synthetic spinel forms octahedra which can be doped to give a variety of colours, or left colourless. The colourless octahedra often show trigons on the faces: in diamond these are also seen but here they do not follow the edges of the face so that the trigon's point does not take the same direction as that of the point of the face. With synthetic spinel octahedra the reverse is the case, with the edges of trigon and face echoing one another. A similar feature can be seen in some colourless octahedral crystals of synthetic flux-grown corundum.

Should you be faced with a desirable transparent octahedron try to examine the interior. A transparent specimen is pretty likely to be synthetic and this could be confirmed by the presence of flux inclusions. Octahedra in red, pink and blue varieties could just be natural but those of any other colour are more likely not to be – colourless ones certainly will not be natural as colourless natural spinel is virtually unknown.

Synthetic flame-fusion spinel normally has more alumina (aluminium oxide) content than the natural material (up to 2.5 times as much) and this accounts for the higher RI and SG at 1.728 and 3.64 compared to the 1.718 and 3.60 for natural spinel. As long as the RI can be taken this makes synthetic spinel easy to test – when you expect it.

A very convincing imitation of moonstone can be made by increasing the alumina content of synthetic Verneuil spinel by up to five times. This material has synthetic spinel properties with RI 1.728 and SG 3.64 rather than those of orthoclase moonstone, 1.54 and 2.57.

In the 1960s and 1970s a number of crystal growers were experimenting with the production of spinel group minerals doped with different elements which on occasion produced unusual and beautiful colours. It is unlikely that these research products will ever turn up in the gemstone or collectors' markets and if purchased through the official channels the cost is very high. Nonetheless a small orange octahedron, for example, may be found and may baffle investigators who, if they are sensible, will look for flux inclusions.

Experimental materials apart, crystal growers in Russia have produced and are producing on a semi-commercial scale, crystals of red and blue spinel. Stones were reported from the early 1990s and GIA issued a report on them in the Summer 1993 issue of *Gems & Gemology*. The report involved the examination of nine red and 12 blue faceted stones ranging in weight from 0.19 to 8.58 ct together with two rough and faceted blue flux-grown Russian spinels known to have a higher iron content than the other samples. The red stones were a vivid medium dark and slightly purplish-red with some specimens showing a very slight orange to brown component in incandescent light: this could not be seen in fluorescent light under which the specimens showed an enhanced purplish element.

The blue specimens showed a saturated medium dark to dark blue with a slight tinge of violet. When the stones were moved about under incandescent light they showed red flashes and in fluorescent light a grey component could be seen.

All the specimens were in the refractive index range for natural spinel while those with the higher iron content gave 1.717. The SG also fell in the natural spinel range, immersion in di-iodomethane (SG 3.32) showing an easy descent. Under LWUV the red stones gave a strong purplish-red to slightly orange–red reaction with a weaker but similar reaction under SW. Some chalkiness on the edges between faces could be seen under SWUV and these also showed a yellowish-orange in some directions. No phosphorescence could be detected after either kind of irradiation so that UV testing cannot separate the synthetic from the natural material.

Testing with the hand spectroscope found that the red material had an emission line in the red between 685 and 680 nm with a broad absorption band between 580 and 510 nm. A general absorption extended from the blue at about 450 nm to the end of the visible spectrum. There was no sign of the group of emission lines in the red as reported from natural red spinel, one emission line only being seen.

The blue specimens absorbed strongly between 635 and 615 nm, between 590 and 560 nm and between 550 and 535 nm. These absorptions are characteristic of cobalt and the expected red to orange through the Chelsea filter was observed. Both red and blue stones gave a strong red to orange transmission luminescence when observed in a strong visible light.

Inclusions in the Russian spinels showed clear evidence that they had been grown by the flux-melt process; residual flux was seen to have formed net-like patterns and particles with jagged edges. Some of the flux inclusions contained gas bubbles and others reflected a greyish-silver; these last could have been detached crucible material. In some of the specimens the larger flux inclusions formed pyramidal-shaped phantoms in very close alignment with the edges of the octahedron (these might be confused with the pyramidal or octahedral inclusions found in natural spinel). The triangular markings on the octahedral faces, mentioned above, were not observed in the red Russian material but growth hillocks could be seen under a magnification of 80×. Dendritic inclusions were observed in the flux material: they showed a metallic appearance in reflected light. None of the samples showed colour zoning.

The red spinels can be identified by the presence of flux inclusions or metallic residues or by the absence of natural solid inclusions. The blue stones show the cobalt absorption spectrum more strongly than the rare cobalt-bearing natural blue spinels and contain flux inclusions similar to those found in the red specimens. Under LWUV they give a weak to moderate chalky-red to reddish-purple fluorescence, the same colour, but stronger being seen under SW. Verneuil-grown blue spinels do not show this effect but appear a mottled blue to bluish-white. Natural blue spinels are inert to SWUV.

Synthetic flame-fusion cobalt-bearing blue spinel has been grown for use as an imitation of lapis lazuli (see below): flecks of gold are inserted to simulate the brassy-yellow flecks of pyrite seen in nature.

Pink synthetic spinel turns up from time to time but should easily be identified by one or other of the standard methods.

Reports of interesting and unusual examples from the literature

Items in this section have been chosen to illustrate points made in the chapter and to bring one-off items to your notice.

Gem spinels from Tunduru, southern Tanzania, show several shades of blue with some lilac and light pink examples. A few specimens showed a colour change. Calcite and högbomite were among the mineral inclusions identified.

A blue cobalt-coloured spinel from Burma contained abundant iron, a trace of manganese and an unusually high content of nickel, though cobalt itself was not found.

A colour-zoned red spinel from Mogok gave two refractive indexes of 1.732 and 1.718. This was found to arise from the stone's having a dark core and a paler surround. The dark core gave the higher RI contained higher amounts of chromium and vanadium.

The magnesium spinels form an isomorphous series with the zinc aluminate gahnite which is occasionally faceted: the colour is a fine dark green. A fine blue specimen found near Jemaa, Nigeria, had RI between 1.793 and 1.94 with an SG in the range 4.4–4.59: the absorption spectrum showed two sharp areas, with no absorption between them, where the absorption is mixed in intensity: these are at 580 and 555 nm. There are also absorption bands in the red and yellow similar to those in spinel and gahnospinel (the name used for material intermediate between spinel and gahnite). The Nigerian material showed markedly red under incandescent light. Crystals of beryl and hematite are reported as inclusions.

Chapter 13

Peridot

Introduction

The gemstone peridot is a variety of the mineral group olivine and has a composition between the two mineral end-members forsterite and fayalite, magnesium-rich and iron-rich silicates respectively. There is now no individual species name, olivine but gem olivine (=peridot) is richer in magnesium than in iron, the content of the latter being an optimum 12–15 per cent to give the finest green. As the iron content increases the colour changes to a greenish-brown and gemstones of this colour are sometimes seen. A very bright green peridot from Hawaii contains some chromium.

Identification

Olivine may be found in quite large crystals and many gems are large as a consequence. There is a distinct cleavage and care needs to be taken when cutting or setting peridot. The hardness is about 6.5 and the refractive index will be in the range 1.654–1.689 with a DR of 0.036 and a specific gravity in the range 3.3–3.4. Pleochroism is weak and there is no fluorescence in either type of UV light or under X-rays due to the iron content.

The same iron content is also responsible for a very useful absorption spectrum which shows three bands at 493, 473 and 453 nm. The first two show a darker centre with diffuse edges. An additional band at 529 nm can be seen in certain orientations and in large stones a band at 653 nm. If in a dark brownish-green peridot-like specimen an additional band at 463 nm is seen the specimen is likely to be the rather uncommon mineral sinhalite, dark specimens of which were once taken to be olivine. Sinhalite is in fact a borate rather than a silicate and will be discussed below.

Probably the easiest way to identify peridot is by examining the absorption spectrum and calculating the birefringence. The green of olivine is not very like the green of other gemstones, certainly not like that of emerald, except for the chromium-bearing material. As always, it all depends on what the customer is told and even if the colour of a particular specimen looks nothing like emerald, for example, the gemmologist will have to test the specimen to make sure. There is no synthetic olivine in significant sizes.

The colour of peridot varies considerably, some East African specimens showing an intense green to brownish-green while nickel as well as chromium and iron has some influence. If standard tests (spectroscope, refractometer) are not available a study of the quite complex inclusion scene will help in identification.

One of the best known of the inclusions has been given the name 'lily pads' or 'lotus leaves': they appear as negative crystals surrounded by delicate droplets or decrepitation haloes. Sometimes the negative crystal may be filled with glass or the crystal may be chromite. Peridot from Mount Kyaukpon in Myanmar contains considerable numbers of biotite crystals which often show a rectangular form and parallel orientation. The classic peridot from Zeberged (Island of St John) in the Red Sea often contains fluid inclusions reflecting its hydrothermal origin. Arizona peridot very frequently shows the lily-pad inclusions. Hawaiian material often contains blebs of glass. The occasional star stone has been reported, the six-rayed star arising from biotite platelets and needles.

This interesting inclusion scene makes it reasonably easy to distinguish peridot from specimens of other green gemstones but to use them on their own as locality pointers may be hard to justify. Peridot from different areas does seem to show distinctive optical and physical properties: yellowish-green Kenya specimens have the composition Fo_{90} Fa_{10} and SG 3.45 while stones from Myanmar usually show a lower SG at 3.22. The RI ranges of Kenya and Myanmar material respectively are 1.650–1.686 and 1.654–1.689. Birefringence is very similar in both instances. Colourless olivine will be rare (very Fe-poor) but examples have been reported from Sri Lanka and it is worth knowing that they do exist.

Peridot has a somewhat oily appearance which adds to its attraction and is not very like any other green gemstone. Its high birefringence of 0.036 will separate it from green sapphire whose DR is 0.008. If the spectroscope is brought into service first the observer will find that the absorption bands in both species may at first appear to be similar. However, careful study should make the test effective. The bands in green sapphire are at 471, 460 and 450 nm, the last one much the strongest. In peridot the bands are at 493, 473 and 453 nm with a darker centre in each band whose edges tend to be vaguely rather than well defined. As already mentioned, an absorption band at 529 nm can be seen in one direction but even without this the two spectra should be distinguished from one another without too much difficulty.

In any case, the gemmologist can always turn to a measurement of the refractive index, though we prefer to try the spectroscope and microscope first with most coloured stones. There is not a great resemblance between green sapphire and peridot but tourmaline, with its great range of colours, provides a greater challenge. Tourmaline's high birefringence is not as great as that of peridot, at 0.018 against 0.036, and the refractive index is slightly lower. The characteristic fluid inclusions in tourmaline are nothing like the inclusions in peridot.

Peridot could only be mistaken for emerald by those unfamiliar with either species: emerald's birefringence and inclusions as well as its much lower refractive index and specific gravity (mean figures 1.58, 2.71) separates them.

The most likely cause of confusion is with the variety of synthetic spinel which is grown and doped with the stated aim of looking like peridot. This material will show no double refraction and will have the higher RI and SG of 1.728 and 3.64. Between crossed polars synthetic spinel will show the unmistakable anomalous double refraction ('tabby extinction') which will not be seen if the specimen is peridot.

Glass will show no birefringence and will show the usual swirly interior with gas bubbles – it will also show anomalous double refraction.

With its high birefringence, zircon may provide some cause for caution. Green zircon, in particular, may not always show quite so high a birefringence in some specimens as they may be the nearly amorphous (non-crystalline) low (metamict) zircon. Only crystalline substances can display birefringence. If the green zircon is fully crystalline its birefringence will be above that of peridot, something near 0.01 with an RI of 1.81. In practice the refractive index and birefringence of many zircons varies a great deal for the reasons given above and the refractometer liquid used today will allow readings only below that figure – almost all zircons will give a negative reading (no shadow-edges seen). This means that some experience of judging birefringence should be cultivated.

Some other transparent green species may cause momentary confusion with peridot. Some colours of chrysoberyl may be mistaken for the browner-green olivines and this may also apply to the rarer mineral sinhalite. Taking chrysoberyl first, the RI of 1.74–1.75 is higher than that of peridot and the SG of 3.71 is higher than peridot's 3.34. Instead of the quite strong multi-band absorption spectrum of peridot, chrysoberyl shows a strong band at about 444 nm. The birefringence of chrysoberyl is 0.009 compared with 0.036.

The borate mineral sinhalite was long believed to be a dark green to brown variety of olivine with whose refractive index range it coincides. If a suspected sinhalite is examined with the spectroscope bands in the same position as those in olivine will be seen: they are joined, however, by an additional band at 463 nm. Sinhalite's RI is in the range 1.665–1.712 with DR 0.038 and SG 3.47–3.50. The hardness is 6.5–7.

Reports of interesting and unusual examples from the literature

Items in this section have been chosen to illustrate points made in the chapter and to bring one-off items to your notice.

Peridot from the Nanga Parbat region of Pakistan Kashmir shows RI in the range 1.648–1.689 with DR in the range 0.035–0.038, SG 3.29–3.37. The normal spectrum of Fe^{2+} was observed. Feathers of small liquid inclusions and growth structures could be seen under magnification: there were also rope-like inclusions of the ludwigite–vonsenite solid solution series (magnesium–iron borates).

Peridot from the Black Rock Summit lava flow, Nye County, Nevada, USA, showed RI in the range 1.650–1.699 with a DR in the range 0.033–0.038, SG in the range 3.351–3.388. One notable inclusion was described as resembling a printed circuit board though its identity was unestablished. Lily-pad inclusions were virtually absent and chrome diopside crystals were 'strung like rosary beads' and there were also smoke-like veils and black metallic spheres identified as pyrrhotite and pentlandite. Faceted stones are a darker green than most gem peridot.

Peridot from Lam Dong province, Vietnam, shows properties consistent with those of peridot from other localities. Inclusions too are consistent, lily-pads having been recorded.

Peridot reported from Damaping, Hubei Province, China, has RI in the range 1.652–1.600 with a DR of 0.034–0.036. The SG is 3.34–3.36. The lily-pad inclusions, fingerprint-like healing cracks and fine growth inhomogeneities have all been observed.

Chapter 14

Zircon

It should be said from the start that zircon and (cubic) zirconia have nothing in common but the element zirconium. Zirconia is the oxide and zirconium the element. Zircon, described in this chapter, is zirconium silicate ($ZrSiO_4$) and the gemstone which appears in so many colours from such places as Sri Lanka.

Chemically and structurally zircon is of particular interest since it reflects some of its own geological history to the observer. The point should also be made that virtually all zircons reaching the public, apart from fancy colours which normally reach only the connoisseurs and collectors, have been heated after mining and that disclosure, as with aquamarine, is not required. We shall see later on why zircon's properties vary so much but the hardness is usually between 7 and 7.5: stones, especially those that have been heated, are notably brittle and damage to facet edges is a feature of zircons that has been kept together in a parcel – for this reason zircon used to be kept in twists of paper and if these twists are encountered during a valuation or auction their contents are far more likely to be zircon than any other stone.

We can delay no longer in introducing something of zircon's geological history. A collection of zircons of all colours will be very likely to contain some 'high' and some 'low' specimens.

High zircon

High zircons are fully crystalline and give the higher RI, DR and SGs of the zircon family. These will be: RI 1.926–1.985 (often around 1.925) (remember that the limit of the contact liquid used on the refractometer is well below these values so high zircons cannot be tested with this instrument). The DR is very high at 0.059 and, like diamond, zircon has a high dispersion so that colourless stones could be confused with diamond. In fact, the dispersion of zircon is quite close to that of diamond at 0.039 compared to diamond's 0.044. Generally we have not mentioned dispersion values save in those cases when it helps in identification or is a major feature of the species concerned. The SG of high zircon is in the range 4.67–4.70.

Low zircon

Over geological time low zircons have lost their crystalline structure (in the most extreme cases) due to the continuous bombardment of the crystal structure by alpha particles emanating from included uranium and thorium. This results in a change of mass over geological time and thus the SG is lower than that of high zircon, in the range 3.95–4.20 but generally around 4.00.

The RI is much lower too, in the most extreme cases being around 1.78–1.815 with no double refraction or very little. Some specimens are virtually isotropic. The lower refractive index of low zircon could be seen when di-iodomethane with RI = 1.81 was used as the refractometer contact fluid but with the present liquid of 1.79 the shadow-edge will be blurred and hard to distinguish. It may look as though the unknown specimen is singly refractive but fortunately other features are available for testing. Most low zircons are a rather attractive quiet green (sometimes brown or orange) with distinguishing internal features which we shall meet later but not all green zircons are low – the correct name is *metamict* and the process of gradual disintegration of the crystal structure is *metamictization*. The hardness is 6.5.

As might be deduced the process of metamictization leaves some specimens neither at the top of the high end nor at the bottom of the low. These intermediate zircons will usually have an RI in the range 1.83–1.970 with a birefringence between 0.008 and 0.043 and specific gravity 4.08–4.60.

Zircons of whatever colour seem to have a particular lustre which has often been called oily or greasy – it can be seen very easily on crystals. The adjective sub-adamantine has occasionally been used. The colours cover virtually the complete visible spectrum though some of the heated stones develop colours not usually encountered in the unheated crystals. The colours of the heated stones seen in commerce are yellow or golden, bright blue and colourless, the latter more than able to substitute for diamond. The original crystals are not found in these colours but are commonly reddish-brown.

Metamict zircons are not usually heated but if they are they may rise to the intermediate level. The absorption spectrum will usually give them away.

Zircon has a most interesting range of absorption spectra. High zircons and especially some of greenish-brown colour from Burma show a multi-line spectrum with as many as 40 fine lines crossing the entire visible spectrum. This is a uranium absorption spectrum although uranium is not the direct cause of the stone's colour. The strongest of the absorptions is at 653.5 nm and this will be seen in almost all zircons of whatever colour and wherever they originate. The colourless, golden-yellow and bright blue heated zircons will show this line, often with another one at 659 nm.

In the absorption spectrum of low zircon only a vague broad band near 653.5 nm may be seen while in others a prominent narrow band at 520 nm may be seen. If the low zircons (those with a specific gravity less than about 4.02) are heated to 800°C a series of anomalous absorption bands will develop: these are at 687, 668.5, 652.5, 589, 574, 560.5, 473.5 and 451 nm. This effect is rare in natural zircon. We should imagine that such stones might be found collectable by gemmologists.

A number of zircons may show no absorption spectrum: these are usually red, sometimes orange or pale brown. Most will turn up as mineral specimens rather than as fashioned stones.

Zircon shows very little pleochroism though some of the heated blue stones show fairly strong colours. Colourless zircons give a yellowish glow under LWUV light or X-rays: when the response is seen examination with the spectroscope will show two groups of bright lines in the yellow and in the blue. The response may be due to the presence of one or another of the rare earth elements (lanthanides). Should the gemmologist wish to see this effect do not try it with a customer's stones! The irradiation will cause the stones to discolour to a muddy brown and they will need heating to restore them to their previous state. A dull red heat should do the trick and if the heating is done in the dark on a hot plate the observer will note the phenomenon of thermo-phosphorescence. While this is visible the spectroscope will show the same bright line spectrum already described. As zircons are heated their refractive index will rise but we imagine that few gemmologists will want to crack the glass of their refractometers with a hot specimen, especially when it will almost certainly give no reading!

Heat treatment of low zircons causes the SG and RI to be altered. At an SG of about 4.2 further heating usually causes the SG to be lowered. Stones with an SG above 4.2 when heated show a rise. Stones heated to temperatures in excess of 1100°C are more likely to revert to the normal constants.

Fancy colours

Zircon like spinel and corundum turns up in a number of fancy colours (other than blue, yellow and colourless). Many of the fancy zircons are very attractive and Sri Lanka is the major producer. Somewhere along the line there has been some astute promotion if the family-based-producing groups can be induced to pass along some of the unusual colours as well as those which they know from experience will sell satisfactorily.

In our experience the red zircons from Sri Lanka sell for the highest prices among the fancy colours but orange and brown stones can be very attractive. Red specimens will not show the absorption at 653.5 but their birefringence will separate them from the darker red–orange garnets (pyrope, almandine, spessartine and hessonite).

Orange zircon may look very like the magnificent orange isotropic sphalerite (zinc blende) which has a lower specific gravity at 4.09 compared to zircon's mean 4.69. Sphalerite is worth looking out for as, though soft and with six directions of cleavage, it has a very high dispersion of 0.156, considerably higher than that of zircon (and of diamond).

Brown zircon is unexpectedly attractive too and with the recent acceptance of diamonds of this colour there may be the temptation to place a brown zircon with small brown diamonds (surely there would be very little to gain from this practice, but it does happen). Once suspected the high double refraction would give the zircon away. In general commercial practice the zircon would probably get quite far along the road.

The faceted yellow and orange sphenes (the latter very rare) make beautiful specimens for collectors. With similar lustre, high birefringence and dispersion (0.105–0.135 and 0.051) the distinction from zircon could be difficult to make but sphene contains mineral inclusions (frequently pyrrhotite and lepidocrocite in growth tubes) which are uncommon in zircon which also shows far weaker pleochroism than sphene. The RI at 1.885–2.050 is higher than that of zircon though the SG of 3.52–3.54 is lower. There is strong pleochroism.

Brown dravite tourmaline would look more like brown zircon if it had more dispersion: in any case its birefringence is lower at 0.018 compared to 0.059 and it is much more markedly pleochroic. The birefringence of sinhalite is quite high, at 0.038, but the refractive index should be easy to obtain (mean figure 1.685). The absorption spectrum is quite unlike that of zircon with bands in the blue at 493, 475, 463 and 450 nm.

Natural materials that may be confused with zircon

Distinguishing zircon from other natural gem species must take into account both the wide range of colours possible and the wide variation in the properties of zircon. Clearly the greasiness or oiliness of some of the darker shades (not the blue, golden-yellow and colourless faceted stones) should alert the observer and the absorption spectrum, which should be an early, if not the first choice, of testing feature with coloured stones, will be very helpful. When magnification is available (if the spectroscope is available a nearby lens or microscope is likely) the strong birefringence can hardly be missed, apart from in metamict stones.

Looking at other gem species with high birefringence and taking colourless stones first we have to remember that colourless zircon cannot be tested on the standard refractometer. Since zircon has a high dispersion (0.039 at its greatest compared to diamond's 0.044) it will be likely that such a stone will be tested carefully just in case it turns out to be diamond. Highly dispersive colourless stones should perhaps be kept away from the refractomerter! Colourless zircon will show the absorption band at 653.5 (best seen by reflected light) and this will be sufficient both to distinguish it from diamond and to prove its identity. Should the stone be tested for diamond first and LWUV light called into play, remember the dangers of discoloration in colourless and blue (heated) zircons.

We should mention at this point that zircon is often said to 'fade' by those who have got an idea that something can happen to zircon's colour. The colour does not fade but the stone gradually develops a dark brownish, muddy stain. This happens only with heated colourless and blue zircons, the golden and yellow varieties seemingly remain unaffected.

Colourless zircon

The high birefringence of colourless zircon (which is never metamict) will rule out such possibilities as *glass*, *synthetic corundum* or *spinel*, or, of course, *diamond*. While this property, together with its high dispersion, may cause confusion with synthetic rutile, also highly birefringent and dispersive (figures are 0.287 and near to 0.28 to 0.3 respectively compared to diamond's lack of birefringence and dispersion of 0.044), *synthetic rutile* always displays a yellowish cast when otherwise colourless, a feature not seen in zircon. The heated and annealed synthetic rutiles in shades of yellow, blue and black, are much more likely to be mistaken for zircon though the dispersion of rutile is so pronounced that anyone familiar with the two species (and with diamond) will not easily be taken in.

Blue zircon

The blue heated zircon (no blue zircons have been found in nature so far) could be confused with *benitoite* (you are more likely to think of benitoite if you live in California and collect gemstones). Benitoite is also highly birefringent (0.047) and dispersive (0.046). It will show an intense bright blue fluorescence in SWUV light and under X-rays but if you think zircon may be a possibility for the stone you are testing don't go to the UV source first (or second – see above).

There is little possibility of blue zircon being taken for blue *sapphire* as the appearance is quite different – if you're familiar with both species. Sapphire will virtually always show characteristic inclusions while those in zircon are often indistinct. In general zircon shows few inclusions that immediately say 'zircon'. Cracks which have occurred post-formation and healed fractures are probably the most typical features and the heated stones show very little to help the investigator. The cracks may be surrounded by yellow to brown films of liquid. Metamict zircons do show a recognizable inclusion pattern which we shall discuss later.

Testing the stone which looks most like blue sapphire will give the corundum RI and SG. You will not have the same success with blue zircon as it reads well above the limit of the standard refractometer (negative reading). With suspected blue zircon/blue sapphire try the spectroscope first: if zircon, at least the 653.5 nm absorption should be seen and a blue sapphire ought to show at least the 450 nm band. If the sapphire is a flame-fusion synthetic this absorption band may not be present but neither will there be any absorption in the 653.5 area.

A greater danger is that a blue synthetic *spinel* will be mistaken for blue zircon since this particular colour is aimed at by crystal growers. The dispersion will be lower (0.20) and, more importantly, there will be no birefringence. Spinel shows no absorption at 653.5 nm and though the band at 632 nm is close it will be broader than anything seen in zircon. Blue synthetic spinel usually shows orange through the Chelsea filter and will show 'tabby extinction' between crossed polars. The cracks sometimes seen in this type of synthetic spinel appear to meet at near-right angles (the natural spinels do not seem to show this) and they should not be confused with the healing cracks and their yellow–brown surround seen in the blue zircons.

As far as natural blue spinel is concerned the rather dark colour and lack of marked dispersion, as well as the lack of any birefringence, should make them easy to test. Some natural blue zircons owing their colour to cobalt will show orange to red through the Chelsea filter but those coloured by iron will not, but a green colour instead, though the multi-band absorption spectrum is nothing like the absorption at 653.5 nm shown by blue zircon.

Blue *zoisite* (tanzanite) looks nothing like zircon and in any case its strong pleochroism could not be missed: its double refraction is only about 0.009. *Aquamarine* lacks the fire (dispersion) of blue zircon and again has a low birefringence of 0.006. Blue *tourmaline* is usually rather dark and although it shows a high birefringence (0.018) this property is still noticeably less than that of zircon. It is possible that some of the smaller bright blue Paraíba tourmalines may be confused with zircon on colour grounds but their dispersion is not high and their refractive index easy to establish.

Blue *topaz* and blue zircon could quite easily be confused on colour grounds although zircon has a much more noticeable dispersion (0.039 compared to that of topaz (0.014) which is distinctly low). Topaz is normally quite easy to test on the refractometer and will show RI readings (in the range 1.62–1.64) when zircon shows none. Blue topaz will show no absorption in the visible and will not display any significant fluorescence other than a weak yellow–greenish-yellow under LWUV light and a weaker but similar effect under SW.

Yellow zircon

Golden and yellow zircon show the same strong birefringence and high dispersion as the colourless and blue stones but are not, like them, prone to discoloration. From the colour point of view the yellow zircons could easily be taken for yellow diamond though magnification will quickly settle matters when their high birefringence is seen.

Yellow zircon is bright but in small sizes the colour difference between Cape diamond or even canary yellow diamond is less obvious. Cape diamonds will not absorb in the 653.5 nm 'zircon' region and yellow zircon will show no absorption in the 415 nm Cape region.

Yellow sapphire is birefringent but much less so than zircon. Stones from Sri Lanka may show apricot-yellow fluorescence under LWUV light and X-rays while iron-rich stones from Australia and elsewhere will show the three iron bands at 471, 460 and 450 nm in the blue.

Green zircon

Leaving green zircon until last, any milky green stone should be tested with the spectroscope to see what absorption bands may be present. If the stone is a zircon in which the process of metamictization is far advanced the only absorption seen will be the band near 653.5 nm though in some green low zircons the band at 520 nm in the green may also be visible. We saw above that these stones can be heated to give an anomalous absorption spectrum.

It is understandable that the gemmologist who in most cases will see very few of these stones may be deceived by the absorption spectrum and will wish for further confirmation of a possibly tentative identification. Fortunately the microscope will help here: in metamict zircon angular tension fractures echo the structure of the original crystal and thus show as systems of fissures at angles of 57.5° to one another (this is the angle at which two major crystallographic directions in zircon meet). Tension fissures may also occur as disc shapes in metamict zircon and systems of parallel stripes are also known.

Though cat's-eye zircon and some star stones turn up from time to time (they are usually of Sri Lankan origin) the effects arise from random orientation of inclusions allowing the lapidary to bring out the phenomena.

A synthetic zircon has been grown hydrothermally by Bell Laboratories of Murray Hill, New Jersey, USA, but this was for its experimental rather than its ornamental properties. The crystal we have seen was doped with vanadium to give a very attractive purple colour with strong pleochroism.

Reports of interesting and unusual examples from the literature

Items in this section have been chosen to illustrate points made in the chapter and to bring one-off items to your notice.

Transparent orange, orange–yellow or brownish red zircons from the Primorye placer deposits in Russia have been reported to turn colourless on heating.

Chapter 15

Moonstone and the feldspar minerals

Moonstone whose appearance is almost impossible to describe to anyone else has (1) to be a member of the *feldspar* group of minerals and (2) to fulfil the general conception of moonstone. The name traditionally given to the particular lustre of moonstone is adularescence (or schiller), the first name deriving from a European place name where moonstone was found.

The feldspar group of minerals is one of the largest on Earth and after the most recent classification contained 16 distinct species: the older gemmology texts are now dated in this respect, the number of separate species with which the gemmologist is likely to have to deal having reduced to five – albite, anorthite, microcline, orthoclase, sanidine. All previous 'species' are now varieties of some of them. Normally this book does not deal with the wider study of mineralogy but in the case of the feldspars some explanation is necessary before we look at individual feldspar gemstones.

We have not yet finished with the feldspars as a mineral group! The minerals *orthoclase*, *sanidine* and *microcline* all contain potassium (K-feldspars) (alkali feldspars) while the others belong to the sodium–calcium feldspars which form a solid solution series similar to that shown by individual species of the garnet group. We shall see how all this works as we consider the different species.

Orthoclase

We are no nearer to defining the particular lustre of moonstone (the noun lustre in this context signifies an optical effect reaching the eye as light is reflected from sub-surface inclusions). In the case of the majority of moonstones, an orthoclase (K-feldspar) host contains inclusions of albite, one of the sodium–calcium species. It is from the albite that light is reflected in such a way as to give the soft, sometimes bluish appearance characteristic of moonstone. Other colours are known as we shall find below.

Looking at the inclusions causing the moonstone effect in orthoclase we find that the finest schiller (bluish) is found when light is reflected from fine plates of albite. The thicker the included albite crystals the whiter the sheen becomes. The body colour of the orthoclase may be coloured, also from included mineral material: reddish colours, for example, are caused by the iron oxide goethite. The best blue sheen is usually found in moonstone from Burma and Sri Lanka.

The orthoclase moonstones have RI in the range 1.520–1.525 with a double refraction of 0.005 and specific gravity in the range 2.56–2.59. Properties do vary widely and these figures are general. Again generally speaking orthoclase moonstones from Sri Lanka and those from India tend to be found respectively at the low and high ends of the scale. The adularescence shown by some orthoclase moonstone may be concentrated into a band which gives a cat's-eye effect and sometimes a four-rayed star.

Orthoclase moonstone is 6 on Mohs' scale of hardness and also shows two directions of easy cleavage. Owners have been surprised to find their moonstone cabochons (the best way to show adularescence) have separated into parts which resemble the dissociated segments of an orange. Indications of the cleavage may also be noticed on the surface of the polished stones, where they show interesting and characteristic structures to which the name 'centipedes' has been given. The structures show distinct combined markings at right angles to one another (the name orthoclase indicates cleavage directions at right angles).

The moonstone effect in non-feldspar materials

Before going on to the other members of the feldspar group (including other moonstones) we may consider examples of moonstone-like materials which are not members of the feldspar group. The best imitation of moonstone (there is no synthetic moonstone) is provided by a white synthetic spinel which in this case has extra alumina added to it (it is grown by the Verneuil flame-fusion method). The extra alumina may amount to as much as five times that of natural spinel so that the RI may be as much as 1.727 and the SG 3.64 compared to orthoclase moonstone's 1.525 and 2.56–2.59. Spinel shows no double refraction. Glass of various kinds is often found imitating moonstone but lacks the surface markings and never achieves true adularescence: in any case there should at least be traces of gas bubbles.

Occasionally there are reports of a whitish heat-treated amethyst being passed off as moonstone but we should have thought that since amethyst usually occurs with colour in the tips of the rhombohedra terminating the crystals which are otherwise white or grade into it, that to use the white material without bothering to heat it would be more likely. The RI of such a piece would not be too far from that of orthoclase moonstone but they should not turn up very often. It is much more likely that translucent chalcedony would be used.

Moonstones other than orthoclase are most likely to be albite which has a refractive index in the range 1.530–1.540 and specific gravity 2.63, higher than orthoclase. These figures are taken from Madagascan material which is sometimes faceted. Many albites show quiet colours of yellow, pink, green, grey, and reddish, mainly if not entirely from inclusions. The moonstone effect may also be present.

The name sanidine is now used to denote a potassium–sodium (alkali) monoclinic feldspar but to the gemmologist it is known as a very attractive transparent material very like smoky quartz in appearance. Sanidine sometimes shows the moonstone effect: it has RI in the range 1.518–1.534 with a double refraction of 0.005 and specific gravity 2.52–2.62. These figures are lower than those for smoky quartz which will give RI 1.544–1.553 and 2.65.

Other feldspars

Moonstone is not the only gem variety of the feldspar group minerals. Orthoclase is also found as beautiful transparent yellow crystals which are sometimes faceted and which can be quite large on occasion. The finest specimens come from Madagascar and have a refractive index in the range 1.518–1.539 with DR 0.008 and SG between 2.55 and 2.63. The hardness is 6–6.5. Yellow orthoclase may show a weak blue fluorescence under LWUV and orange under SW, white to violet under X-rays though it is hard to believe that such tests will be found necessary for identification. The absorption spectrum is probably the best test to begin (and end) with: there are bands at 420 and 448 nm. It is unlikely to be confused with yellow sapphire or topaz, whose refractive indices and specific gravity are much higher. Yellow apatite and the rare gemstone brazilianite are much more like orthoclase in colour.

Yellow *apatite* very often shows a rare earth absorption spectrum with two groups of fine lines, in the yellow and in the green: it has a higher RI than orthoclase, at a mean 1.63. Brazilianite shows no distinctive visible absorption spectrum and RI in the range 1.60–1.62, well above the values for orthoclase.

The yellow varieties of *beryl* may be mistaken for orthoclase though the reverse would be more likely. Beryl's refractive index with a mean 1.58 is higher than that of orthoclase and yellow beryl shows no distinctive absorption spectrum in the visible. The paler yellow tourmalines have a much higher birefringence (0.018) than moonstone's 0.005 and doubling of back facets and inclusions are very easy to see under magnification.

Another species of potassium (alkali) feldspar which provides some ornamental material includes *microcline*, sometimes known as amazonite or amazon stone. This has a most distinctive blue–green colour with whitish streaks: crystals are often well formed and collectable in their own right – probably more specimens are kept for their form rather than for anything that could be fashioned from them. Nothing looks like microcline and it is not synthesized. It may be thought to resemble jade if you have never seen jade. Microcline has an RI 1.539 at its highest, well above nephrite's 1.60 and jadeite's 1.66. It may show a yellow–green fluorescence under LWUV and green under X-rays. At 6.5 it is harder than orthoclase.

So far we have looked only at the alkali feldspars though albite has been mentioned as another type of moonstone. The other feldspar group forms a continuous solid solution series between the sodium aluminium silicate (albite) and the calcium aluminium silicate (anorthite). This group used to be called the plagioclases but the name has now been dropped during reclassification. The individual species in the series provide the gem varieties albite, oligoclase, labradorite, sunstone and peristerite. Some of these names are not in common use but the gemstones are of some importance and include some rarities.

Albite is white as the name suggests and provides a moonstone effect on occasion. It will give a refractive index in a range very close to that of orthoclase moonstone and there seems no reason to differentiate between them: as long as 'moonstone' is a feldspar it doesn't matter to gemmologists which species it is.

Oligoclase is sometimes found as colourless to pale green transparent crystals but these are not of ornamental significance. The name *sunstone* describes a sodium–calcium feldspar with its colour derived from inclusions which may be hematite or goethite (both are oxides of iron with slight differences in chemical composition) or of copper. Most sunstone, especially the material found in Norway, is flecked so profusely with reddish-golden platy inclusions that the entire stone appears to have one single body colour. Sunstone may chemically be anywhere in the series between albite and anorthite but does not itself possess species status.

The same is true for *labradorite* (varietal name) which gemmologists tend to associate solely with dark, almost opal-like material within or against which a play of colour can be seen. While this certainly does give a well-known ornamental stone, labradorite is more often found as transparent colourless to yellow crystals which are quite often fashioned and which may also contain inclusions by which it may earn sunstone status. Magnetite and zircon may be found as inclusions but their presence, while interesting, is insignificant as a clue for testing.

To make all clear both oligoclase and labradorite can provide sunstone material but neither have individual species status. Gemmologists normally do not need to know where in the series a particular specimen comes, only to know that it is a feldspar group member.

While sunstone and clear transparent sodium–calcium feldspars present no problems of identification and are neither synthesized nor imitated, labradorite may easily be taken for black opal by those unfamiliar with either material. It is not advisable to take the refractive index of opal since its porosity will allow the refractometer contact liquid to enter and cause discoloration but an examination of opal's individual colour patches will show that they are smaller, brighter coloured and relatively similar sized. Should a refractive index be found in ways not involving contact liquids then opal's much lower value of 1.45 is well below the mean 1.56 of labradorite, which has a specific gravity of about 2.70 compared to the 2.10 of opal.

Transparent garnet-red feldspars

In recent years a magnificent transparent garnet-red sodium–calcium feldspar variety has been mined in Oregon, USA. This is perhaps the most costly of the feldspars apart from the largest and most exceptional moonstones. Its mean refractive index and specific gravity of around 1.56 and 2.70 are well below the figures for similarly coloured red garnets (about 1.75 and 3.66). While the coloured feldspars in general show little or no pleochroism it is quite marked in the deep red material. This might be helpful in testing – the colour seen will not be very distinctive (a lighter and a darker red): some doubling of back facets and inclusions may be seen (the birefringence will be about 0.007) while garnet shows no birefringence.

Sunstone is not really well or deliberately imitated though a glass often known as *goldstone* is very common. This is made by incorporating cuprous oxide during manufacture. The copper is reduced during annealing to produce small flakes of metallic copper which act as the hematite or goethite do in the sunstone feldspars. Under magnification the familiar swirls and gas bubbles characteristic of glass can be seen.

Reports of interesting and unusual examples from the literature

Items in this section have been chosen to illustrate points made in the chapter and to bring one-off items to your notice.

Moonstone with a darker than usual body colour is found in Sri Lanka and the name 'smoky moonstone' has been used. Specimens show a blue or white sheen against the dark background. The iron content appears to be greater than in colourless moonstones.

Moonstone with a royal blue sheen has been extracted from a mine near Patna, Bihar, India. Most of the best specimens are 1 ct or less. Its properties indicate a high calcium feldspar.

Chapter 16

Rhodochrosite, tanzanite and rhodonite

Two other gem materials which have gained in importance in recent years do not present too much difficulty in identification. They are rhodochrosite and the tanzanite variety of zoisite: rhodonite with which rhodochrosite can easily be confused is also described here.

Though both occur in transparent and opaque forms, larger and especially orange–pink transparent faceted specimens are much more likely to be rhodochrosite than rhodonite which is usually pinkish-red when transparent. Most rhodochrosite is found as masses and it is from these that the well-known carving material is produced, the pink rhodochrosite often combining with white veins of calcite.

The hardness of rhodochrosite is 3.5–4, RI in the range 1.695–1.840, DR for transparent material 0.201–0.220. The SG is 3.4–3.6. There is a perfect cleavage and the spectroscope will show absorption bands at 551 and 410 nm with weak lines at 565 and 535 nm. Rhodochrosite as a carbonate will effervesce in warm acids.

It was not until the 1960s that transparent rhodochrosite was found in the Kalahari area of South Africa though transparent specimens were already known from Colorado and Argentina as well as Peru. The Kalahari crystals can be magnificent both in form (scalenohedra) and in colour which is a superb orange–pink. Very deep red material may show a dull red to violet fluorescence in SWUV light (Argentine and Colorado specimens in particular).

The transparent rhodochrosites show so high a birefringence that confusion with other orange stones (spessartine or hessonite, for example) is unlikely. The garnets of this colour will also show readings on the gemmological refractometer which cannot be used for rhodochrosite. One point worth making is that many crystals from the Kalahari mines (Hotazel, N'Chwaning, for example) are so fine that they are unlikely ever to be faceted but they will be on display (commanding very high prices) at the major gem shows.

Rhodonite is also found in masses which can be carved but the veining is more likely to be black, from undifferentiated manganese oxides. It has a hardness of 5.5–6.5, RI in the range 1.723–1.737 with DR 0.012. The SG is 3.68–3.70 (readings for the transparent material from Broken Hill, New South

Wales). There is a perfect cleavage in one direction and the spectroscope will show a broad absorption band at 548 nm with a strong line at 503 nm, a weak band at 455 nm and sometimes lines at 412 and 408 nm.

Rhodonite when clear enough to produce faceted stones could be confused with several different transparent red species. As stones are almost always small, the spectroscope will confirm the presence of manganese which does not colour the major red species.

The gemstone *tanzanite*, named after the country of origin, is a variety of zoisite, itself a member of the epidote group of minerals. It was not until the 1960s that a transparent variety of zoisite was found and then discovered to be capabie of achieving a magnificent blue colour after heat treatment.

The crystals when found show little hint of the blue that can be attained when they are heated. They have a hardness of 6–7, RI in the range 1.692–1.700 with DR 0.009 and SG 3.35. The pleochroism is very notable and distinguishes tanzanite from iolite, from very deep blue beryls and light blue sapphires – in any case the colour is never quite like the blue of these species. In tanzanite the pleochroic colours may vary from deep blue to purple to green or from reddish-purple to blue to yellowish-brown. The spectroscope may show some faint absorption lines in the red with a broad absorption centred at 595 nm and other bands at 528 and 455 nm.

Tanzanite sizes can be quite large with one stone recorded at over 120 ct. A green transparent zoisite has been found in northern Pakistan but is not heated to give a blue colour.

Imitations include a synthetic forsterite (with higher DR) which is described in the synthetics section, a doped glass with a high SG which has been offered for sale under the name U.M. Tanzanic, a YAG with the name purple coranite, a synthetic corundum named blue coranite and a calcium phosphate glass with no trade name. Tanzanite's pleochroism will easily distinguish it from these materials.

Chapter 17

Turquoise

Turquoise is a common, relatively inexpensive but popular gem/ornamental material owing its colour to copper and iron. Connoisseurs will heatedly deny that the colour resembles the advertising of a well-known bank – this colour is more greenish blue than blue, the green arising from a higher iron than copper content. Anyone with a set of saleroom catalogues extending over many years and including the vital European and American issues can trace the desirability of the different colours of turquoise – the material offered 30 years ago inclines far more to blue than to green and this is not just a photographic technicality.

The mineral turquoise is essentially a hydrous phosphate of copper and aluminium, forming a series with chalcosiderite which is a phosphate of copper and iron. This is how the iron comes into the story. The ferric iron substitutes for aluminium in the turquoise structure.

Turquoise is 5–6 on Mohs' scale of hardness and has a naturally waxy lustre. The refractive index for the masses in which turquoise occurs (though small crystals are known) is 1.62 and the specific gravity falls in the range 2.6–2.9: the very fine material from Iran is 2.75–2.85.

With the spectroscope the two copper absorption bands, though hard to see, can often be detected deep in the blue at 460 and 432 nm. In treated stones (which will show a darker colour) the bands are easier to see and this is a useful identification feature. Under LWUV light turquoise may show a greenish-yellow fluorescence but most specimens remain inert under SWUV light and under X-rays.

If turquoise is presented as a cabochon with a flat back the refractive index should be easy to establish though the action of the contact liquid on the stone should be minimized as far as possible.

Turquoise from different places can sometimes show features pinpointing its origin. One example is turquoise from the Lavender Blue mine in Nevada, USA, which shows a black patterning reminiscent of a spider's web with minute specks of turquoise. The finest specimens still come from the Nishapur area of Iran, stones frequently containing characteristic brownish veining, probably of limonite. Intense dark blue turquoise has been found at Bisbee, Arizona, USA, and mines at least 3000 years old have operated in the Sinai Peninsula of

Egypt at Serabit el Khadim and Magharah. Turquoise has been reported from China and is described as 'porcelain turquoise' if the hardness is more than 5 on Mohs' scale. Some at least of this turquoise fluoresces a weak greenish-yellow under LW and SWUV light.

Much of the turquoise found in the south-western United States is pale and powdery: it is routinely stabilized (and darkened in the process) by impregnation with plastic material of one kind or another, probably epoxy resins. Examples are given below.

Identification

Turquoise is admittedly hard to test because of the numerous treatments and imitations around. There is, however, a synthetic turquoise which was manufactured by Gilson in 1972. This is not an imitation as it has the same chemical composition and crystal structure as natural turquoise. It is not of a very deep colour and has a specific gravity close to 2.7. Examination of the surface with the microscope shows entirely characteristic angular blue grains in a whitish matrix – magnification of about 50× is necessary so the 10× lens will not help in this case. The Gilson material does not seem to show the absorption band at 432 nm.

The Gilson product was sold in at least two different varieties. Medium blue stones have been sold under the name 'Cleopatra' and darker blue ones as 'Farah'. The best survey of turquoise simulants, including the Gilson product, can be found in *Gems & Gemology*, 15, 226–232. No iron is present in the Gilson turquoise which is manufactured from pure materials.

One of the simulants has been marketed as 'Turquite' and has been manufactured by Turquite Minerals of New Mexico. Chemical investigation showed that Turquite was poor in aluminium but rich in sulphur, silicon and calcium. An imitation turquoise made by the Syntho Gem Company in California was found to be similar to the Gilson product and this was also true of another imitation made by Adco Products, also of California. Only the Gilson turquoise has the same crystal structure as natural turquoise.

If the investigator is presented with a suite of turquoise jewellery an exact match between all the stones should be a warning: they may be surface treated or even synthetic material but imitations are more likely.

It may be possible to test for turquoise imitations by placing a drop of Thoulet's solution (iodide of potassium and mercury, unlikely to be easily available) on the surface of a doubtful specimen. The surface of an imitation will turn brown under the drop while turquoise is unaffected.

Dyed blue magnesite (magnesium carbonate) can make a quite convincing turquoise imitation. As a carbonate it will effervesce with warm dilute HCl: it has a refractive index of 1.51–1.70 (carbonates have a high birefringence) and specific gravity of about 3.0. Dyed howlite has a refractive index of 1.59 and specific gravity 2.53–2.59. Ivory has sometimes been dyed to imitate turquoise but the lines of Retzius should always be visible in one direction of viewing. Ivory's specific gravity of 1.80 is much lower than that of turquoise so specimens will feel very light in comparison.

'Viennese turquoise' is a precipitate of aluminium phosphate consolidated by pressing and coloured blue by copper oleate, and 'Neolith' a mixture of copper phosphate and bayerite, an aluminium hydroxide. Both of these products will show a yellow colour when touched by a drop of dilute hydrochloric acid.

Fading or discolouration of turquoise can arise from skin contact, the colour often changing from blue to green.

Paraffin wax is often used for impregnation, the stone first being dried then soaked in warm melted wax in a process taking several days. Dyestuffs used to deepen the colour include different copper compounds and Prussian blue. In *Gemstone Enhancement* (1994) Nassau comments that Egyptian buyers prefer green to blue turquoise so that blue material found there is altered to green to increase sales.

Turquoise can be coated with the aim of improving a dull-looking stone or to protect a layer of dye. Nassau (1994) describes a method of coating in which the surface of turquoise beads was painted blue, black paint added to imitate matrix and the whole bead then coated with clear lacquer.

If treatment is suspected the microscope should be used. Colour concentrates in cracks and drill holes, some dyes being removable with ammonia on a cotton swab. The thermal reaction tester may cause some coatings to show local melting while plastic coatings will give the characteristic pungent smell. Disclosure is not needed for turquoise which has been stabilized without the addition of colour.

The Zachery treatment of turquoise may or may not improve the colour but does help it to take a good polish. Porosity decreases with this treatment so that the harmful effect of skin oils is reduced. No polymer impregnation is involved and only chemical analysis can identify the treatment using different non-gemmological techniques. Treated stones show more potassium than untreated ones.

Turquoise of a powdery consistency can be consolidated with resins: the colour is darkened and the absorption spectrum is easier to see, as already mentioned. Glass, plastics, enamel, porcelain and dyed chalcedony or marble may imitate turquoise but all these can be distinguished. Other natural imitations include variscite with an RI and SG of 1.56 and 2.2–2.5; odontolite has RI 1.57, SG near 3.0; microcline (amazonite) feldspar has RI 1.51–1.53, SG 2.54–2.63; lazulite has RI 1.60–1.66, SG near 3.08. Chrysocolla has RI 1.57–1.63, SG 2.0–2.45.

Chapter 18

Lapis-lazuli

Introduction

The finest lapis-lazuli comes from Afghanistan: the material is a rock rather than an individual mineral. It is composed of a number of minerals of which the most important are lazurite (which gives the colour), haüyne (also a bright blue), sodalite (a darker blue) and nosean. These four species (feldspathoids) are members of the sodalite group and lazurite, haüyne and sodalite also provide facetable gemstones in their own right. Calcite, diopside and pyrite are usually found in lapis-lazuli, a high content of the former lowering the value considerably. On the other hand the bright brassy yellow flecks of pyrite enhance the general appearance of lapis-lazuli.

As lapis-lazuli is a rock rather than a single mineral the properties vary: the refractive index (which can be taken only on flat-backed cabochons – much lapis is carved) will be near 1.50 and the specific gravity 2.7–2.9 (sometimes higher if there is a high pyrite content). Some lapis may show orange fluorescent streaks in LWUV light. This is commonest in Chilean lapis which contains a good deal of calcite and some specimens may show pink under SWUV light. A drop of HCl placed on the specimen may produce a rotten eggs smell but we would hope that the acid test is used infrequently: if the acid is placed on the calcite part it will cause effervescence.

Imitations

As a rock it could be hard to imitate lapis successfully but specimens are large enough for fine blue cabochons and small carvings to be obtained without using any material which is not of the best colour. For this reason all the imitator has to do is to produce the colour and this is very well achieved by a variety of synthetic spinel said to have been placed on the gem markets in 1954 (we mention this in case any reader knows the date of manufacture of their lapis jewellery).

The material is grown by flame-fusion, doped with cobalt and fused at a temperature of not less than 2000°C. The name 'synthetic sintered spinel' was used for a while.

The spinel's specific gravity of 3.64 is much higher than the 2.7–2.9 of lapis but few will want to make SG tests today when other methods are available. It is easier to examine a suspect under a strong light with the Chelsea filter. As cobalt is the cause of the colour it will cause the specimen to appear a strong red. Some lapis specimens may show a dull brownish-red with this test but there should be no confusion. In some instances small flecks of gold are added to simulate pyrite (fool's pyrite!) but no attempt is made to imitate the calcite for obvious reasons. Many examples of the spinel will have flat backs (of cufflinks, for example) and they will enable an RI reading to be obtained – it will be around 1.725 compared to the 1.50 mean figure for lapis.

The spectroscope will show cobalt bands in the red, yellow, green and blue regions where lapis-lazuli shows no absorption in the visible. If a thin enough piece of the spinel imitation can be obtained it will show a distinctive reddish-purple colour.

Just as soil in the garden appears darker after rain, so a light-coloured specimen of lapis can be darkened by impregnation with plastic material. Chilean material is usually chosen for treatment since it diminishes or removes the whitish calcite areas. The thermal reaction tester will produce a pungent 'plastic' smell.

Gilson lapis-lazuli

Pierre Gilson, known for his manufacture of fine quality synthetic emerald, brought out an imitation of lapis-lazuli in the 1970s. Several varieties were offered, some containing added pyrite. The material was found to be porous and for that reason has a lower specific gravity than natural lapis-lazuli, at about 2.46. When a piece was drawn across an unglazed porcelain streak plate it left behind a strongly coloured blue powder: this streak test with true lapis-lazuli leaves a much weaker blue. When this test is carried out on the Gilson material a distinct sulphurous smell was noticed: a similar smell but much less distinct occurs with the natural material but in this case a greater pressure on specimen and plate needs to be applied.

If tested with acids the Gilson material reacts more strongly than natural lapis-lazuli. The composition is ultramarine (an intense blue pigment) with hydrous zinc phosphates.

Enhancement

Natural lapis-lazuli is frequently dyed or waxed to improve the colour and lessen the effect of any calcite. Specimens should be examined under magnification to see whether or not dyestuffs have concentrated in cracks. Dye can often be removed with an acetone-soaked swab (if prevailing health rules allow it): nail polish remover will probably be in most gemmologists' reach and will do as well.

Natural material may occasionally be dyed to give what is thought to be the lapis-lazuli colour. One of the commonest examples (though it is not really very common) is howlite which has a specific gravity of 2.45–2.58, well below the 2.83 of lapis.

Unstained howlite from California gives a brownish-yellow or intense orange fluorescence under UV light and a specimen dyed to resemble lapis-lazuli showed the same fluorescence colour in a part of the stone from which the dye had been removed. Stained dolomite (magnesium calcium carbonate) has also been reported as an imitation of lapis-lazuli. It will react with *warm* acids (dilute HCl) by effervescing. It has an RI of 1.50–1,68 and SG 2.85 (near lapis-lazuli): it would not be easy to carve as it cleaves so easily and signs of this would be apparent.

Reports of interesting and unusual examples from the literature

Items in this section have been chosen to illustrate points made in the chapter and to bring one-off items to your notice.

Plastic imitations of lapis-lazuli (and malachite) have been used as inlays in watch-face material. Watches with natural lapis faces will feel much heavier. Unpolished edges between each link of the inlay will appear uneven and grainy where plastic will be smooth.

An imitation of lapis-lazuli has been described whose colour could not be removed by acetone even though the beads of the necklace were reported to have stained skin and clothing. In the end denatured alcohol (found in most scents and colognes) did the trick!

In 1986 GIA described a lapis-lazuli necklace whose beads were violet–blue and very deeply coloured. They fluoresced a patchy red under LWUV light while under SW light, only a few gave the usual chalky-green response of natural lapis-lazuli. The acetone swab did not remove so much of the colour as expected but paraffin treatment was confirmed by the specimens sweating when the thermal reaction tester was brought near. Some of the beads showed a purple dye in cracks, visible under magnification, and the necklace showed a definite brownish-red through the Chelsea filter. This was a brighter colour than had been reported previously for natural lapis-lazuli. It is likely that the beads had been paraffin treated and then dyed after the seal created by the paraffin treatment had been removed. The dye was strong enough to cause virtually all the colour.

GIA in 1991 reported an imitation of lapis-lazuli which closely resembled the natural material, showing an even, dark violet colour and with randomly distributed pyrite grains. Gemmological testing gave a specific gravity of 2.31 compared to the 2.83 of natural lapis-lazuli and a refractive index of 1.55. The specimen did not react to LWUV but gave a weak chalky-yellow fluorescence under SW. The pyrite inclusions were proud of the surface thus showing that they were harder than their host. Also visible were some random, shallow whitish areas. The specimen transmitted more light from a fibre-optic guide than would be expected from natural lapis-lazuli. The specimen became virtually invisible when viewed through the Chelsea filter using the same source of light. When the filter was used with reflected light the piece showed a slightly dark reddish-brown colour. The thermal reaction tester produced a weak acrid smell with a slight melting and whitish discolorations. It is possible that some kind of plastic binder may have been used; the specimen turned out to be barium sulphate with a polymer binding agent, the identity being proved by X-ray diffraction analysis.

A dyed blue calcite marble was described by GIA in the same issue (Spring 1991) as a possible lapis-lazuli imitation. The item was a single-strand necklace with uniform beads which were believed to be lapis-lazuli. The RI was

1.4–1.6 with birefringence which suggested a carbonate. When tested with a 10 per cent solution of dilute hydrochloric acid, effervescence could be seen. The specimen could not have been magnesite since it will not react with a 10 per cent solution of dilute HCl at room temperature. An acetone-soaked swab removed the colour, and when the specimen was examined along the drill hole a yellow underlying colour could be seen. X-ray diffraction analysis proved the piece to be a dyed calcite marble.

Jewellery that stains the skin or clothes of the wearer is not likely to command much in the way of sales. A necklace of opaque blue beads with yellow metal spacers tested by GIA and reported in 1989 had been found to stain the skin: dyed lapis was suspected. The RI was not that of lapis-lazuli and not all the beads looked the same. There was no fluorescence under X-rays. Some of the beads were found to be dyed calcite while others were dyed jasper of the type once known as 'Swiss lapis' (see below).

For a time a synthetic sodalite was grown in China and offered for sale as lapis-lazuli: samples were heavily twinned and included. The specimens were colourless when grown, but later irradiated to give the blue colour.

Acrylic substances have been used to coat a number of different ornamental materials. Wax and paraffin are the most commonly used vehicles for the coating of lapis-lazuli. GIA in 1992 quoted an article in which aerosol sprays were recommended for surface improvement. One of those recommended gave a transparent colourless surface which was tested and found to give a glassy coating on fashioned pieces of lapis-lazuli and jadeite. Four light coatings were applied, the surface then showing a concentration of glassy material in irregularities and carved recesses. The coating was easily removed with a razor blade and also with acetone.

A glass imitation of lapis-lazuli was noted by ICA (International Coloured Stone Association) in 1991, a bead necklace and a single loose fashioned stone having been tested. The material was said to be opaque and predominantly a medium blue with darker blue portions distributed in a marbled pattern. The RI was measured at 1.62 and there was no response to LWUV light though a faint powdery blue could be seen under SWUV light. A uniform distribution of very small, highly reflective and transparent, slightly brown, flake-like spots could be seen under magnification, some of the spots having triangular outlines. These were presumed to be intentional imitations of pyrite crystals.

Another imitation of lapis-lazuli took the form of a pair of scarabs reported by GIA. The colour was more like the dark blue of sodalite than that of lapis-lazuli and it was found to be concentrated in fractures. X-ray diffraction showed the material to be a dyed feldspar.

In 1993 GIA reported 'howlite lapis' in the form of large violet cabochons. They were considered to be a quite convincing imitation of lapis-lazuli as they contained white, dye-resistant veining quite like the calcite so often seen in natural lapis-lazuli. When examined the materials were found not to be howlite (no strong birefringence) but the presence of a dye was confirmed by acetone-soaked swabs. Magnesite was proved by X-ray diffraction analysis. Some of the dye came off when the pieces were washed in a mild soap solution.

A lapis-lazuli imitation for which X-ray diffraction analysis appeared to give the pattern characteristic of a phlogopite-rich ceramic was found when a thin section was examined to be predominantly a strongly birefringent mica-like material with high-order interference colours. There were also small singly refractive zones coloured dark blue which appeared black between crossed polars. Further testing with a scanning electron microscope with an energy-dispersive spectrometer showed that the specimen was largely composed of crystals with a roughly rectangular outline and a lamellar structure, characteristic of mica. The spectrum showed that magnesium, aluminium, silicon and potassium were present as major elements, indicating that it was probably the mica, phlogopite. Also present were grains of an unidentified silicate of calcium and magnesium, perhaps diopside, as well as the mineral lazurite.

Attractive gold veining in a specimen of lapis-lazuli proved to be cement with pyrite dust.

A glass imitation of lapis-lazuli is made from a blue aventurescent material in which the spangles are triangular copper crystals.

The material long and unfortunately known as Swiss lapis is a dyed jasper and gives the optical and physical properties of quartz (SG 2.65, RI about 1.55).

Chapter 19

Pearl

Introduction

Pearl is the most important of the organic gemstones, that is, those which have once been living organisms or which have formed parts of living organisms. The normal tests used on the inorganic gemstones are not in general appropriate for organic ones: they are more fragile and easily damaged. Analysis of their chemistry is well beyond the normally non-graduate gemmologist and their optical properties are quite different from those valued in the other major gem species. Many of the tests which are simpler or at least easier to understand are well described in the various editions of Anderson's *Gem Testing* and summarized with modifications below.

Pearls are also plentiful everywhere; they are imitated and cultured, they are often coloured and pieces of jewellery in which they feature usually include a number of pearls of which just one small individual may have been substituted.

Origin of natural pearls

The origin of natural pearls is still under discussion but this does not concern us here. What the pearl tester (and this means in a laboratory) needs is the ability to distinguish between natural, cultured and imitation pearls.

Natural ('oriental') pearls can be distinguished by concentric circles reflecting successive stages of growth and in general these layered structures should not be mistaken for anything seen in cultured pearls. In the formation of a pearl, cells of the organic material conchiolin form a network in each layer and aragonite crystals (aragonite is the orthorhombic form of calcium carbonate) are found in each cell of the conchiolin framework. The concentric circles can sometimes be detected when the interior of the drill hole is examined. A 10× lens is sufficient for this.

While a very large number of natural pearls have been sectioned there has been no sign of an obvious initiator of pearl growth. For the tester this will be of little consequence and time should not be spent looking for a nucleating agent in a pearl known to be of natural origin.

Cultured pearls

The variety of nucleating agents in cultured pearls (in which growth is artificially initiated and regulated) is amazing, ranging from small statues of the Buddha to grains of sand. Probably mother-of-pearl beads are the most common nuclei in cultured pearls and their presence can be detected by a number of simple methods available to gemmologists. There is little point in describing these objects here.

When the mother-of-pearl bead nucleus of cultured pearl has a covering of pearl sufficiently thin, the sheen of the mother-of-pearl bead can be seen beneath the covering and a striped effect may be seen when the pearl is viewed against a strong light. Cultured pearls are often said to have a waxy lustre because of the greater translucency of the coating compared to that of a natural pearl (natural pearls from Venezuela are reported to be especially translucent). The drill holes are often larger in cultured pearls and those accustomed to drilling with the broacher apparently feel that cultured pearls are softer than natural ones.

When the lens or microscope is used to examine the drill hole, cultured pearls will show a definite boundary between nucleus and overgrowth. The presence of a black line of conchiolin often marks the boundary. Growth lines cannot be seen below the boundary whereas in the natural pearl a series of growth lines continues to the centre of the pearl, sometimes with a yellow to brown colour gradually darkening. A pearl necklace whose nature is not obvious may be held against a strong light and rotated so that each bead is closely examined. It is certainly likely that if the pearls are cultured, some will show the signs of a mother-of-pearl bead nucleus at the crucial angle. Marks similar to varicose veins are sometimes seen below the surface and this is characteristic of cultured pearls.

The instruments of pearl testing

Pearl testing with instruments is covered by a multitude of textbooks, all of which describe the endoscope, a name familiar to many who have undergone exploratory hospital treatment with an instrument which operates on the same principle. Today the endoscope is not used in pearl testing so all that needs to be said is that it was designed to distinguish natural from cultured pearls by tracking the progress of an intense beam of light either around the concentric circles of a natural pearl or to the surface of a cultured pearl where it appeared like a cat's-eye on the upper surface. The only type of cultured pearl that could be tested by the endoscope contained a fairly large bead nucleus.

Pearl-testing endoscopes were manufactured as one-offs and there seems only to have been one model with around 25 instruments made. As far as we know there are few if any of these endoscopes working for their living though there may be some still in working order.

The endoscope made use of a hollow needle in which two mirrors placed back to back reflected the light from a carbon arc source. The needles were expensive to make and the rapid improvements made in X-radiography made the results, which were often uncertain, less likely to be regarded as truly diagnostic. In any case pearls had to be drilled for endoscope testing.

Today direct radiography (taking photographs with X-rays rather than light) is the preferred method of pearl testing and this is not likely to be successful with the amateur who in any case should not have access to X-ray equipment. The pictures achieved are in fact as hard to interpret as medical X-rays and skilled investigation is needed. In fact, all pearl testing should be carried out under laboratory conditions and this text is not intended as a pearl-testing manual as much as one designed more specifically for other materials.

Nonetheless some useful points can still be made. Where a bead nucleus is used it is not always easy to distinguish it from the pearl overgrowth unless conchiolin is seen as a black line separating the two. The growth layers in a natural pearl cannot always be seen on an X-radiograph though they may have been seen with a lens and this is one area in which the combination of two testing methods is needed. In some instances the presence of a bead may be suspected from a circular area on the radiograph but there will usually be a broken margin somewhere and a faint hint of other structures in the area of the 'bead' which make it certain that the pearl is natural. This 'bead' will be more transparent to a strong source of light than the area around it while the nucleus of cultured pearl is more opaque than its pearl coating. These distinctions are very fine and a trained eye is needed to detect them – it cannot be done with certainty over the counter.

If a pearl should not respond to these tests X-ray diffraction methods may be used. Details of the procedures, which are not generally available to non-laboratory or non-specialist personnel, can be found in a number of textbooks. X-ray fluorescence can also be useful since the bead used in many cultured pearls contains manganese and this will give a greenish fluorescence under X-rays. This can be seen in very many cultured pearls. Freshwater pearls may show an equally strong fluorescence but the glow is yellowish rather than greenish and showing much more clearly on the outside of the pearl. Some at least of the non-nucleated pearls from Lake Biwa and similar pearl farms show a strong overall fluorescence under X-rays.

Should there be a drill hole in a cultured pearl with a mother-of-pearl bead nucleus a useful observation can sometimes be made with X-rays. The drill hole (which does not need to penetrate the whole pearl) should face the observer who may be able to see that the glow from the drill hole is brighter than that emanating from the surrounding pearl – this means that the bead is fluorescing and the glow does not have the coating to pass through.

Imitation pearls can be shown up by X-rays since they appear either as opaque discs (when the pearl is solid glass) or show an opaque circle with a nearly transparent centre (the hollow glass wax-filled beads).

The standard gemmological tests are not always appropriate for pearl testing and I have left them until the tests which are standard today have been described. Nevertheless there have been and still are some simple tests, apart from the ones described above, that could be mentioned, but the reader is warned that some of the tests involve the use of liquids now proscribed on health and safety grounds so that while they are simple, laboratory conditions are essential. If the liquid bromoform (SG 2.9) (proscribed) is safely available, it can be diluted (proscribed liquids again) until it suspends a clear piece of calcite (SG 2.71). In this the majority (more than 90 per cent in fact) of cultured pearls will sink and the majority of natural pearls will float (about 80 per cent). Some of the cultured pearls are fervent sinkers and go down quite fast. The SG of Arabian Gulf pearls would have an upper limit of about 2.74 while the majority (more than 60 per cent) of cultured pearls of ordinary quality show a higher value. It is less easy to define a range for Venezuelan and freshwater pearls but these will usually show a lower specific gravity than cultured pearls. Australian pearls approach the SG of cultured pearls but only about 25 per cent of them reach a higher value than 2.74. Black pearls show an SG of about 2.65, quite low for a pearl.

Non-nucleated cultured pearls

Non-nucleated cultured pearls are grown in a variety of ways but are beyond the scope of this book which is concerned solely with identification. Up to now we have looked only at those cultured pearls that have a mother-of-pearl bead as a nucleus which are probably still in a majority. Nonetheless the increase to a significant proportion of the pearl markets of the non-nucleated variety (in which there is, as the name suggests, no bead nucleus) ensures that we need look at how to test them.

Some of the earlier pearls were produced at Lake Biwa in Japan where production of non-nucleated pearls takes place on a large scale using freshwater mussels. The pearls have a characteristic oily lustre though they are bright too. The size is usually less than five grains. The pearl grain is 0.25 ct.

Those handling large quantities of this type of pearl claim that they can be identified without too much difficulty: radiography shows small, elongated irregular hollows or conchiolin patches at the centre of the pearl. Nothing like this has been observed in natural pearl where conchiolin patches are usually circular and may be banded.

Freshwater pearls may often contain small amounts of manganese which as we have seen above will fluoresce strongly under X-rays. This is particularly the case with the Lake Biwa pearls which show a stronger fluorescence than any other: they also give a strong and persistent phosphorescence. This is a quick test though, as is pointed out in *Gem Testing*, phosphorescence is not a proof of cultured pearl.

Non-nucleated pearls produced in Australia tend to show structures similar to those found in Lake Biwa pearls and are mentioned above. As in the Biwa pearls the structures are found near the centre, these being in general larger than those shown by the Biwa pearls. Australian non-nucleated pearls often form misshapen drops and buttons.

The name *Keshi* is given to pearls resembling a grain of rice or, by extension, a very small pearl. They are sufficiently insignificant for testing methods not to be specially devised for them.

Blister pearls

Pearls may be found partly adhering to inside of the shell of the mollusc in which they have grown. They have long been known as blister pearls and can be of fine quality. They can be cultured, too, the earlier ones being known at the time of their arrival on the markets as Jap or Japanese pearls. They send a warning to the investigator by their very symmetrical domed surface though the shell backing is not always left on.

The symmetrical dome is not part of the original pearl but a covering made from mother-of-pearl placed over it. More recently the mother-of-pearl placed over the pearl is lined with wax, backed with a small bead and given a domed mother-of-pearl base. These cultured blister pearls are known as Mabe pearls. Under X-rays the transparency of the wax filling is very distinctive.

Pearls are harvested from a number of other sources which include abalones, giant clams and the great conch. The last two yield pearls which do not have the surface structure of overlapping platelets from which the characteristic pearly lustre is derived. Pearls from the giant conch (*Strombus gigas*) are often called pink pearls and they may sometimes occur in orange. While they have a somewhat ceramic appearance they show a bright sheen when viewed at certain angles and also a flame-like surface patterning which can be seen under magnification.

The pink pearls could be confused with coral but they have a higher specific gravity in the range 2.83–2.86 compared to coral's 2.69. Pearls from the giant clam (*Tridacna gigas*) have a similar SG to conch pearls but are usually white and show no flame-like structure. Clam pearls is a name sometimes used for shiny dark brown or black pearls which lack the characteristic pearly lustre.

Imitation pearls

Imitation pearls get their lustre from a substance (*essence d'orient*) which was once made from fish scales, this being applied to the outside and more rarely to the inside of a glass bead. The bead may be of other materials such as mother-of-pearl and these pearls are graded according to the number of applications of the coating they receive. When tested by rubbing against the front teeth (dentures will not do) imitation pearls feel smooth compared to the definite grittiness of natural and cultured pearls. The coating will often be seen to have worn or be broken in places, especially around the drill hole.

The specific gravity of imitation pearls varies but they are usually within the general range 2.85–3.18, above the range for natural and cultured pearls. Some poor quality beads have been found to have an SG as low as 2.33. Beads filled with wax have an even lower SG of around 1.55 or below. Coating has also been applied to plastic and mother-of pearl beads which will also show a distinctly lower SG than any real pearl.

Black pearls

Black pearls are really a very dark grey, bronze, gunmetal or blue–black and will often have had their colour deepened by a judicious application of silver nitrate. Pearls with a really intense black will almost certainly not be true pearls and may even be hematite, this mineral giving a red streak (powder) on an unglazed porcelain plate and having an SG of about 5.1 which will certainly be noticed in a necklace. The streak test is not recommended for any supposed pearl.

The presence of silver nitrate staining may be spotted by the dye having penetrated between the layers: the pearl needs to be drilled for this to be seen most clearly. This penetration will cause the nitrate to be reduced to a black metallic silver which will show as white lines on an X-ray negative.

Natural black (unstained) pearls give a rather hard-to-see red fluorescence under LWUV light (or between crossed filters where the specimen is illuminated by blue light and viewed through a red filter) while those stained by silver nitrate will remain inert.

Cultured black pearls are now well established on the market though at one time any cultured black pearl would have been stained.

Natural pearls may sometimes show a strong bluish-white fluorescence which may be seen under either type of UV radiation. Under X-rays cultured pearls often give a greenish-yellow response whereas natural pearls rarely respond. Non-nucleated pearls do not give predictable fluorescence under X-rays, Lake Biwa specimens giving a strong greenish-yellow response while salt-water pearls with the same structure appear not to respond.

Reports of interesting and unusual cases from the literature

Items in this section have been chosen to illustrate points made in the chapter and to bring one-off items to your notice.

Repeated drilling of pearls for the purpose of eliminating signs of tissue nucleation (in non-nucleated pearls) was reported in *Gems & Gemology* during 1996. X-radiographs showed them up. Predrilled bead nuclei have also been found: they are inserted into freshwater mussels which open less widely than other molluscs, forcing the pearl farmer to use a special tool for opening them. The drilling is to help with this activity.

Imitation blister pearl has been cut from the central whorl of the nautilus shell, the specimen being probably a Mabe pearl or Nautilus pearl, both of which are common names for this shape. Under magnification parallel transverse ridges were visible, an effect not seen in true blister pearls formed by any of the pearl-producing animals. Sections of nautilus shell have been called Coque de perle.

An interesting investigation of a set of grey saltwater cultured pearls was reported in *Gems & Gemology* during 1986. Three supposed pearls were 10 mm rounds, two of them in earrings, the other in a brooch. While this colour is associated with cultured pearls, natural or irradiated, these specimens, under X-rays, showed an opaque centre inside a shell of nacre of normal thickness. Visible down the drill hole was a white central material with properties similar to those of French pearl cement which appeared slightly soluble to the immersion liquid. What was the nucleus and where had it gone? The drill hole was about four times the diameter of the normal drill hole. GIA felt that the nucleus could have been dissolved after the pearl had formed around it and that it could have been made of some type of plastic. In blue to grey saltwater-cultured pearls the colour comes from a coloured bead nucleus whose colour shows through the pearly outer coating. Some of these accumulate a large deposit of dark conchiolin around the bead nucleus before deposition of pearl begins – this is thought to be due to some abnormality in the growth conditions. If dyed, traces of the dye will be found in the vicinity of the nucleus and under the pearl material and will be visible through the drill hole. If these pearls have been irradiated, the freshwater shell bead nucleus will darken under the colourless pearl layer.

Cultured pearls in which the nucleus appears dark when viewed down the drill hole may have been irradiated as a report in *Gems & Gemology* (1988) suggests. In the case described a well-matched rope of 9 mm grey saltwater-cultured pearls was found to show not only dark nuclei but enhanced orient (pearly lustre) as well as improved body colour.

Examples of the wax-filled glass imitation pearls are not so common as may be supposed. *Gems & Gemology* (1998) describes a fine necklace of uniformly sized 9 mm cultured pearls. All but one fluoresced under X-rays in a dark room.

Pushing a pin through the drill hole of this pearl showed that the core was soft, a small portion of it melting under low heat. The pearl surface was nacreous and showed the same structure as the other pearls in the necklace. Wax-filled pearls are light and do not hang well – the same is true of plastic imitations.

A pair of earrings tested by GIA contained drop-shaped apparent pearls which turned out to be glass coated with essence of orient, the liquid used to give the impression of the surface of a pearl. These were accompanied by smaller natural pearls so the glass imitations may have been replacements. It is not certain when the manufacture of wax-filled imitation pearls was succeeded by the introduction of coated ones.

Dyed black cultured pearls either fluoresce a dull green or remain inert to both types of UV radiation. Natural black or grey pearls fluoresce a brownish-red, this being seen best under darkroom conditions. Under X-rays, pearls treated with silver nitrate to darken their colour show silver concentrated in the area of the conchiolin (silver is opaque to X-rays). In 1990 GIA reported that a number of grey to black and brown cultured pearls said to have come from Tahiti showed a distinct yellow fluorescence under LWUV light. From three of these pearls examined with X-rays two showed a good contrast between the nucleus and the nacreous layer while the third had a very thin nacreous layer. On the surfaces of two of the beads a dark fingerprint-like pattern could be seen.

It is not necessary for a pearl to have been drilled for it to have colouring material applied to it and the coloured area may be only a thin surface stain. In this case X-ray fluorescence tests showed that tellurium was present in all three specimens.

The Fall 1990 issue of *Gems & Gemology* describes cultured pearls with green coloured bead nuclei. The beads were said to be made of powdered oyster shell which had been bonded with a type of cement, then dyed and fashioned into spheres. Beads were a dark greyish-green and approximately 8 mm in diameter. An aggregate structure was indicated by the surface. Embedded in the spheres were near-colourless transparent and opaque white grains.

The SG was determined at 2.74 and the RI was just below 1.50. These properties are shared by a number of carbonates and this was proved by the effervescence of the specimens when brought into contact with a 10 per cent solution of hydrochloric acid. Some green dye was removed by an acetone-soaked swab. X-ray diffraction proved the sphere to be calcite and the infra-red spectrum showed that a polymer was present as a bonding agent.

The pearls with these bead nuclei had a notably transparent nacreous layer and when the drill hole was examined the thin nacre over the green nucleus could be seen. Some of the plastic bonding material melted when the pearls were drilled and this can also be seen inside the drill hole. As the green nuclei were not made from freshwater pearls there was no fluorescence under X-rays. The conchiolin layer could be seen on an X-radiograph and so could the differences between the X-ray transparencies of the nucleus and the nacreous layer.

An assembled blister pearl had the normal white mother-of-pearl base but the top was made from a dark purplish-brown nacre with a very high degree of lustre. Under magnification the pearl showed an uneven distribution of colour in the nacreous layer and with strong lighting and higher magnification darker brown areas were visible. When the patches of colour were touched with a cotton swab soaked with acetone or 2 per cent dilute nitric acid they gave a dark stain which proved that dye had been used. The pearl fluoresced a dull reddish-orange under LWUV light: natural black pearls usually show a brownish-red to red fluorescence.

A number of white Mabe-assembled blister pearls were noted at the 1991 Tucson Gem & Mineral Show where they were selling at low prices. Most were between 15 and 20 mm in diameter and showed very strong pink overtones. Some showed a spotty, uneven colour distribution which could have resulted from enhancement. A plastic dome had been fixed to the pearl, with a very thin nacreous layer on it. This could be removed from the plastic dome which was found to be coated with a very fine highly reflective material. This layer was a lacquer rather than the customary pearl essence.

Treated black Mabe pearls started to appear on the market in 1992. In one sectioned specimen GIA noted that the nacre top gave a dull reddish-orange fluorescence under LWUV light which could have indicated treatment. The top was an evenly coloured dark purplish-brown but a cotton swab soaked in 2 per cent nitric acid removed no dye. The pearl had three components: a white mother-of-pearl base, the dark purplish-brown nacre top, averaging about 0.5 mm in thickness, and a dome-shaped core which gave a granular appearance in reflected light.

When viewed through the microscope the core was seen to be made up of translucent white fragments in a whitish mass. A very sharp black demarcation line separated the core material and the nacre top. The core reacted with a 10 per cent solution of hydrochloric acid and the thermal reaction tester caused the whitish mass to melt and at the same time give off a characteristic epoxy resin smell. When the nacre top was examined by EDXRF it was found to contain calcium with trace amounts of silver and bromine.

Non-nucleated pearls, though rice-grain shaped in the early days of their production, can now achieve round and near-round shapes. GIA in 1994 described a necklace of freshwater tissue-nucleated pearls with beads up to 7 mm, a size popular with the manufacturers of the bead-nucleated 'Akoya' cultured pearls in Japan. Under X-rays the pearls in the necklace showed no signs of the voids which normally show on the film as a sign of tissue nucleation.

Evidence may have been eliminated by the drilling of the hole and it is possible that some freshwater tissue-nucleated pearls may not be distinguishable from some freshwater natural pearls just from X-radiographs. It is possible that in some instances saltwater tissue-nucleated cultured pearls have been substituted for natural pearls in old pieces of jewellery with the hope that the larger-than-usual drill hole will eliminate the voids left by the tissue nuclei.

Pearls dyed to imitate American freshwater pearls were reported in 1994 by GIA. X-rays showed them to be early assembled cultured blisters with a characteristic rectangular saltwater shell insert. The pearls were set in a ring.

An interesting composite imitating black pearl turned up as a strand of dark silvery grey-to-black pearls averaging about 9.44 mm in diameter. The surfaces of the pearls looked hazy and had a rubbery texture. Under magnification the pearls showed three distinct sections around the drill holes: the inner core was a colourless translucent bead covered by several thin, silvery, grey-to-black layers with another coating of thicker material forming the outer layer.

The bead core showed a vitreous lustre and conchoidal fracture, and X-ray diffraction showed that it had an amorphous structure. It was thought to be lead glass and as the necklace felt heavy this seemed plausible. The composition of the thin layers was found to be a bismuth oxide chloride, bismoclite.

This has been found to be a constituent of the coating of other imitation pearls, these pearls differing from those coated by the fish-scale (guanine) liquid. The haziness and rubbery effect on the pearl surface were caused by a coating used to strengthen the bead. Strings of black pearls should be examined to see whether different materials have been used, looking especially in the area of the drill hole and looking carefully also at strings feeling unnaturally heavy.

On the developed X-ray film taken of a black pearl coloured by treatment, a white ring could be seen surrounding the nucleus, proving treatment by a metallic compound which was opaque to X-rays. EDXRF analysis showed that a silver compound often used to stain pearls was present. Magnification showed that the treatment had damaged the nacreous layers with some portions showing iridescence as one effect of the damage. When a pearl is treated with silver salts the entire surface absorbs some of the dye. Characteristic dimpling is seen on many pearl surfaces and when treatment is applied the dimples will become coloured as well as the rest of the surface. The presence of a white area in the centre of one dimple showed that there was one place on the surface which the dye did not reach.

Pearls with circumferential depressions have been called circle pearls and can often be found among cultured black pearls. The cause may be either rotation of the nucleus in the pearl or wrinkles formed on the pearl sac, from unknown causes. This effect has not been reported from natural black pearls.

Good quality cultured pearls from Indonesia are produced from blister pearls grown in the mollusc *Pinctada maxima*. Resin has recently been used as the bead nucleus so that they resemble natural pearls with a low specific gravity. A deceptive sheen is caused by a thin nacreous layer.

Abalone 'Mabe' pearls made with blister pearls cultivated in the abalone shell showed a fine cellular structure in the blister portion. A strong yellow fluorescence under LWUV light showed its derivation from abalone.

An undrilled pearl of 29.22 ct showed a brown colour over most of its surface but had a white circular area at the base. This area was nacreous while the dark brown area was non-nacreous. The pearl fluoresced a strong orange–red over the brown areas and a characteristic blue–white over the nacreous basal area. This effect has been reported in other partially coloured brown, black or grey pearls, including the famous Hope pearl and is usually taken to indicate that the pearl is natural.

A barrel-shaped black pearl showed both a solid black appearance and distinctive banding. The dark areas glowed a very strong red under LWUV light (as seen in other black pearls from the Gulf of California) while X-radiography showed a large dark central core which proved the pearl to be natural.

Raman spectroscopy has been found useful in separating naturally coloured from dyed culture pearls: tests are described in the October 2001 issue of the *Journal of Gemmology*. Irradiation treatment is not detectable by Raman methods, however.

In recent years Chinese freshwater cultured pearls have taken an important place in world markets. It is possible to differentiate between these and natural and bead-nucleated pearls by X-radiography. Details of the pearls are described in the Summer 2001 issue of *Gems & Gemology*.

Chapter 20

Amber

Amber at the time of writing is plentiful in the shops and a good deal of it is genuine, specimens from the Dominican Republic being especially prominent. Amber is a fossil resin and differs in this respect from the many imitations which are made from contemporary resins – copal is probably the best known of them. No tree which produces resin of the amber type exists today (the tree was *Pinus succinifera*).

Amber specimens will be less than 250 million years old and many will contain plant and animal remains. As a general rule, the better-preserved these appear, the stronger the possibility that they may not be what they seem and that their host is a contemporary resin into which a contemporary insect or plant section has been introduced. Some of these specimens are grotesque.

Amber on the market may come from several different areas: though locality information is not required there may be some commercial value in stating that a specimen is from the Dominican Republic whose amber is known to be attractive. In general this amber is bright and a greenish-yellow to brown (it is very difficult to describe the colours of amber with any degree of accuracy): amber from the Baltic shores is usually a lightish yellow and can be transparent though some material contains small air bubbles in profusion. Browner colours are common in amber from Romania and Sicily and some specimens are reddish-brown to black. Cracking is often found in Romanian amber.

Brown is the predominating colour in Burmese amber and specimens very often contain animal remnants. In general amber when touched with the thermal reaction tester will emit an aromatic smell while plastic imitations will give a pungent smell. Some ambers fluoresce, the Dominican Republic material often showing greenish colours in strong sunlight and always responding with a green fluorescence to ultra-violet stimulation.

Amber has a hardness of 2.5–3 on Mohs' scale and a specific gravity in the range 1.04–1.10, the bubble-filled material showing a lower SG than most specimens. The refractive index is usually around 1.54 but use of the contact liquid is not advised. At about 180°C amber will soften. It is true that when amber is vigorously rubbed for a moment or two small fragments of paper will be attracted to it but several amber imitations behave in the same way. If an amber-like specimen does not show this effect it cannot be amber.

Pressed amber

An early imitation of amber is known as Ambroid and was made by heating and compressing fragments of Baltic amber. The heating at around 200–250°C (amber melts at 180°C) softens the pieces which are passed through a fine steel sieve to form a block which has very much the same appearance and properties of normal amber. Pressed amber can be recognized by its characteristic flow structure with amber globules forming lines. Elongated bubbles may also be seen. Some pressed amber may appear treacly and be free or almost free from bubbles and flow structure. This material has a specific gravity of 1.06 which is slightly lower than that of normal amber.

Amber's specific gravity can be roughly ascertained by flotation in a brine solution (50 g salt in 250 ml water). Between crossed polars block amber shows extinction (dark) bars crossing the specimen on rotation: in the same situation pressed amber shows strain birefringence with some light invariably transmitted. Amber and pressed amber show interference colours but those in pressed amber are more patchy than general. Discoid spangles can also be found in pressed amber. Pressed amber gives a brilliant chalky-blue fluorescence under LWUV light and while fluorescing the material may show granular structures.

Amber which is cloudy may be clarified by heating in rape seed oil which enters the air spaces which cause the cloudiness. Material treated in this way often shows characteristic cracks which have traditionally been called sun spangles.

Locality clues

There is some property and colour variation in amber from different locations: Burmese material is brownish and some specimens are reddish. Some Romanian amber has an SG in the range 1.05–1.12 and hardness 2–2.5. Some amber from the area of Gdansk, Poland, has SG near 1.02 and hardness 1.5–2.

Amber from the Dominican Republic covers a wide range of colours and contains no succinic acid. Some specimens may show a surface blue colour and clear yellow to yellowish-brown in transmitted light: all colours fluoresce strongly in blue or green colours even in daylight though the effect is reported to diminish with age.

Inclusions in amber are accepted by the general public as a sure sign of natural origin, this being particularly the case where animal remnants are involved. In general, a well-formed insect which is not dismembered should be cause for disquiet since most animal remains are in poor condition, individual insects, for example, often leaving a leg behind in their haste to avoid encroaching resin. Flies make up about 54 per cent of all trapped insects. A number of amber specimens contain bubbles some of which may contain possible quartz crystals.

Contemporary resins

Of the contemporary resins copal is probably the best known. Amber itself is soluble to a small extent in ether or chloroform but the contemporary resins are much more soluble and this remains one of the best distinguishing tests. Shapes resembling nasturtium leaves are probably due to some form of treatment.

Copal is the best known of the natural resins that may be mistaken for or offered as amber. The source of copal is still in existence and fossilized specimens may be anything from 100 to 1000 years old. It may be transparent or translucent, yellow or brown. It will burn with a smoky flame and give off a smell of resin: the RI and SG are similar to those of amber but the surface feels tacky. Copal may contain insect and plant inclusions.

Copal from New Zealand comes from the Kauri pine *Agathis australis* and is known locally as Kauri gum, which can be found up to 100 m below the surface. It may be yellow to reddish-yellow or brown and insect inclusions are very common. It contains no succinic acid and the specific gravity is 1.05. It has a low melting point, causing stickiness to develop during the polishing process. Dammar resin is clear and a pale yellow and has a melting point of about 140°C compared to the 187–343° of copal in general. Manila copal from the Philippines is sometimes very dark, even black: it has an SG of 1.072 and a melting point of 190°C for the harder and 120°C for softer specimens. Pontianak from Borneo has an SG of 1.068 and a melting point of 135°C.

In general the ether test is the most effective way of distinguishing between amber and copal: the latter's surface yields easily to a knife blade and will often appear crazed.

Plastics

The commonest imitations of amber are plastics, in particular bakelite, casein and celluloid. In all cases the specific gravity is higher than that of amber and the thermal reaction tester will produce the familiar acrid smell. Perspex has an SG near 1.18 while some polystyrenes give values closer to those of amber near 1.05. Some of the polystyrenes will give the high RI figure of 1.58 compared to amber's 1.54 but in practice the testing of amber with the usual contact liquids should be ruled out as damage to the specimen is likely.

One of the polystyrenes, Distrene, is soluble in benzene but this is a dangerous substance, to be used only under laboratory conditions and certainly not by amateurs. The string of an amber bead necklace makes little difference to the result of specific gravity determinations – this applies also to pearl. If the specimen under investigation is large and has places where the eye rarely reaches, a test with a knife blade will show that amber chips while most plastics peel.

As with many of the organic materials used for ornament, successful testing of amber depends more on the experience of the observer and on the outcome of several investigations rather than one. It is true that some of the tests will appear destructive and it should be emphasized that such tests are carried out on the smallest possible areas of a specimen and any chipping should not be visible to the unaided eye.

Reports on interesting and unusual examples from the literature

Items in this section have been chosen to illustrate points made in the chapter and to bring one-off items to your notice.

In 1993 a 'lac' (gum from the insect *Coccus lacca*) bead was said to have been produced from a natural resin from the New Delhi area of India. Beads were opaque and showed a swirly texture with a yellow to orange–brown colour. Marks very like the surface crazing seen on some fossilized resins were noticed: the refractive index was found to be 1.51 and the specific gravity 1.67. Cutting a section showed that the interior was medium dark brown with streaks of yellow like the exterior colour which turned out to be a bright yellow opaque layer with a brownish-orange transparent coating. The swirly appearance arose from intermixing of the two layers: dark brown specks of colour and shallow hemispherical cavities were seen, the cavities probably arising from gas bubbles breaking the outer layer.

Is alcohol harmful to amber? This was the subject of a letter to *Gems & Gemology* in the Summer 1992 issue and the topic has been the subject of argument for many years. The example quoted cited an amber specimen which was polished then washed with a soap to remove the polish. The amber was then given a drying rinse in denatured alcohol and it was reported that this part of the treatment caused hazing to develop in areas below the surface. The hazing could be removed only after some time was spent in scraping, filling and sanding. Cosmetic and other sprays may also dull amber surfaces as pointed out by Fraquet, *Amber* (Butterworth-Heinemann, 1987).

A large transparent to opaque amber imitation was coloured orange, white, yellow and brown. On one side there was a thin white coating over a brown area in which broken patches showed a third layer which was orange–brown. On the other side was more of the white coating with what seemed to be stamped impressions. There were numerous gas bubbles and on the same side another break showed a dark yellow interior. Those familiar with slag from plastics factories said that the piece was not unlike this material but neither an RI nor an SG could be obtained. With the thermal reaction tester a pungent smell associated with burning plastics was produced and the infra-red spectrum gave a curve closely matching that of polyvinyl chloride.

Natural resin in plastic has been used to simulate amber, the material thought to have been produced from Baltic amber. GIA laboratory staff found a resemblance in this material to examples of pressed amber as veil-like grain boundaries could be seen with the unaided eye. The terms pressed, reconstructed, reconstituted and synthetic amber have been applied to a material probably produced in Poland. Publicity stated that the amber-like substance began with small fragments of amber recovered from the Baltic shores or from mining and that pieces after grinding are embedded in fresh tree sap, after which they are polished and set. The material has been found in a number of different forms but differs from pressed amber in the presence of very clearly defined irregular transparent to semi-transparent yellow–brown fragments in a lighter-toned transparent yellow groundmass.

The refractive index was found to be 1.56 and the specific gravity 1.24. Between crossed polars a strong anomalous birefringence with strain colours could be seen. The body of a cabochon fluoresced a moderate greenish-yellow under LWUV light and the included fragments a moderate bluish-white.

The body fluoresced a faint yellowish-orange under SWUV light while the included fragments remained inert. The thermal reaction tester produced an acrid smell but an included fragment which reached the surface gave an aromatic smell. Fourier-transform infra-red spectroscopy showed peaks consistent with natural amber when an amber-like fragment was tested, while the matrix gave results consistent with an unsaturated polyester resin. GIA who examined the piece concluded that the cabochon was a plastic with fragments of a natural resin, probably amber, embedded within it.

The name 'Polybern' has been used in factories in the traditional amber-producing areas of Poland and Germany: it refers to small amber chips embedded in a synthetic resin.

It is interesting to note that amber (and plastic) worry beads used in the Middle East are generally found in multiples of 33, thus 66, 99. Some of the beads have been found to be made from amber pieces embedded in a plastic bead frame and show a central line where two adjacent pieces of plastic meet.

It is possible that Raman spectroscopy may be able to distinguish between amber and its imitations and to identify them.

Chapter 21

Ivory

Ivory can be considered as an ornamental if not a gem material. Ivory's composition is mainly hydroxyl apatite with collagen. It can be softened by nitric and phosphoric acids. Ivory has been used for the most intricate and delicate carvings as well as for simple bead necklaces. The original creamy white tends to yellow with age but this need not necessarily be considered unattractive. At its best ivory has a recognizable surface glow, due to its fine texture – a quality that can be best seen when ivory is compared to bone by which it is often simulated.

Though some of the best ivory (in which trade is now illegal in many countries) comes from the tusk of the elephant, the tusks of a number of other animals also furnish workable ivory and some of the differences can be distinguished. A list of these animals would include the walrus and the narwhal – these form a category which has been called marine ivory. Hippos, hogs and sperm whales also provide ivory.

Fossil ivory means material from the tusks of the woolly mammoth, the remains of which can be found in Siberia and elsewhere. This material is not fossilized and the quality can be very acceptable commercially: the properties are consistent with those of contemporary elephant ivory.

Hippopotamus teeth can also be of fine quality, even exceeding in fineness the best elephant ivory, but is more dense.

Ivory has a hardness of about 2.5, a refractive index of about 1.535 and specific gravity of about 1.70–1.90 (elephant ivory). Clearly the RI will not be taken unless no other means of identification seems possible since the contact liquid will harm the specimen. The specific gravity of marine ivories will be in the region of 1.80–1.95.

The one characteristic which enables the investigator to distinguish ivory from its simulants is the presence of *lines of Retzius*, seen in all elephant ivory (including ivory from the mammoth) but not in other ivories. The lines of Retzius resemble the engine-turning seen on some ornamental cigarette cases. Elephant ivories contain a network of slender tubes extending from the base towards the tip in a generally longitudinal direction. Because of this transverse tusk sections show a series of arcs and spirals and these are diagnostic for ivory.

Under LWUV light all types of ivory show a whitish to violet-blue fluorescence which appears brighter from walrus and hippopotamus ivories than those of the elephant. Two types of ivory, hard and soft, used to be distinguished and may still be among connoisseurs of antique ivories. The hard ivory is harder to cut and is brighter while the soft ivory cracks less readily, with a better resistance to temperature swings.

Ivory from African countries comes from the African elephant *Loxodonta africanus*. The best is probably from Cameroon: this material has a transparent/translucent appearance without the mottling shown by many other ivories. Sudan ivory shows light and dark concentric rings when sectioned while ivory from Zanzibar (Tanzania) and Mozambique is generally soft. Ivory from Ethiopia is also soft and sometimes shows a bark or skin.

African ivory tends to be whiter and harder to work than ivory from the Asiatic elephant *Elephas maximus*. This material, found on elephants from Myanmar, Thailand and India, yellows more easily and is easier to work.

Fossil ivory from the woolly mammoth, *Elephas primigenius*, is darker than most other ivories. Hippopotamus ivory from *Hippopotamus amphibius*, with notably fine grain, is found in the rivers of central Africa. Walrus ivory comes from *Odobenus rosmarus* and has a finer texture on the outside than on the inside. Ivory from the narwhal and sperm whale is coarse with a texture closer to bone.

Though this is not a test to be recommended ivory will be found to chip when tested with a knife blade where plastic imitations are sectile and will peel (but who would even think of using a knife blade on a specimen suspected of being ivory?).

Tusks are not the only source of ornamental ivory: molar teeth of elephants have been used as knife handles when sliced and the canines of warthogs may be carved.

Hornbill ivory comes from the beak of the helmeted hornbill bird *Rhinoplax vigil* and especially from a variety found in Indonesia. This material has a hardness of 2.5, refractive index 1.55, specific gravity 1.28–1.29. There is a greenish or bluish-white fluorescence under UV light. Nothing happens when the material is tested with acid but when the thermal reaction tester is brought close the smell of keratin can be detected. Hornbill ivory pares easily and is usually found in two colours, carmine or yellowish.

Vegetable ivory

Vegetable ivory is not a true ivory. There are several varieties but the one commonly described is the seed of the ivory palm *Phytelephas macrocarpa*, found in Colombia and Peru. The fruit contains a nut called corozo nut which may be as large as a hen's egg. The composition is close to pure cellulose and the hardness is 2.5 with a refractive index near 1.54 and specific gravity 1.40–1.43. A peeling shows characteristic torpedo-shaped cells in roughly parallel lines. The fluorescence is similar to that of the other ivories, but weaker. Both types are inert to X-rays.

The doum palm, *Hyphaene thebaica*, is found in Egypt and west central Africa. It has a nut with a hardness very close to 2.5: the RI is 1.54 and the SG 1.38–1.40. Peelings show a structure similar to that of the corozo nut though when viewed end-on polygonal-shaped cells may be seen. All types of vegetable ivory take dye well and in some circumstances they may be mistaken for coral. They will not effervesce with acids.

Careful examination of the marine and other non-elephant ivories will show longitudinal striae on some polished surfaces and when peelings are examined, immersed, with a microscope (high magnification is not necessary) fine tubes should be seen, crossed by the heavier growth lines. These should be found in all types of dentine ivory and it has been reported that these tubes show crinkling in hippopotamus ivory while in walrus ivory they are straight and have a larger diameter. Fine branching tubules with prominent growth bands are characteristic of sperm whale ivory while narwhal ivory shows fine branched tubules set in a granular matrix. These findings are reported in greater detail by Brown in *The Australian Gemmologist.*

Simulants of ivory

Bone

Bone is the commonest natural simulant of ivory and in fact the composition is not dissimilar. Bone has a higher specific gravity than ivory at about 2.0 and shows no lines of Retzius. On the other hand bone shows a feature known as *Haversian canals* which permeate the whole specimen and which can be seen as long white lines (they are not often white since dirt will accumulate in them): seen end-on they show as groups of black dots (the black is also dirt).

It should be remembered that bone also fluoresces violet–blue under UV though some authorities contend that the fluorescence of bone is slightly whiter than that of ivory. Bone is inert to X-rays.

Taken from the long bones of various animals, in particular the ox, bone has a significant grease content and this needs to be removed before working. Some types of washing powder accomplish this very well.

Deer horn

Deer horn has a structure similar to that of bone and shows the Haversian canals less prominently. It has been used for ornamental objects rather than jewellery. It may be imitated by plastics.

Chapter 22

Jet

Jet has never quite lost its popularity and the work of independent modern designer jewellers has brought it back into prominence once more. In England it is possible to visit one of the world's few jet centres, at Whitby in Yorkshire, and specimens can be found if the searcher is fortunate enough to find the right place in which to look. There are one or two other minor deposits but they are of no significance in what can still be called the jet trade.

Jet is a fossilized wood (this does not mean old packing cases as seen occasionally floating along the Thames, but plant material of similar origin to brown coal). The Whitby deposits lie in the hard shales of the Upper Lias from which they are still mined though most material is found on beaches in the north-east of England.

The scanning electron microscope has in recent years shown that jet is a form of fossilized Araucaria, with a woody structure. X-ray emission spectroscopy of Whitby jet has demonstrated the presence on the surface of a high amount of silicon and of significant levels of aluminium, sulphur, potassium, calcium, iron and copper. In material found in the area of Villaviciosa, Spain, the surface shows a high amount of aluminium, very little sulphur but some iron and copper. This material which has been worked into ornaments at Oviedo is reported to break up with sudden changes in temperature – this may be due to the sulphur content. Jet from sources in Turkey shows significant amounts of aluminium, silicon and sulphur as well as some potassium and calcium.

Jet has a hardness of 2.5–4 and a specific gravity of 1.30–1.35. The thermal reaction tester when brought close (but not touching) will produce a sooty flame and a smell like that of coal (a smell which is not often encountered today). Jet gives a brown streak.

Jet is commonly imitated though the natural material is not seriously hard to obtain. Black is always fashionable and though jet has a general appearance which is hard to mistake, familiarization takes time and encounters with a large number of specimens. Black onyx can immediately be distinguished by its much higher specific gravity of around 2.6: plastics are much more serious imitations and will give the usual acrid smell if the thermal reaction tester is brought close.

The artificial rubber vulcanite (we have had difficulties in obtaining this and it is either quite rare or there is so much about that testing is never carried out – the first seems the more likely possibility) smells strongly of rubber with the thermal reaction tester.

The range of objects made from jet is very extensive: though its nineteenth century use in mourning jewellery is virtually or completely obsolete such pieces are sought by collectors and this helps to keep up the prices.

Reports of interesting and unusual examples from the literature

Items in this section have been chosen to illustrate points made in the chapter and to bring one-off items to your notice.

Jet found in Siberia has been identified as sapropelic coal but contains no trace of wood structures (saprope). It appears to be formed from unconsolidated mud or mire with a jelly-like consistency and contains plant remains, chiefly algae, which have putrefied in anaerobic conditions on the beds of shallow lakes or seas. Sapropelic coal originates from these muds and includes the varieties cannel, boghead and torbanite. The jet-sapropelites differ from the coals in their compact structure, low ash content, conchoidal fracture, brownish colour and brown streak. It is resistant to air and water and can be dissolved by benzene, ether, toluene or chloroform.

In general black ornamental materials are hard to identify whether worked or rough but jet's low hardness means that the edges of fashioned pieces will be much less sharp than those of, for example, black onyx. Observers of black opaque ornamental materials should look carefully for the type of fracture: the backs of cabochons as well as edges can give clues here. By using reflected light grain boundaries can sometimes be detected, as well as differences of relief or reflectivity. In the end X-ray powder diffraction or EDXRF techniques may have to be used.

In *Jet* (Butterworth-Heinemann, 1987) Helen Muller lists a number of simulants of jet: they include *cannel coal* which shows a black rather than a brown streak and appears with a silvery grey sheen in some directions. *Horn* gives a grey streak and powder while *bog oak* is a very dark brown rather than black, taking a less high polish than jet. The black glass known as *French jet* has a hardness of about 6 (higher than that of jet at 3.5–4) and gives no streak.

Chapter 23

Coral

Ornamental coral is very popular and some of its colours are most attractive, especially a rosy pink which seems inseparable from coral. As far as Europe is concerned, coral means the south of Italy and in particular Sicily where a good deal of coral is worked at Torre del Greco. These corals are red or pink.

The type of coral used as ornament is distinct from the type of coral that builds reefs. They all belong to the phylum *Coelenterata* and include the species *Corallium rubrum*, *C. japonicum* and *C. secundum*. Like all forms of coral the ornamental varieties are secreted by marine coral polyps with the precious material being harvested from the horny or calcareous internal skeletons of simple polyps, which give a rigid support to the colony on the outside.

Corals from the *Corallium* genus grow as arborescent colonies whose twigs turn upwards and forwards. They are pigmented with red spicules and growth rates are about 1 cm a year for *C. secundum* and *C. japonicum*, and about 0.3 cm a year for *C. rubrum*.

Red precious coral when sectioned shows a radiating structure formed by dark red bands radiating from the central canal to terminate at convexities on the external surface of the axis. The areas between the radial structures are lighter in colour and fibrous in structure. Some dark circumferential growth rings are superimposed on this pattern.

Coral used in jewellery is usually *C. nobile* or *C. rubrum* with a hardness of about 3.5–7 and refractive index near 1.65 and 1.49 (these are the figures for calcite, the crystals of which form an integral part of the coral). The specific gravity is between 2.6 and 2.7. Coral contains calcite and will effervesce with acids.

Most red or pink coral will show an absorption band at 494 nm. Under UV light most specimens show only a dull red (some imitations made from uranium glass give a strong green fluorescence and some plastics give an orange–red).

Coral from the Ryuku Islands south of Japan may be white, blood-red or pink, and a black coral, *Antipathes spiralis*, is found off the coast of Cameroon – this has been given the name Akabar or King's coral. Black coral (which comes from a number of places) is an arborescent coelenterate and in cross-section looks like a tree with ring patterning.

Two types of black coral are recognized: ‘true’ black coral is derived from Antipatharian coral found off Hawaii and the Great Barrier Reef, Australia. ‘False’ black coral is derived from Gorgonian sea fan coral from shallow tropical or sub-tropical waters. Both types have been formed from concentric lamellae of a horny material deposited round a central canal. The axis of the false black coral is derived from a protein named antipathin.

Black coral

The true black coral shows fine linear structures radiating from the central canal to the external surface. Such structures are not found in false black coral. Sections of true black coral show strain birefringence between crossed polars. Interference fringes always surround the radially arranged spines of true black coral but are never seen in false coral.

Black coral from the waters around the island of Maui in the Hawaiian group is used in jewellery and some specimens show a vague RI near 1.56 which suggests that it is composed of conchiolin rather than calcite.

Both black and golden-yellow coral have distinctive structures, the golden material showing a spotted or pitted surface while the black coral shows concentric rings.

No coral imitation will show the characteristic ridged structure of the true material. Glass is one of the commonest imitations though plastics can be more convincing. The pink conch pearl, which can resemble coral, does not show the ridges but has a higher SG of 2.84 compared to coral’s 2.68. The ornamental plastics will sink in a glass of distilled water much more slowly than coral.

Gilson coral imitation

A coral imitation was made by Gilson in shades of reddish-pink: the material is a calcite from a French source. The Gilson coral has an RI near 1.55 compared to natural coral’s 1.49 and 1.65. The SG of the Gilson specimens is about 2.44 (compared to 2.6–2.7 and the hardness is similar to that of the natural material at 3.5. The Gilson imitation, like natural coral, will effervesce with acids.

Some of the Gilson coral shows a brecciated structure while natural coral presents a wood-grain appearance. Gilson coral leaves a reddish-brown mark on a streak plate while natural coral gives only a white mark.

Reports of interesting and unusual examples from the literature

Items in this section have been chosen to illustrate points made in the chapter and to bring one-off items to your notice.

A slab of imitation coral reported in the Summer 1990 issue of *Gems & Gemology* was found to be barium sulphate. The slab weighed 21.26 ct and was an orange–red colour, transmitting a fair amount of light while appearing opaque by reflected light. The polished side showed a fairly even orange–red with a waxy lustre.

When the slab was examined under oblique lighting and magnification, an irregular whitish-pink veining and black metallic inclusions were visible. The RI was found to be about 1.58 and the SG 2.33 with a hardness of 2.5–3. X-ray powder diffraction analysis proved its mineralogical nature.

Also in *Gems & Gemology*, for Spring 1984, was the description of a supposed blue coral. The 17 mm round bead had a blue core with a near-colourless transparent coating containing gas bubbles. The coating could easily be indented with the point of a pin and the thermal reaction tester produced the characteristic acrid smell of a plastic.

The most likely substance to be a problem with coral identification is the pink conch pearl which has a higher specific gravity of 2.84 compared with the 2.68 of coral.

Golden and black coral are popular, some of the golden specimens showing a surface which has been called 'plucked chicken'.

Chapter 24

Shell

Trade in most varieties of shell is now, quite rightly, limited in the interests of conservation (a more important pursuit than ornament) but many examples from jewellery of the past are still around. Shell with a play of colour is particularly valued with the *pearl oysters Pinctada maxima* and *P. margaritifera* providing mother-of-pearl. *Pinctada margaritifera* may reach 20 cm in size and is valued more for its shell than for its pearls.

Abalone: Paua shell (Haliotis)

Some dark forms of shell can produce an effect similar to the cat's-eye though there is no real likelihood of serious confusion with chrysoberyl: they may be used for small items like buttons. Abalone is the name given to the Paua shell which has coloured bright green and blue nacre: it is found in New Zealand. Abalone is the name used in America while the taxonomic name is *Haliotis*. Sometimes they produce pearls.

Sea snail

The sea snail (Turbo sp.) produces material which could be mistaken for pearl as the upper surface shows a near-pearly lustre, the back being yellowish and non-nacreous. The name 'oil pearls' has been used and in some places they have been called Antilles pearls. *Coque de perle* is cut from the central whorl of the nautilus shell and resembles a blister pearl with a thinner skin. The hollow interior may be filled with wax or cement.

Operculum

The gastropod *Turbo petholatus* lives in a shell with a 'door' or lid (operculum) which is used for ornament. This is because it shows attractive patterning with a generally eye-like form. The usual size range is around 12–25 mm in diameter: it is rounded with a slightly domed upper surface. This surface may be green at the apex, turning to yellow and white on one side and reddish-brown to dark brown on the other. On the reverse side are growth lines in spiral form. The hardness is about 3.5 and the specific gravity in the range 2.70–2.76. The gastropod comes from the tropical seas of Oceania, Melanesia and Polynesia.

Cameos

The cameo is made from two different shells, the commonest being the *Helmet shell* (of *Cassis madagascariensis*) from the seas surrounding the West Indies. Cameos made from the Helmet shell are in white relief against a brown background. Italy is the major cutting centre for material from Madagascar. Cameos can also be fashioned from varieties of quartz.

Conch shell

Cameos may also be fashioned from the shell of the giant conch found off the West Indies and the coast of Florida. Carvings may be white on a rose-coloured background or the other way round and some specimens have been reported as fading after exposure to bright light. In 1982 Mitchell reported that in nearly all cases the brown, orange–brown or pink layer used as the background surrounding the white cameo proper shows a fine grain with lines virtually parallel in one direction and the white cameo material shows fine lines running nearly at right angles to those in the background. The various layers do in fact show different grain directions and this helps to strengthen the cameo as the grain is a direction of weakness. Cameo imitations made from chalcedony and other materials do not show graining.

Tortoiseshell

One of the most important examples is tortoiseshell which is taken from the Hawk's-bill turtle, *Eretmochelys imbricata*, which is found in a number of tropical seas. The best examples have a rich brown colour on a translucent yellow background. The dorsal carapace has 13 large scutes (plates) these being fringed by marginal scutes which may number 26.

Gemmologists will not usually encounter whole plates so there is no real need to know much about the biological structure of tortoiseshell: what is needed are means of distinguishing tortoiseshell from its imitations. Small chips of tortoiseshell may be moulded together after softening.

Tortoiseshell is formed from keratin and will give a burning hair smell when approached by the thermal reaction tester. It is also sectile. A less dangerous test (for the specimen) is to examine the item under the microscope. Tortoiseshell will show mottling of colour and this can be seen under magnification to be made up of spherical dots. In plastic imitations (which are very common and usually of casein which smells of burnt milk when heated) the mottling shows colour in patches and swathes. Under UV light the clear yellow parts of the shell give a bluish-white colour while most imitations give a yellow fluorescence. Tortoiseshell is usually fashioned by ivory workers who will also repair damaged pieces. Tortoiseshell is inert to X-rays.

Chapter 25

Natural and artificial glass

By far the commonest imitations of gemstones (even crystals) are glass. Examples may be solid or form a section of a composite and the gemmologist can be sure that in any group of specimens submitted for examination the likelihood of glass turning up is very high.

This should not be surprising as glass is not only cheap to make but can be very attractive in itself: some types of glass have a high dispersion and appear almost diamond-like to the inexperienced eye (sometimes to experienced ones if the conditions for examination are difficult). Colour is very easily introduced into glass and it can be most convincing, especially in the paler colours.

For the gemmologist, suspicion is the best first test and once that state of mind is achieved glass should not be hard to detect. Trouble often arises when the specimens are small and when they occur in large quantities spread over a piece of jewellery. Over the centuries jewellers and jewellery repairers have slipped the odd piece of glass into a jewel when replacement of a natural gemstone suddenly became necessary – this has happened before on more than one English coronation!

Nomenclature

As always, nomenclature problems soon confuse the casual or amateur tester. 'Crystal' as a noun or adjective usually means 'cut glass' in the world of giftware (which is often found with jewellery in large department stores). Glass is one of the only two known solid inorganic substances (opal is the other) which are not crystalline. The term 'paste' is now dropping into disuse – at one time it had a definite meaning, but it always means glass.

Glass is quite difficult to define though the term 'super-cooled liquid' serves in general. For the purposes of this book the chemical and structural nature of glass does not need lengthy discussion – we want to be able to recognize it in the field. However, it may be worth knowing that most glasses from which imitations are fashioned are silica glass (old windows are crown glass, a lime-soda silica glass). The addition of lead oxide to glass gives added powers of dispersion but softens the glass at the same time as well as increasing the likelihood of tarnish and the development of a yellowish colour: lead also raises the SG. The glass containing lead oxide is called flint glass.

Dispersion can also be increased by the addition of thallium oxide and opacifiers can be used to make glass translucent to opaque. Colouring agents (metal oxides) are used for specific imitations. Glass is made by melting in ceramic crucibles in kilns to high temperatures and the cheaper imitations are moulded though the better-class ones are faceted on a lap in the same way as natural gemstones.

Properties

In general glass is relatively soft with a hardness between 5.5 and 6 on Mohs' scale. It is easily scratched and is brittle but shows no cleavage. Examination of facet edges shows rounding when the stone has been moulded and frequent conchoidal (shell-like) fractures are highly characteristic of glass. The well-known bubbles which inhabit glass can be seen to be well rounded with bold edges, this being due to the difference between the refractive indices of the gas content of the bubble and the glass host. Due to incomplete and imperfect mixing of the ingredients swirl marks are very common and easy to see when the specimen is examined (from the side) under dark-field illumination.

Glass is a poor conductor of heat and feels warm compared to most natural (crystalline) gemstones. The tongue test (touching the specimen with the tip of the tongue) gives a good idea of the comparative warmth of glass but this test is not recommended. Apart from the danger of swallowing a customer's specimen, previous tests may have involved the use of chemicals which have not been adequately removed from the surface. Always wash the specimen after testing.

In general the refractive index of the glass most commonly used as gemstone imitations will be between 1.50 and 1.60 (though there are many exceptions): only pollucite and rhodizite (see Chapter 29) are, like glass, singly refractive and fall in this range. Most other gemstones and especially those imitated by glass will be birefringent: glass on its own never shows doubling of inclusions or of back facet edges. Specific gravity values cover a wide range, from 2.3 to 5.0. Hardness also varies though many glasses are around 5–5.5 on Mohs' scale.

Uses in gemstone imitation

There are many examples of the ingenious use of glass in what have come to be called phenomenal gemstones: *Cathay stone* is an excellent imitation of chrysoberyl cat's-eye in which bundles of glass rods are heated and drawn in such a way that light reflected from them when they have been incorporated in a host glass of different refractive index gives an excellent eye. Colouring agents are added to give the honey-yellow chrysoberyl colour. This material is very effective when set and the high constants are not likely to suggest glass in the first instance. The manufacturers, Cathay Corporation of Stamford, Connecticut, USA, give RI 1.8, SG 4.58 and hardness 6.

Glass has been used to imitate star-stones, sometimes by having six rays engraved on the back or by using a foil on which 'rays' have been scratched. A white opaque glass has been made to imitate star-stones: in one example the molten glass is pressed to give a cabochon with six ridges giving an imitation of the rays of a star. On completion of this process the stone is covered with a thin layer of deep blue glaze which hides the ridges, making it look as if the star was just below the surface.

Opal can be imitated by translucent lime glass with added fluorides or phosphates. Calcium compounds which precipitate in the glass complete the illusion.

The name '*Victoria stone*' (kinga-stone) or the less commercially attractive 'meta-jade' was given to a partly crystallized glass designed to imitate jade though there is not much resemblance. It is possible to create a chatoyant effect in this material by using fibrous inclusions in parallel bundles. *Iimori stone* is also used as a jade imitation but is clear to translucent where Victoria stone is opaque. Goldstone is made by including cuprous oxide with the glass ingredients and reducing (oxygen) during annealing. This allows small crystals of copper to precipitate and give the sunstone (feldspar) effect.

If the glass is blue it may be and has been used as a rather unconvincing imitation of lapis-lazuli though the glassy effect is nothing like that of the rock. A green glass was reported to have been manufactured from fused material from the eruption of Mount St Helens in Washington, USA. This material was found to contain 10 per cent at most of this ash and many specimens contained none of it. I (MO'D) am indebted to my friend the late Alan Rowlands of Calgary, Alberta, for specimens of this glass in other colours.

Glass may also be recognized by its behaviour between crossed polars on the polariscope: it shows apparently irregular birefringence (due to strain), the field being neither completely dark (as an isotropic material ought to be), nor alternately dark and light four times during a complete rotation. Sometimes (glass is full of surprises and this is just one of them) an apparent birefringence may be shown between crossed polars but there is usually something else in the specimen that hints that all is not what it seems. This sounds facetious but glass does give itself away.

The gemmologist usually encounters glass when it is pretending to be something else. When it is imitating blue zircon it will not show the strong birefringence or the reliable refractive index of natural red spinel (1.718) when spinel is imitated. It will not show the three-band iron absorption spectrum of peridot or the 'horsetail' inclusions of demantoid garnet. It will show inclusions of its own as we have seen above.

One type of inclusion peculiar to glass that is very likely to catch out the gemmologist is devitrification which is mentioned elsewhere in the book. Crystals of some of the compounds which have been used in its manufacture tend to 'come out' from the mix and these can easily be mistaken for crystals of an included mineral. Other tests will show them for what they are.

Properties of glass vary widely with composition and it is not really vital for the gem tester to know exactly what type of glass is involved in a masquerade. However, the specific gravities and refractive indices of some coloured glasses are given in a table by Anderson in *Gem Testing* (reproduced in the 10th edition of 1990, page 310). Among the entries of interest are a colourless glass with an RI of 1.47 and SG 2.30. This is a borosilicate crown glass. A yellow calcium crown glass has an RI of 2.43 and SG 1.498 and a colourless light flint glass with lead gives 1.54 and 2.87.

Other examples include an emerald-green fused beryl glass with RI 1.516 and SG 2.49 and a yellow flint glass, with lead, showing RI 1.77 and SG 4.98. The dense glass used in the refractometer has an RI of 1.962 and an SG of

6.33 – this is an extra dense flint glass. This and some of the other very dense glasses are not likely to be used as wearable gemstones but there are serious collectors of synthetic and imitation productions who will need to know about them.

We have already mentioned the fused beryl glasses which would have shown RIs close to those of natural beryl had they been crystalline and thus denser. The composition is the same, however. Some of the lead glasses have refractive indices in the topaz range (one glass cited in the table had RI 1.633 and was yellow, though its SG was 3.627, above the topaz limit). It is worth noting that glasses which in general have a higher dispersion than most natural gemstones give a sharper shadow-edge reading in white light on the refractometer. These glasses are the lead glasses with RI readings over 1.60. If a spinel refractometer is used the reverse effect is seen with the lead glass specimen giving a shadow-edge with a colour fringe and the natural stone a sharp colour-free shadow-edge. This is a good way in which to assess (not measure) dispersion since a specimen with a low dispersion will show a notable colour fringe to the shadow edge on a glass refractometer while a specimen of a high dispersion will not. White light is of course necessary for any colour to be seen and the reverse effect, as above, is seen when the spinel refractometer is used.

Dispersion cannot be accurately measured with the standard refractometer as these instruments are calibrated for sodium light.

A fused quartz with refractive index near 1.46 and specific gravity 2.21 (natural quartz gives 1.54–1.55 and 2.65) is another example of a glass with similar composition to a natural material having lower constants because of its amorphous nature. Both the fused beryl and the fused quartz can be coloured by various dopants (elements not forming part of the regular chemical composition of the specimen but added in minute quantities during the growth process to gain some end, usually a desired colour).

Cobalt is added to give blue and chromium to give green. The cobalt-doped glass shows bright red through the Chelsea filter and shows a characteristic absorption spectrum of three strong broad bands in the red, yellow and green at about 655, 580 and 535 nm. The centre band is usually the narrowest of the three (it is the widest in cobalt-doped synthetic flame-fusion spinel).

Red glasses may be coloured by selenium and then show a single broad absorption band in the green and some pink or red glasses may owe their colour to rare earths and show a characteristic fine-line absorption spectrum. These are not very common but when pink rather than red they may first attract attention by their colour change from pink to a pale slaty blue.

If a specimen looks fairly normal to the eye but will not give a reading on the refractometer when you might expect to obtain one with no difficulty it will most likely be glass. The difficulty in obtaining a reading from the table facet of many if not most faceted or moulded glasses usually means that the table facet being tested is not sufficiently flat for optical contact to be made with the instrument. There are ways of getting round this, e.g. the spectroscope or the microscope may be used.

The natural glasses

Although the natural glasses could have been placed with the rarer species, for convenience they can be examined at this point. Both natural and artificial glasses are cooled quickly from the molten state so that a crystalline structure has no time to develop.

Obsidian

The best known of the natural glasses is obsidian, which forms from the cooling of volcanic glass. The cooling is very rapid so that no crystals have time to form, thus producing a glass. Obsidian may contain up to 77 per cent silica with 10–18 per cent alumina.

Most obsidian is black but other colours are known, mostly yellow, red, greenish-brown and grey. Snowflake obsidian found in the United States is black with white flecks but this is only one variety among many; fanciful names are rife.

Examination of obsidian may show signs of incipient crystallization and small vesicles occur in most specimens. These are found in cavities of varied shapes formed by the entrapment of gas bubbles during solidification. Strain may cause a slight strain birefringence. The hardness is about 5, the refractive index is usually between 1.48 and 1.51, and the specific gravity is in the range 2.33–2.47. Obsidian has a very marked conchoidal fracture which gives sharp edges.

Basalt glass, a natural glass containing 50 per cent silica, has an RI between 1.58 and 1.65 with SG 2.70–3.00.

The name *marekanite* has been given to an obsidian showing a network of curving cracks (perlitic cracks) with disintegration along the curves giving bead- or ball-like pieces which take the name marekanite from a type locality in Siberia. Peanut obsidian has a spherulitic texture with radiating feldspar fibres: it is found in the Sonora region of Mexico.

Surprisingly, obsidian is sometimes imitated by bottle glass which has a higher RI and SG and is also more transparent.

The name tektite has been used for natural glass of presumed meteoritic origin. The material attracts collectors and some specimens have been used in jewellery. Specimens of dark green glass (moldavite) were found near the Moldau river in Bohemia during the eighteenth century though the theory of an extra-terrestrial origin came much later.

However, a recent study (McCall, 2001) suggests that tektites are not meteorites after all although they are associated with meteoritics. The new theory holds that tektites are mostly projected melt from terrestrial rocks at a site of impact. Tektites are composed of silica and alumina and have a hardness of 5 (this figure was established for a bottle-green moldavite), refractive index of 1.48–1.52 and specific gravity 2.30–2.50. These figures vary with colour and place of origin but the interiors will always show the glass characteristics.

When collected some tektites show a characteristic flattening of one of the surfaces, this being thought to be due to their flight through the atmosphere from an unknown origin. McCall's theory, quoted above, explains this flattening equally well.

Libyan glass

The material known as Libyan glass or silica glass is nearly pure silica. It is an attractive transparent pale lemon–yellow with a hardness of 6, RI 1.462 and

SG 2.21. The dispersion is low. McCall (2001) believes the origin of Libyan glass to be impact related and that the glass came originally from two craters both just over 100 km from the strewn field.

The natural glasses are usually on view at gem and mineral shows and there should be little difficulty in identifying them. It is less likely that one of them will be offered in place of a similarly coloured natural crystalline substance except through a genuine mistake. Few natural glasses are completely colourless: devitrification signs are more likely to be found in obsidian than in the green moldavite.

Glass is usually the main material in composites which are described in Chapter 28. It is also used to fill surface-reaching fractures: gemstone enhancement is dealt with in Chapter 33. The ingenious material Slocum stone has been described in Chapter 5 on opal.

Reports of interesting and unusual examples from the literature

Items in this section have been chosen to illustrate points made in the chapter and to bring one-off items to your notice.

The Winter 1993 issue of *Gems & Gemology* described a material with the name Junelite which had been produced in a range of colours. Publicity material quoted an RI of 2.0 and a specific gravity of 4.59. On examination specimens were found to show an RI above 1.81 and an SG of 4.44. Anomalous birefringence could be seen between crossed polars and X-ray diffraction proved that it was amorphous.

A glass egg weighing 198 g was found to contain gas bubbles and swirls thus proving that it was glass. However, the specimen showed a very strong greenish-yellow fluorescence. GIA tested for radioactivity and found a level ten times above normal background. This rather lurid yellow–green colour is characteristic of uranium glass and any stone of this colour should be tested for fluorescence and/or radioactivity.

A convincing chatoyant glass with a more natural eye than those seen in other imitations turned up in the form of an 11 ct cabochon. The eye was intersected by a series of evenly spaced dark lines. The microscope showed a honeycomb structure in which the individual cells had hexagonal outlines – the cells could be seen when the specimen was examined under magnification from the side perpendicular to the top of the stone. Individual fibres were found to be thicker than those in most imitation cat's-eyes and their edges were not transparent: this caused the eye to be less sharp. RI readings taken on the dome of the cabochon were 1.48 and 1.62 (chrysoberyl would give a mean reading of 1.74).

A light green emerald-cut stone gave a single RI of 1.529. Between crossed polars it appeared to be singly refractive with strong anomalous birefringence. The stone gave a very weak dull yellow fluorescence under LWUV and a very weak chalky greenish-yellow under SWUV. No absorption bands could be seen with the hand spectroscope. While the stone was identified as glass it looked very like light green beryl or unheated aquamarine. The stone (hefted by hand) seemed to have a specific gravity close to that of beryl. Testing with heavy liquids (now available only under laboratory conditions) showed the SG to be about 2.50, too low for beryl.

Glass is of course widely used as an imitation of rough gem material. I (MO'D) was once offered a green emerald-coloured hexagonal crystal, reported to be emerald of African origin, which had suspiciously smooth and evenly coloured faces and which was coated with yellow flakes of a supposed mica. No emerald is coated with this kind of mica, biotite being dark brown to black: under magnification and a strong light large gas bubbles could be seen inside. The owner, who had told me that his life savings had been spent on this piece, seemed unaffected by my verdict on it!

A black bead necklace from Asia was found to consist of glass with dendritic intergrown fluorite. A 'new' gemstone emanating from Namibia turned out to be blue glass with tridymite inclusions.

A transparent smoky natural glass occurs at Superior, Arizona, USA, the material being rhyolitic in composition and shows RI 1.480, SG 2.33 with anomalous birefringence in rough pieces only.

While the fibre-optic glass imitation Cathay stone has imitated tiger's-eye and cat's-eye chrysoberyl it has been manufactured in colours other than yellow and brown. Pink, purple, blue and gun-metal grey (this one resembling hawk's-eye quartz) were reported in 1996.

Rainbow obsidian from Jalisco, Mexico, owes its play of colour to oriented rods of augitic pyroxene. Interesting patterns, including a double heart with purple centre and green surround have been cut.

A faceted obsidian from northern Chile had RI 1.540 (close to the usual upper limit for obsidian) and the SG was 2.36. Three faceted examples, one completely transparent, the other two milky, all remained bright during a complete rotation between crossed polars, the cause being a profusion of anisotropic (doubly refractive) inclusions. Chemical analysis showed a high silica and alkali content which confirms a rhyolitic origin. Among the inclusions were blue columnar hexagonal crystals with a very strong pleochroism of deep bluish-violet and colourless.

These were confirmed as cordierite. Long colourless rods with striations parallel to the main axis were found on analysis to be sillimanite.

Blue faceted beads sold as quartz turned out to be artificial glass. They showed rod-like inclusions identified by GIA as synthetic wollastonite. Gas bubbles were present but scarce.

Chapter 26

Metals and ceramics

Metals are not often used to imitate gemstones and apart from gold, silver, platinum and copper they do not come before the jeweller or gemmologist for testing. Metals apart from imitations are really outside the scope of this book but one or two examples need to be mentioned.

There are substitutes for hematite which in its natural form gives a red streak (powder) on an unglazed porcelain streak plate. Hematite is often cut for use in the cheaper forms of jewellery and the substitutes do not give the red streak.

One imitation has for some time been known as Hemetine and is a mixture of stainless steel with chromium, lead and nickel sulphides. It will be attracted by a pocket magnet whereas hematite is not attracted. Hemetine gives a black streak and has an SG near 7.0 compared to hematite's 5.1.

Granular hematite forms the main constituent of a material seen in the gem-cutting industry: this has a bluish-grey colour with a dark brown streak and has an iron content of around 69 per cent. It has a specific gravity of 2.33 compared to hematite's at just over 5.

A titanium dioxide substitute for hematite gives a yellow–brown streak with a body colour of steel-grey; the hardness is about 5.5 and the SG about 4.

Powdered lead sulphide, perhaps with added silver, has been reported as a possible hematite substitute. This had a hardness of 2.52–3 and was easily fusible. The SG is 6.5–7 and the material is notably brittle.

Ceramics

Ceramics are made from finely ground powders of inorganic substances which can be sintered, heated or fired with or without pressure being imposed. The result is a fine-grained polycrystalline solid which can be completely clear. The Gilson turquoise and lapis-lazuli as well as the Adco, Syntho and Turquite imitations of turquoise are made in this way. The material Yttralox, made since 1966 by General Electric, has been used as a diamond simulant. It is a stabilized yttrium oxide, Y_2O_3 which can be made isotropic by the addition of 10 per cent ThO_2. Some specimens may be radioactive. Hot pressing gives the final form: it has a hardness of 7.5–8, RI 1.92 and dispersion 0.039. The SG is 4.84.

Porcelain may sometimes be encountered in the guise of turquoise but can easily be detected by the presence of gas bubbles immediately below the surface. The SG is usually around 2.3. When chipped the surface may be seen to be a glaze.

Faience simulating turquoise has been known since ancient times in Egypt where is has been found as a quartz-rich material with a blue glaze which may arise from copper or cobalt, the appearance varying according to which element is present. The method of manufacture is unknown though current investigations suggest that three different methods were used.

Chapter 27

Plastics

Many of the cheaper imitations of gemstones simulate such materials as amber, pearl and jet and plastics are very cheap and often quite effective substitutes for the natural material. In general the natural materials will always show a higher specific gravity than plastics which average 1.05–1.55 (1.55 is a common value). Plastics (when their nature and that of the contact liquid are not incompatible) show a refractive index of 1.5–1.6 and a hardness of 1.5–3.0. These values can be raised by the addition of fillers.

Most plastic imitations of gemstones are made by the injection moulding process: they will commonly give off an acrid smell when touched by the thermal reaction tester. They are sectile but this of course is a destructive test.

Celluloid

Celluloid was one of the earliest plastics to be manufactured: in its original form it was highly inflammable and there may still be pieces of jewellery ready to burst into flames but the advent of safety celluloid prevented such items from becoming too common. Acetic acid prevents flammability.

Celluloid is made by heating the lower nitrates of cellulose and camphor at 110°C under pressure and is rendered safe by treating with acetic acid. The older forms of celluloid gave a specific gravity in the range 1.35–1.80 (rising towards the higher figure by adding fillers): the hardness is 2 and the refractive index between 1.495 and 1.520. Safety celluloid (cellulose acetate) has an RI in the range 1.49–1.51 and SG 1.29. On heating it will give off a smell of vinegar. Both types of celluloid are sectile: the hardness is 2.

Casein

Casein is made from the protein part of milk and hardened by the addition of formaldehyde. Running ahead a little, casein will give off a burnt milk smell when the thermal reaction tester is brought near. Casein can be dyed to give a range of quite attractive colours and has an RI of 1.55 and an SG in the range

1.32–1.39 (most commonly 1.33). A drop of concentrated nitric acid will develop a yellow spot on the surface but this may very well be thought likely to spoil an otherwise acceptable specimen. Nonetheless, if rough pieces are available and you are equipped to use a dangerous chemical, the test could be useful. Casein is sectile.

Bakelite

Bakelite is a phenolic resin and can be made in a variety of colours though it tends to yellow with age and is not very tough. It may be transparent and has a rather high RI range for plastics, at 1.61–1.66. The specific gravity is 1.25–1.30. A test for bakelite is quoted in *Webster's Gems* but is unlikely to be available today except in laboratory conditions: nonetheless, here it is: chips of bakelite are covered by distilled water in a test tube. After boiling, a pinch of 2-dibromoquinone chlorimide is added. The liquid cools and a drop of a very dilute alkali solution added. Phenol is indicated if the solution turns blue and the piece must be bakelite. The translucent amino plastic which is a modification of bakelite is translucent and can be dyed in different colours. The hardness is near 2, the RI 1.55–1.62 and the SG about 1.50.

Perspex

Perspex is an acrylic resin which is made as a clear, glass-like material for use as cheap beads and for the cores of imitation pearls. It has an RI of about 1.50 and SG 1.18, low for a plastic material. Polystyrene resins have been moulded to form faceted stones: these have an RI of 1.59 and SG 1.05. These resins are sectile and also easily dissolved by some organic liquids such as toluene or bromoform (now with restrictions on their use) or by di-iodomethane.

Perhaps the one important gemstone (apart from pearl and other organic materials) that could be successfully imitated by one of the plastics is fire opal but this has a higher specific gravity, at 2.00, than the plastics could supply. Such imitations would feel suspiciously light and hang unevenly in a necklace. Amber will show a lower SG of about 1.08 than almost all plastics and will to some extent float in a brine solution when most plastics sink. The density of the solution can easily be established by trial and error using pieces of plastic and amber as indicators.

Chapter 28

Composite stones: doublets, triplets and soudé stones

Many gemstones are not the solids they appear but are made up of two or three separate sections, classed as doublets or triplets respectively. The sections may be of different substances or the same substance: they may even be the same as the material to be imitated, as in the diamond doublet. The intention is not always to cheat the customer but sometimes to aim at preserving fine-quality gem material which would otherwise be too thin or fragile to stand alone. The opal and ammolite doublets and triplets are good examples.

Garnet-topped doublets

The best-known doublet, which has been around since the nineteenth century, is the garnet-topped doublet (GTD) which may be found in a number of different colours, including red, though green and blue specimens appear to be the most common. The best clue to doublet/triplet spotting is to 'think composite' and in the case of a GTD this should be at the forefront of one's mind when any rather oddly coloured transparent stone is submitted for testing. Even before any routine test is carried out, the presence of red flashes where no red ought to be should alert the observer when the stones are not red or cobalt-bearing blue.

The GTD is usually said to have been devised to defeat the jeweller's file (used as a test for soft glass) but this cannot have been widely used as customers might have become attached to their glass jewellery. Bearing in mind that most GTDs are glass (apart from the G) it is at least possible that the relative wear on the table facets of glass became a drawback and that the harder garnet, fused in very thin slices to the glass table, helped to prolong the life of the stones.

Examination of a GTD (if the red flashes are considered insufficient evidence for identification) is most easily carried out under magnification for which the microscope should be used (this is because many GTDs are too small to hold safely in tongs). The slice of almandine garnet is fused to the glass table and will probably show needle-like rutile crystals. An almandine absorption spectrum (or at least the persistent 505 nm band) is not too difficult to see. Any red flashes in a stone which is not red can be concentrated if the stone is placed table down on a white surface. The rest of the stone will show the characteristic swirls and gas bubbles of glass as well as conchoidal fractures on facet edges. Stones which appear from their colour to be a known species may fail to show the RI of that species as the reading for almandine is above the limit of the contact liquid (1.79).

Though on the refractometer the almandine shadow-edge will not be seen, the absorption spectrum of the stone imitated will not be visible either so that a 'peridot' or 'blue sapphire' will not show the spectrum anticipated. Since almandine is a member of an isomorphous series it is possible that the slice has a lower RI than the darkest almandine (see Chapter 11 on garnet) and thus could be confused with a corundum reading. If the slice is cemented to a red glass the gemmologist could just be deceived into thinking that the stone is a ruby. The spectroscope will sort this out – provided the tester thinks about using it. It should also be remembered that the garnet is a thin slice fused quite unsymmetrically to the glass table and not a large section cemented to the remainder of the composite at the girdle. It is possible that some portion of the doublet will fluoresce a greenish-white under SWUV light and the almandine slice remain inert on account of its iron content. The resulting effect will indicate a composite.

Diamond doublets

The diamond doublet is very rare (and the authors would like to obtain one!): it may have been manufactured in the first instance to make a single stone from two pieces not worth using singly. In any case, the diamond doublet will show the usual adamantine lustre which is hard to mistake and if the point of a pencil be placed on the table facet its reflection will be seen a little way down inside the stone. While immersion will show up the two sections it may also cause them to become separated, hardly the desired end of the exercise.

Lasque diamonds, a feature of some Indian jewellery, are thin parallel-sided diamond plates whose top is often faceted. The shadow of the table edge may be seen reflected from the lower surface. This is not intended as a deceit but merely as a way of using diamonds which are found in this form. Nevertheless it is easy to see how suspicions may arise.

Diamonds showing internal interference colours should be magnified to ensure that the colours do not indicate that thin films of air are trapped between two portions of a doublet or if they denote cracks or cleavages (or colour enhancement by filler).

Opal doublets and triplets

Almost any combination of material on material may be found in doublets but the opal doublet and triplet (this is merely a capped doublet) is easily the most common. To ensure that thin but fine quality opal layers are not wasted either the rock (matrix) on which they are found is left in place or they are cemented to another hard base of a different material. Capping the doublet thus formed both protects the opal and magnifies it at the same time as the cap forms a convex lens. Doublets will usually show flat backs and a low dome and the lens will show whether or not the colour patches from the opal layer extend downwards into the base. If they do, then the stone will not be a composite. The cap of a triplet may be of plastic, glass or even rock crystal though the latter is rare: plastic caps are easily scratched. Synthetic opal may also be made into doublets and triplets but the same testing criteria apply.

Opal slices may form the top of doublets whose base may be black onyx or black glass.

As well as opal doublets the occasional jade doublets turn up: one example was reported to be made from a thin layer of green jadeite on a thicker layer of white jadeite, the whole being cut as a tablet.

The number of possible doublets is large: some examples include pale emerald on green glass, aquamarine with natural inclusions on green glass and a highly transparent chrysoberyl cat's-eye on a darker base of unknown composition. Natural sapphire on synthetic sapphire or ruby may be deceptive since there may be natural inclusions in the upper part of the stone and a characteristic absorption spectrum from the base, though Verneuil flame-fusion-grown blue sapphires do not usually show the blue sapphire absorption band at 450 nm. Curved growth lines should be seen with the correct adjustment of lighting conditions: SWUV should cause the synthetic sapphire to fluoresce greenish. If ruby forms the base LWUV will show the usual bright red fluorescence and this will contrast with the top which will usually (if red) be inert.

A specimen with a quartz top glued to a green-dyed base with a green cement was offered as green jadeite but would not show the jadeite absorption band at 437 nm nor any chromium elements.

In *Synthetic, Imitation and Treated Gemstones* (Butterworth-Heinemann, 1997) which has been drawn on extensively for this and similar sections, it was noted that rock crystal with its hardness, clarity and lack of cleavage makes an ideal upper portion of a triplet or doublet when the base could be cheaply made of glass. The use of a closed setting can make things difficult and although the rock crystal may show the near-unique quartz interference figure it is usually the colour of the complete stone which will alert the observer to something being wrong.

While some of the older textbooks ponderously cite every possible SG such a test will not be carried out in practice save in a gem testing laboratory. In commerce all the buyer/seller wants to know is whether or not a particular specimen is what it is claimed to be: if it is not what it seems its true identity is a matter of indifference.

In *Webster's Gems*, fifth edition, revised by Peter Read and others (Butterworth-Heinemann, 1994), examples are given of rock crystal on different bases giving sapphire blue, purple and yellow. The blue and purple specimens gave a cobalt absorption spectrum and this was fortunate since there was a close resemblance to amethyst and an RI reading (which would, admittedly, be obtained with some difficulty from a domed top) would have given a reading consistent with amethyst.

The yellow specimen, looking like a yellow synthetic spinel (these are not too common, giving a vivid yellowish-green colour and strong yellow fluorescence under LWUV light), fluoresced from the base only, the fluorescence spectrum showing uranium bands. Uranium glass in green or yellow occasionally turns up in jewellery. In the doublet described the cement layers showed up as a bright line surrounding the girdle. This type of doublet can be found in reverse with a quartz base and a glass top.

Another type of doublet reported is a white topaz with the tip of the pavilion made from natural blue sapphire. In many natural blue sapphires, particularly those from Sri Lanka, the lapidary leaves just such a spot of colour in the culet area in order to turn the whole stone blue when viewed through the table so that this imitation could be thought to be dangerous. In fact the report stated that the effect was not successfully imitated.

A rare type of doublet may be filled with a coloured liquid (this was reported as long ago as 1896 in Bauer's *Edelsteinkunde* and later). The material may be glass or rock crystal, hollowed out with the walls of the cavity highly polished. The coloured liquid was placed in the cavity and the stone completed with a base of the same material as the crown. What was the liquid and did it degenerate? The cost of making such specimens must have been quite high.

A red doublet reported in the literature consisted of two pieces of colourless glass joined by a coloured cement which was said to show an almandine absorption spectrum. This sounds odd: the RI was reported to be 1.52 and the SG 2.48.

Soudé stones

The name soudé is used most commonly in the emerald context and denotes stones whose top and bottom may be colourless quartz, the colour being supplied by a green layer at the girdle. This layer may be a green glass or green gelatine. Do not rely on the Chelsea filter for identification since these stones show red, though not very strongly. Over time the gelatinous materials can discolour to yellow. Today there are better quality soudé stones around though reports do not come out very often.

Some of the later examples remain green under the Chelsea filter but the quartz readings will be sufficient for identification. The joining layer may sometimes be a coloured lead glass which would give a slightly higher SG. Such a glass seems to have been used in a composite in which the top and bottom were synthetic spinel. These stones were made in France from 1951 and show a refractive index of 1.728 from colourless synthetic spinel and also fluoresce a sky blue under SWUV light. If immersed in a liquid with a higher RI than water (1.33) the assembly is easily seen. Unfortunately most if not all of the available liquids are usable only under laboratory conditions.

Under the Chelsea filter the green soudé spinels remain green while imitations of blue zircon and amethyst imitations also show green. When a solid synthetic spinel imitates these two species it will usually show orange through the filter. The Chelsea and other such filters should never be relied upon alone. Other tests should always be made.

Examples of other composites can be found in the species chapters but one very clever example is worth mentioning: this is an imitation of diamond made from stones with a synthetic colourless corundum or spinel top with a highly dispersive strontium titanate base. They are very dangerous in small sizes which do not always get tested and the join is usually above or below the girdle. *Webster's Gems* (1994) describes a mosaic composite in which two colourless substances sandwich a three-colour mosaic in a transparent filter. We have not seen a specimen and cannot therefore comment on its effectiveness.

Reports of interesting and unusual examples from the literature

Items in this section have been chosen to illustrate points made in the chapter and to bring one-off items to your notice.

The Fall 1986 issue of *Gems & Gemology* reported a jadeite doublet of a fine green colour but which consisted of a very thin green layer on the top and another, thicker white layer beneath, the two layers joined by a slightly yellowish cement containing profuse gas bubbles. The dark green upper layer was mottled with many near-colourless veins and with a dark green-to-black jade-like material frequently known as chloromelanite but not a valid mineral species. The white lower layer showed a distinctly crystalline structure. The RI readings obtained were 1.64–1.74 for the green layer (which showed characteristic chromium absorption lines) – a normal reading for jadeite would be 1.66. Both layers were finally identified as jadeite by X-ray diffraction analysis but there has been no explanation of the varied RI readings.

In an expensive necklace 'emeralds' proved to be rock crystal with a green backing. The heart-shaped green stones gave an RI reading of 1.54 and showed two-phase hexagonal inclusions. The stones were in a closed setting.

Fine-coloured carvings purporting to be emerald and showing characteristic inclusions of the natural material were offered for sale by a German firm. Some of the pieces had been carved as cameos. The same firm offered beryl triplets with a saturated slightly greenish-blue colour with the intention of imitating fine blue Paraiba tourmaline.

Star doublets are perhaps less common than the literature suggests. GIA described a transparent red cabochon which had been bezel-set in a man's ring and which gave a six-rayed star. The RI and inclusions showed that the top of the stone was a flame-fusion ruby and round to oval bubbles were visible in the cement layer. The reddish-purple base showed strong hexagonal growth zoning, partially healed fractures and other features which suggested that it was natural corundum of low grade (these have sometimes been called mud ruby). The setting prevented a positive identification of the base but the star in the crown was provided by rutile needles within it.

Faceted tourmaline composites with red–white and green sections were reported in Germany during 1990. The stones were assembled from portions of differently coloured faceted tourmalines glued together. A composite imitating tourmaline cat's-eye was made by cementing a transparent crown to a fibrous pavilion. In both cases the cement layer was said to be clearly visible.

A synthetic spinel and glass triplet imitating emerald (and other colours) was first made as *soudé sur spinelle* in 1951 by Jos Roland of Sannois, France. A report by GIA in 1986 said that the stones were made by sintering coloured glasses to the colourless spinel crown and pavilion. One of the emerald imitations examined by GIA weighed 11.66 ct and was emerald-cut. It showed cross-hatching with curled black bands between crossed polars, a characteristic feature of synthetic spinel.

The RI from the table was interesting in that two shadow-edges were obtained, at 1.724 (strong and characteristic of synthetic spinel) and at 1.682, which was weaker. Some shaded areas were detectable between the two readings. An RI of about 1.682 was obtained from the glass layer which was about 0.55 mm thick and had a hardness of about 4. No absorption bands could be seen with the hand spectroscope but using fibre-optic lighting small bubbles could be seen in the separation plane.

In a direction perpendicular to the girdle the thick glass could be seen to contain swirls and rounded edges. When the stone was immersed in di-iodomethane its composite nature was clearly seen. Under LWUV light the crown showed a strong chalky yellowish-white fluorescence when viewed nearly perpendicular to the girdle with the table closest to the UV source.

The glass layer was inert and the pavilion showed a strong clear yellow fluorescence with no hint of chalkiness. When the culet was placed near to the UV source an opposite effect could be seen with colours reversed. GIA thought that this was due to the glass layer diminishing the amount of radiation reaching those parts of the stone that were not directly facing the source.

There was no response to SWUV light and only a very weak chalky-green fluorescence could be seen under X-rays. There was no phosphorescence.

Chapter 29

Less common gemstones

The choice of species as less common gemstones is arbitrary! Some may become better known over time but on the whole these stones are relatively soft, relatively uncommon or occur as crystals which are more likely to bring high prices than faceted stones. In general it should not be too difficult to find examples of all these species from specialist dealers.

Species which are hardly ever encountered are described in the next chapter. Species described in this chapter may not often be seen in jewellery, if they ever are, but feature strongly in the collectors' market. Some may display properties which may cause confusion with better-known stones and are worth mentioning here as well as in the chapters dealing with them.

We should remember that many gemstones which never reach jewellery are just as beautiful as those that do. Only such factors as low hardness or durability, extreme rarity (no one has ever heard of them) or, of course, the lack of material to display and publish may prevent them becoming much better known and there are instances in which the discovery of excitingly coloured material has elevated a previously obscure mineral to a place among the major gemstones: tanzanite is a good example.

Several considerations make a gemstone rare. Elements and compounds have to come together during some geological event in just such conditions of temperature and pressure that one substance is formed rather than another. Over geological time this happens over and over again but only about 4000 distinct mineral species have been validated. It is not surprising that few of them produce gem or ornamental varieties – those minerals crystallizing fast will be too small while slow crystallizers (the only ones likely to occur in gem sizes) may very well suffer many alterations over time before the lapidary gets to them and in any case may be unattractive or lack durability.

Those species which have escaped these pitfalls without becoming classics are described below. Only general details are given – there are plenty of textbooks which will tell you more – but those points which make them important and those which may cause confusion are highlighted.

Actinolite is a mineral of the amphibole group which has recently been

Plate 33 Ramaura ruby

Plate 34 Ramaura ruby

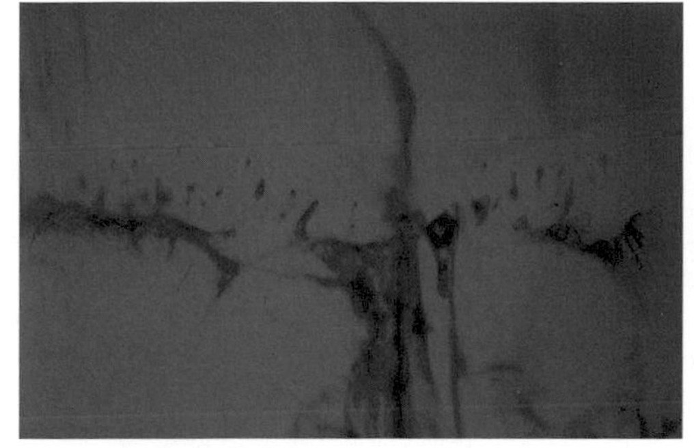

Plate 35 Knischka ruby

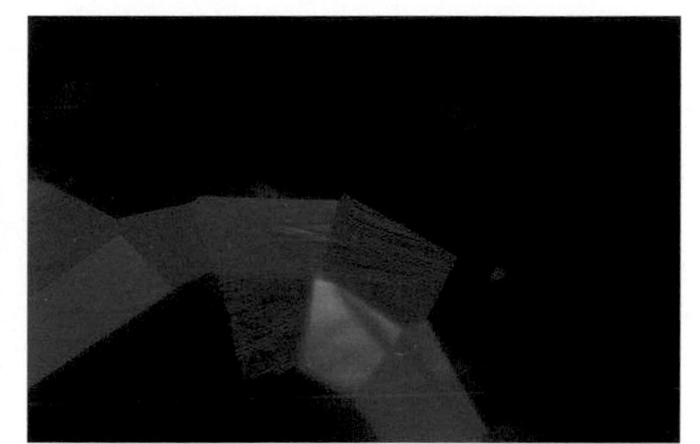

Plate 36 Roiling in natural ruby

Plate 37 Glass infill in synthetic ruby

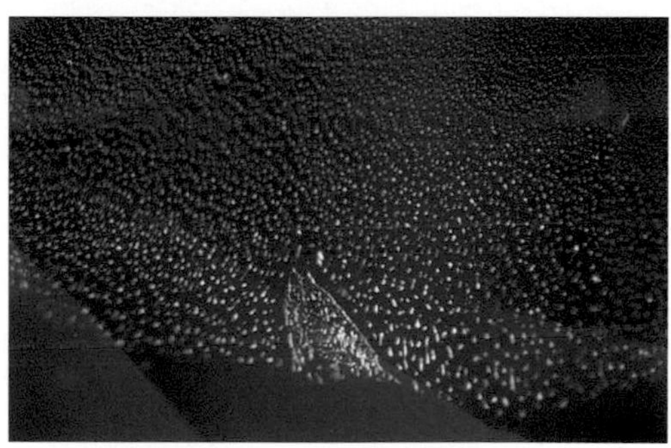

Plate 38 Veiling in early Kashan ruby

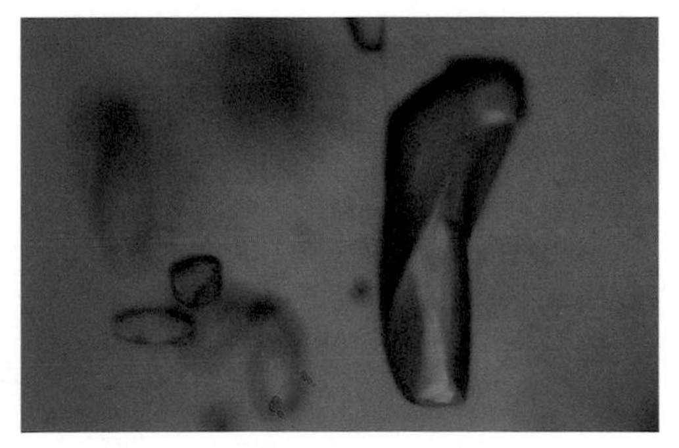

Plate 39 Crystals (?apatite) in Myanmar ruby

Plate 40 Glass imitating a gemstone

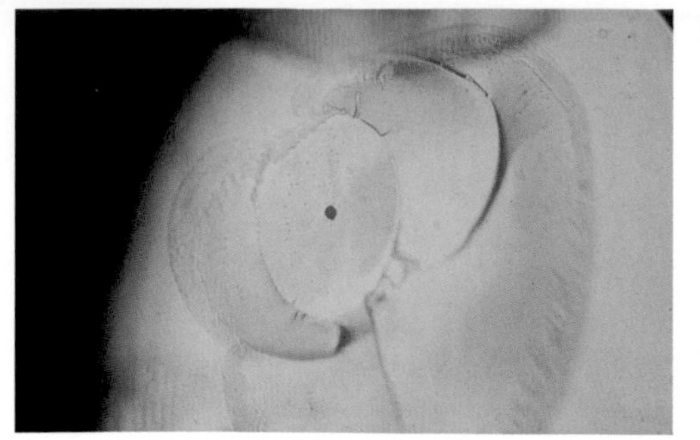

Plate 41 Water-lily in peridot

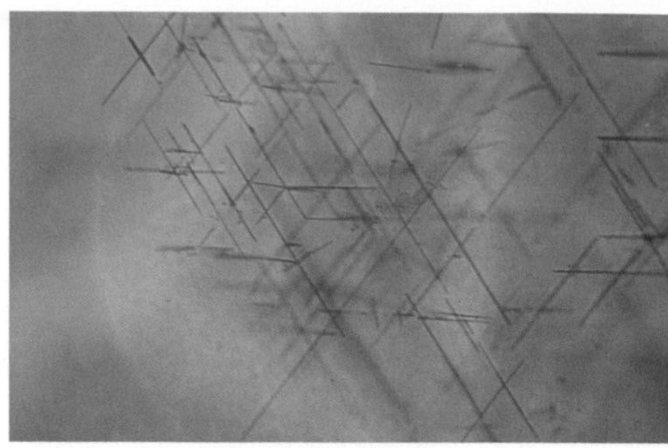

Plate 42 Silk in almandine

Plate 43 Taaffeite crystals and cut stones

Plate 44 Yellow (?Sumitomo) diamond crystal

Plate 45 Silk in natural ruby

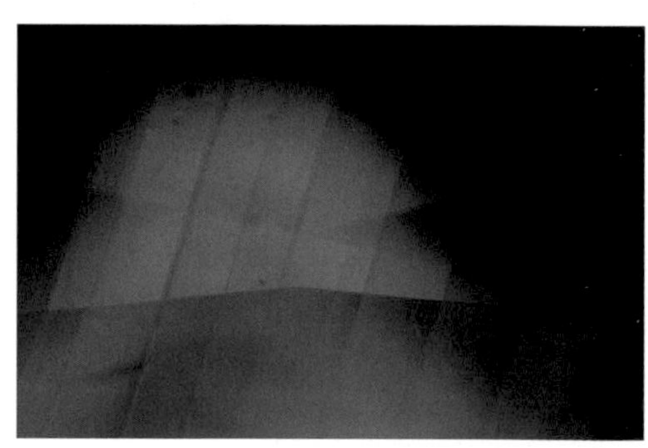

Plate 46 Twin planes in synthetic ruby

Plate 47 Lechleitner overgrowth of emerald on beryl, cracking

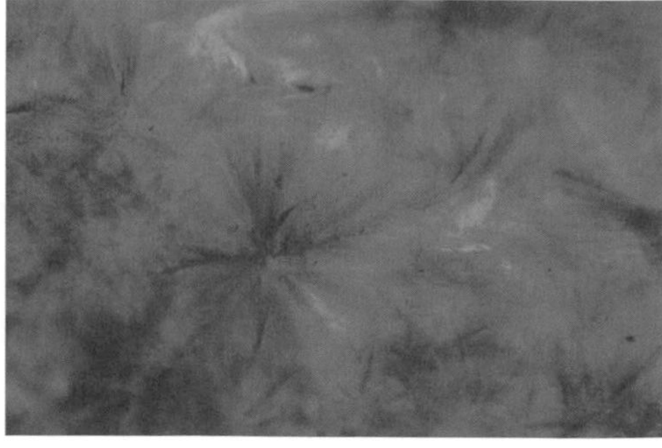

Plate 48 Devitrification in blue glass

Plate 49 Neolith imitation of turquoise

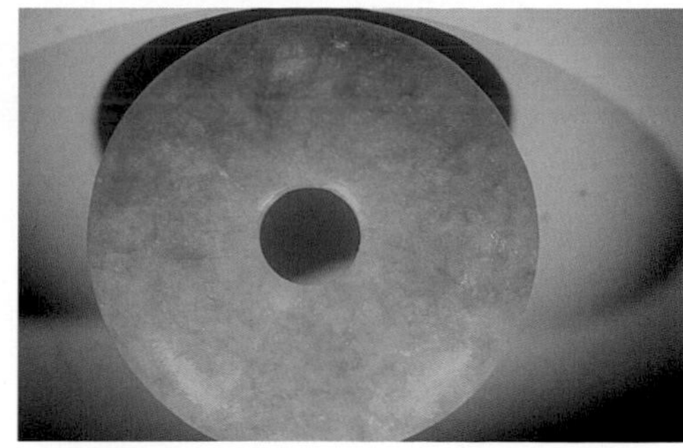

Plate 50 Stained jadeite

Plate 51 Silk and zoning in natural alexandrite

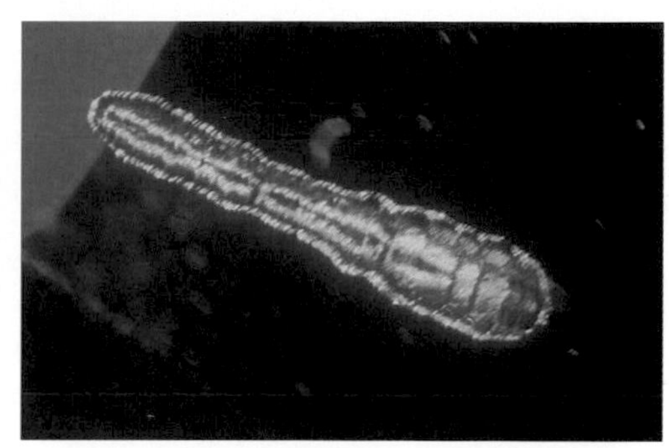

Plate 52 Profilated bubble in synthetic spinel

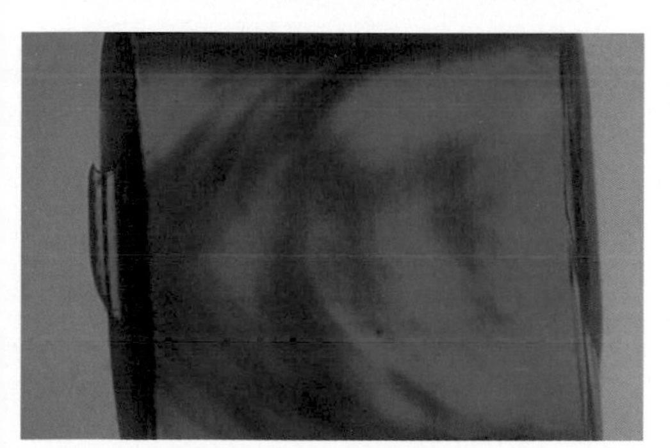

Plate 53 Kyocera blue sapphire-swirls

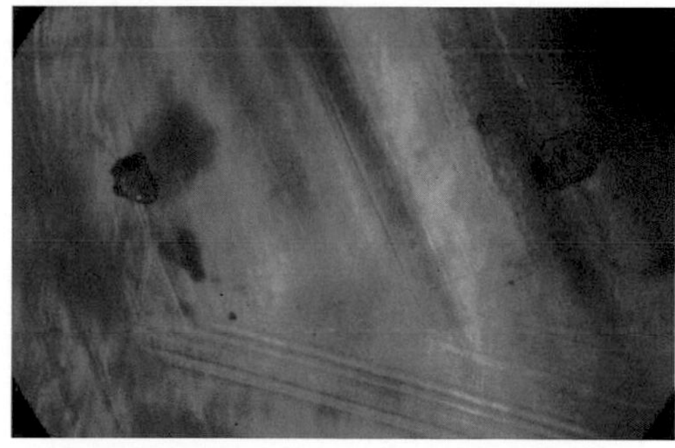

Plate 54 Zircon crystal in natural alexandrite

Plate 55 Stress cracks in moonstone

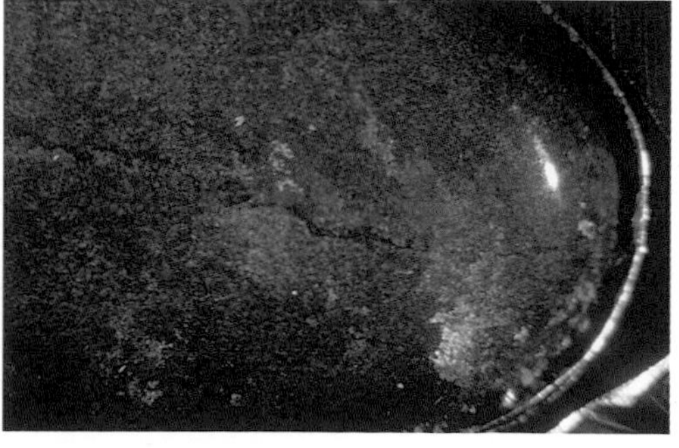

Plate 56 Treated opal (matrix)

Plate 57 Plastic-impregnated turquoise

Plate 58 Synthetic opal

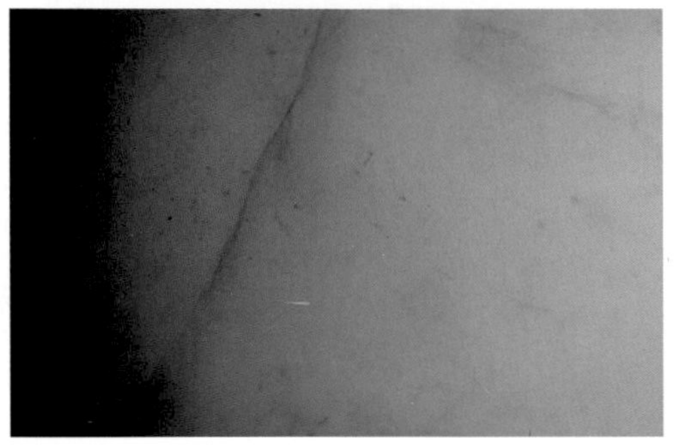

Plate 59 Stained quartzite

Plate 60 Treated colour (yellow faceted stone-corundum or diamond)

reclassified. More than one of its members may provide ornamental material: actinolite is usually green and fibrous, sometimes showing chatoyancy: nephrite is largely actinolite. Green transparent stones from Tanzania and Madagascar among other places have RI in the range 1.619–1.644 with SG 3.05 and a birefringence of about 0.026: pleochroic colours are yellow to dark green and a faint absorption line at 503 nm may be seen. It may be confused with other small dark green stones.

Amblygonite is an attractive transparent lemon yellow – mineralogists will usually hold that most faceted amblygonite is the mineral montebrasite which contains more hydroxyl than fluorine (amblygonite is the other way round). The RI range extends from 1.578 to 1.612 with a double refraction of 0.021 and an SG of about 3.1. It is about 5.5–6 in hardness. Confusion could most easily arise with yellow orthoclase or brazilianite.

An unexpected soft transparent colourless mineral that nevertheless appears from time to time in cut form at gem shows is *analcime*, a member of the zeolite group of minerals. Crystals which show attractive shapes and faces are seen quite often, faceted stones far less as the hardness is only 5.5. There is no cleavage, however, so cutting is possible. RI usually near 1.48, SG near 2.25.

Andalusite may well be taken for alexandrite by those unfamiliar with either species. This is because andalusite, according to direction of viewing, shows a flesh red or a slightly dull green. These colours are nothing like those of alexandrite (in which they do not change with direction but with the nature of the illuminating light). There is a bright green variety of andalusite, coloured by traces of manganese, which gives an absorption spectrum with two groups of fine lines in the yellow and in the blue.

Anglesite when cut (this is not easy due to its softness and easy cleavage) may resemble diamond in its fire but will show yellow fluorescence under both types of UV, not shown by diamond. It will not be encountered outside collections of rarities or at the major gem and mineral shows. RI is too high at over 1.8 for the gemmological refractometer but the high birefringence of 0.017 distinguishes it from diamond and can be seen with the lens.

The name *apatite* is given to a mineral group but collectors will be particularly interested in colour varieties of fluorapatite. Particularly fine are purple specimens from Maine, USA, and yellow crystals from Mexico, the yellow ones often showing a rare earth absorption spectrum with two groups of fine lines, in the yellow and in the blue. These are convenient since yellow apatite is quite like a number of transparent, quiet, yellow materials such as orthoclase and brazilianite. Crystals of the yellow material are easy to obtain and show good crystal form.

Transparent blue apatite has been found in a very bright colour which very closely resembles the Paraíba tourmalines and all stones of this startling colour should be very carefully tested, especially as the physical and optical values are close. Apatite's birefringence is lower than that of tourmaline, however.

Axinite crystals show very strong attractive pleochroism of cinnamon brown, violet, olive green or colourless. Faceted stones turn up at shows but the wedge-shaped crystals are highly characteristic and plentiful enough for the serious collector to come across once in a while. It would be hard to mistake axinite for any other gemstone. Mean RI about 1.70 with DR 0.012, SG 3.30. May show a strong absorption line at 512 nm with others at 492, 466 and 415 nm. The hardness is 6.5–7.

Azurite is found with malachite in intimate massive association but rare transparent crystals from Tsumeb, Namibia, are a superb blue and keenly sought by collectors.

Barite occasionally appears as small, pale yellow, transparent faceted stones which are remarkable for their high specific gravity of about 4.5. Yellow–brown cabochons are fairly common, mostly of French or English origin. The mean RI is 1.637 and DR 0.012 but as most barite encountered will be opaque these values will not easily be obtainable. As much cabochon material comes from England collectors should expect to find examples at British gem and mineral shows. The perfect cleavage makes faceting difficult and those stones which do exist are quite rare for this reason. The hardness is 3–3.5.

Transparent blue *benitoite* when pale could be mistaken for blue diamond as it shows considerable fire. However, its large double refraction of 0.047 makes it easy to see doubling of inclusions and of back facet edges, effects impossible in diamond. Benitoite fluoresces intense blue under SWUV light, again unlike diamond. Beware of anyone offering a large faceted benitoite as most examples are around 1–2 ct. Benitoite crystals are unique and will be sought by crystal collectors. There is only one known locality for this mineral (so far), in California. The hardness is 6–6.5.

Beryllonite occurs in colourless transparent crystals which have occasionally been faceted though it cannot be distinguished from a number of other colourless stones by eye alone. All faceted examples are small and the mineral is rare so cut stones are sought by collectors. The mean refractive index is about 1.55 with a DR of 0.009 and the specific gravity about 2.82. Examples have been reported to give a dark blue fluorescence under X-rays, with phosphorescence, features that will distinguish it from most other colourless stones. The hardness is 5.5–6.

Though crystals of *brazilianite* have been mentioned elsewhere, the large faceted stones can be mistaken for yellow orthoclase and other quiet yellow species. Remarkably discovered as recently as 1944 quite large specimens have been cut. The mean RI of 1.621 is higher than that of orthoclase though the values may overlap with those of yellow apatite: neither brazilianite nor other stones of this type of yellow will show a rare earth absorption spectrum, however. The hardness is 5.5.

Calcite despite its easy and perfect cleavage has been cut into crystal balls (through which objects will be seen doubled if the appropriate direction is used) and even faceted. Most calcite is colourless but the pink *sphaerocobaltite* is very attractive. As carbonates, both minerals show a high birefringence when faceted and crystal fragments (the only examples you will want to test in this way) will effervesce with acids. Let us hope that the fragments tested are not those of a calcite crystal ball! No other gem mineral closely resembles calcite though dolomite (q.v.) may come close. The RI ranges from 1.486 to 1.740 with a birefringence of 0.172–0.190: the SG is 2.71–2.94. The hardness is 3.

While the oxide of tin, *cassiterite*, is found in England, material of faceting quality comes mostly from the Arauca mine in Bolivia. While colourless cassiterite shows even more fire than diamond the mineral has a high birefringence of 0.098, thus easily distinguishing it. The RI is too high to be measured on the gemmological refractometer and the SG is near 6.99. The hardness is 6–7, preventing the lapidary presenting an adamantine lustre and really sharp facet edges. The yellowish, sometimes brown tinge of some faceted stones may remind the observer of diamond.

The attractive blue, translucent, occasionally transparent and orange strontium sulphate *celestine* is found in south-west England so at British gem shows there is a strong possibility of crystals turning up. Some material does get faceted though the cleavage could cause problems. Celestine is soft, 3.5 at best, with an RI in the range 1.622–1.635, DR 0.009–0.012. There may be a bluish-white fluorescence under SWUV light with phosphorescence. Neither the blue nor the orange material looks hard enough to be mistaken for better-known species.

Cerussite is one of the star attractions at mineral shows on account of the fine crystals from Tsumeb, Namibia. Some of them have been faceted to give colourless, transparent, highly dispersive gemstones but specimens are rare due to easy cleavage and brittleness. Nonetheless a faceted cerussite can look spectacular; the hardness is only 3.5 at best, the RI range reaches 2.07 and the SG is very high at 6.55, making specimens feel heavy. By eye alone confusion with diamond might be possible but careful examination in the appropriate direction will show a very high birefringence of 0.274.

The ornamental rock *charoite* is one of the easiest materials to identify by eye alone. It occurs as distinctively purple, banded masses of purple charoite, orange tinaksite and black aegirine and augite. The RI which can be taken on the base of cabochons is near 1.55 and the SG 2.68.

The transparent deep red *chondrodite* is exceptionally rare but is worth mentioning because of its close resemblance to the darker red garnets. It may also turn up when important gem and mineral collections are dispersed (though faceted stones would be found only in the most comprehensive cabinets). Found only in New York state, red transparent chondrodite has a refractive index in the range 1.512–1.646, well away from the garnets, and a specific gravity of 3.16–3.26. Separation from the garnets will also be easy since chondrodite shows a birefringence of around 0.030 where garnet shows none.

Chrysocolla produces a fine blue to green cabochon material which is frequently found mixed with turquoise and pseudomalachite to give a fine greenish-blue. Mixtures with silica are also common and when the silica content is high the hardness may reach 7 and the RI will be close to that of quartz at around 1.55. With low silica content the SG will normally fall in the range 2.00–2.45 and the RI in the range 1.575–1.635. The variety *Eilat stone* can look very like turquoise from which it can be distinguished by the lack of a copper absorption spectrum. It has an SG in the range 2.8–3.2 and will give a yellow stain when touched by a drop of hydrochloric acid – try some other test first.

In the larger gem shows fashioned scarlet-red *cinnabar* has been seen occasionally, especially since Chinese material became available. Fashioning could well be dangerous as cinnabar is a sulphide of mercury but it does turn up and certainly looks very attractive. The colour may fade slowly on prolonged exposure to light (not a recommended test) and the hardness is low at around 2.5. The SG of 8.09 should give a clue to its identity but the very high RI (maximum 3.25) rules out standard gemmological testing.

Clinohumite is related to chondrodite but is a bright orange–yellow rather than dark red; transparent gemstones have come from the former USSR over the past few years and have appeared at gem shows. The refractive index is in the range 1.586–1.614 with birefringence of 0.028, a specific gravity of 2.42 and a hardness of 4.5. Stones may show yellowish or greenish-white fluorescence and phosphorescence under SWUV. There can be a close resemblance to other orange stones, perhaps most closely to spessartine whose absorption spectrum will identify it.

Another bright orange stone, again a gem show rarity though crystals are more keenly sought, is the lead chromate *crocoite*. Faceted stones are always small but most attractive. However, as in other cases, crystals are so fine and well formed that they are unlikely to be cut. The hardness is 3.5–4, the RI in the range 2.29–2.66 (above the limit of the gemmological refractometer) and the specific gravity about 6. A strong absorption band can be seen at 555 nm which should distinguish crocoite from other bright orange stones (but it is one of the brightest).

Fine transparent dark red *cuprite* is also sought for its crystals (from Namibia) from which some faceted stones have been cut. These are too dark and metallic a red to be one of the garnets: the surface may suffer alteration over time and exposure to light so that it may need repolishing. A very fine synthetic cuprite has been made but specimens are very rare. The hardness is 3–4 and the RI 2.848, SG 6.14. Copper is the colouring agent.

Danburite, which has a hardness of 7, RI in the range 1.630–1.636 with a DR 0.006 and which fluoresces a sky blue under LWUV light, may also show a faint rare earth absorption spectrum. The SG is near to 3.00. Most danburite is colourless but a fine pale yellow is found in Madagascar and in recent years a bright orange variety has come from Russia: some of this may have been irradiated. Danburite crystals are described in Chapter 31 dealing with gemstones in the rough but they are favourites with gemmological examiners and not too difficult to obtain. There is some resemblance to colourless topaz though the SG is much lower and the yellow stones closely resemble yellow orthoclase but lack the iron absorption spectrum.

Datolite is colourless and not very often faceted: some specimens may give a blue fluorescence under SWUV but this would not be a diagnostic test. The RI is in the range 1.622–1.670 with a DR of 0.045 and SG about 3. The hardness is 5–5.5. New Jersey has provided some pale green crystals suitable for faceting. *Gems & Gemology* for Fall 2001 reports a 13 ct yellowish-green faceted specimen.

In recent years fine, large, transparent crystals of *diaspore* have been produced in Turkey and some of them have taken the gem shows by surprise when faceted. They are a transparent pale yellow but there are reports of stones showing a colour change from light greenish-brown in daylight to a pinkish-brown under incandescent light. Diaspore looks nothing like alexandrite which is usually dark and does not in fact resemble any other gemstone, the appearance being quite distinctive. The hardness is 6.5–7, the RI in the range 1.702–1.750 with DR of 0.048 and SG about 3.4. Some of the colour-change stones have shown a fine sharp absorption line in the red near 701 nm with other lines at 471, 463 and 454 nm. They have an RI in the range 1.702–1.750 with DR about 0.048 but a higher SG than non-colour-change material, at 3.39. Some of the Turkish diaspore shows three absorption bands in the blue in similar positions to those found in green and blue sapphire.

The mineral *diopside* may provide black cabochons with four-rayed stars, chatoyant stones, dark green or bright chrome-green faceted stones, the range of greens being extensive. The mineral has a hardness of 5.5–6.5 and RI in the range 1.664–1.721 with DR maximum 0.031. The SG will fall in the range

3.22–3.38, higher values corresponding with an increase in the iron content and a corresponding darkening of the green.

The paler green stones show absorption bands at 505, 493 and 446 nm and the chrome-rich (emerald-green) ones absorb in the blue at 508, 506 and 490 nm with a doublet at 690 nm and woolly bands at 670, 655 and 635 nm. In enstatite which closely resembles diopside there is only one line in the blue, at 506 nm.

No other green stone looks really like diopside: the darker peridot has a slightly lower RI and higher DR with a lower SG. Reports of a synthetic chrome diopside grown in Russia never seemed to lead to anything.

Yellow transparent diopside has been found in Burma which also provides green cat's-eyes with a very sharp eye. Chrome diopside is found with and in diamond, together with green enstatite, thus causing some confusion which the spectroscope can sometimes resolve, as described above.

While the mineral shows a range of fluorescent effects most of the iron-bearing green stones show no response to UV.

Dioptase is best known for its groups of very bright emerald-green crystals which are not often cut and even when they are produce very small faceted stones. Added to that there is an easy and perfect cleavage in one direction. Stones have a hardness of 5, RI in the range 1.644–1.709 with DR 0.053. The spectroscope will show an absorption band centred at 550 nm with strong absorption of the blue and violet. The colour is too bright for emerald (which has no significant cleavage) and the RI and SG are higher.

Dolomite forms rhombs rather like those of calcite, sometimes with curved faces and has one direction of perfect cleavage. The hardness is 3.5–4, varying with direction, and the RI is in the range 1.500–1.703 with DR 0.179–0.185. These figures are typical of carbonates and dolomite, like calcite, shows a variety of fluorescent effects. Fine transparent material from Egua, Spain, is sometimes faceted but the high DR means that only calcite could be mistaken for dolomite.

Dumortierite is not a very well-known gemstone but it gives a hard blue or violet material well suited to ornamental use as well as the occasional transparent green or blue faceted stones which are usually small – these are among the rarer gemstones. Dumortierite has a hardness of between 7 and 8, a refractive index in the range 1.668–1.688 with a birefringence of 0.15–0.37 and a specific gravity of around 3.34. Some fluorescence effects have been reported but may not be sufficiently consistent to be reliable tests for every specimen encountered. Dumortierite is not likely to be mistaken for another gem species (nor any other for it) but the small stones are attractive and the hardness makes it adaptable for any jewellery use.

Prominent among the rarer gemstones is the usually radioactive *ekanite* which occurs as a very dark, rather sombre but quite pleasing green. An emerald-green variety has been reported to show a refractive index of 1.590–1.597 (light brown material with this reading is also reported but ekanite usually falls in the range 1.572–1.573, with a specific gravity of 3.28–3.32 and a hardness of 5–6.5).

Properties vary as the mineral is undergoing continuous crystal lattice breakdown, so that 0.001 is about the highest possible reading for birefringence. The green is quite like that of a very dark green sapphire or tourmaline but has other elements in it which make it quite distinctive. The radioactivity can be quite high and specimens acquired should be tested by a gemmological laboratory – collectors have been known to keep their specimens in lead foil.

Enstatite may occur in bright green chrome-bearing crystals from which beautiful, small, transparent faceted stones may be fashioned but it is found more commonly in yellow, green, colourless and brown transparent crystals most of which will show a diagnostic absorption line at 506 nm, often accompanied by another at 547.5 nm. All the enstatite varieties appear to show pleochroic colours of very pale pink to green: four- and six-rayed stars can be found in some Indian material. Enstatite has a hardness of 5–6 and a refractive index in the range 1.650–1.680 with a birefringence of 0.008 and a specific gravity 3.20–3.30. A yellowish-green enstatite from Kenya had RI 1.652 and 1.622 with DR 0.010 and the stone was found to contain both iron and chromium. A colourless enstatite from Sri Lanka had RI 1.658–1.668 with a strong 506 nm absorption band with other fainter bands – the stone showed a faint brownish-pink in one of the directions of viewing. The specimen was found to be rich in magnesium.

Casual observation might have missed the pleochroism which you might not expect to find in an apparently colourless specimen.

The mineral *epidote* gives some beautiful dark green faceted stones with a colour which has been compared to that of a pistachio nut. Epidote like many gem minerals belongs to a mineral group rather than occurring as an isolated species – the zoisite which when heated gives the blue variety tanzanite is a member of the epidote group. The mineral *clinozoisite* is also a member of the epidote group and produces pale yellow or dark green stones. The hardness is around 6–7 and there is a perfect cleavage in one direction. Epidote has an RI in the range 1.715–1.797 with a DR 0.015–0.049 and specific gravity 3.38–3.49. Clinozoisite has RI in the range 1.670–1.734 with DR 0.005–0.015, SG 3.21–3.38.

Epidote group members often show a strong absorption line at 450 nm but this is direction-sensitive and specimens should be checked in more than one direction. Epidote shows pleochroism of colourless, pale yellow, yellow–green/greenish-yellow/yellow–green but no pleochroism can be observed in clinozoisite. None of the group's ornamental species shows any luminescence. Chromium-bearing epidote forms the variety *tawmawite*, a fine chrome-green rock found in Burma, and another ornamental rock, *unakite*, is composed of pink feldspar and green epidote. This is often seen at displays and can be used very successfully in carvings. One red transparent epidote variety is reported from Norway: this has RI 1.715, DR 0.016 and SG 3.32.

Very fine sapphire-blue *euclase* with a close resemblance to blue sapphire used to be found at the now-closed Miami mine in Zimbabwe. Stone sizes are always small and crystals are the more keenly sought (at very high prices) as the very easy cleavage makes faceting difficult. Euclase from Brazil produces green and blue stones with an altogether quieter colour.

The best blue stones are small and can be distinguished from blue sapphire most easily by an interesting absorption spectrum with no bands in the visible – but with an apparently unique transparency region between 500 and 400 nm. Some other blue or greenish-blue euclase has been found to show a chromium absorption doublet in the red at 705 nm with two vague bands in the blue.

Euclase has a hardness of 6.5–7.5 (this may vary within an individual crystal) and RI in the range 1.650–1.676 with a DR of 0.020 and SG 3.10. The colour of the quieter stones could be confused with that of other gem species but RI tests are not too difficult with euclase.

Very fine, small, transparent, rose-red stones of the uncommon mineral *eudialyte* have a hardness of 5–5.5, RI in the range 1.597–1.600 with DR 0.004 and SG 2.8–3.0. The colour is probably due to manganese and absorption bands have been reported at 870, 800 and 745 nm (outside the visible spectrum). The ornamental specimens of this species come from Kippaw, Quebec, Canada, and will therefore be found at North American gem shows. The generally small size of faceted material should alert the collector to the mineral's rarity.

Fluorite, which gets a mention elsewhere in the book, forms large crystals, often interpenetrating and in a great range of colours. Faceted stones are cut with some difficulty as fluorite has an easy octahedral cleavage (four directions), is soft (4) and has a notably low single refractive index of 1.43. The specific gravity is 3.18. While the different colours could imitate other stones fluorite is easily tested. Rainbow-like markings inside a stone indicate incipient cleavage and care should be used.

Rare, attractive and of particular interest in North America is the pink to orange–red translucent to opaque *friedelite* which, if fashioned at all, will most likely be found as cabochons. The interest arises from friedelite's presence at the renowned mineral locality of Franklin, New Jersey, which constantly provides material for shows. Friedelite may show reddish fluorescence under both types of UV and has a hardness of 405, RI in the range 1.654–1.629, DR 0.030 and SG 3.04–3.07. This is a beautiful material and could be taken for cornelian when sufficient brown is present – though this is unlikely as friedelite is rather scarcer!

Madagascar, which at the time of writing (early 2002) seems to be producing fine specimens of most gem species, is also the home of *grandidierite* which, in a distinctive and unusual quiet green, is found mostly in cabochon-quality material though recently some small transparent faceted gemstones have appeared on the collectors' market. The hardness is 7.5, the RI in the range 1.590–1.639 with a DR of 0.037 and specific gravity 2.97–3.00. Stones show a strong pleochroism of dark blue–green, colourless and dark green. There seems to be no likelihood of confusion with any other species.

Gypsum should be regarded rather as an ornamental mineral than a gemstone in the usual sense as it forms large masses from which pieces of considerable size can be fashioned. The three directions of cleavage assist neither the carver nor the lapidary and the hardness is the standard 2 on Mohs' scale. The name alabaster is often used for gypsum which has an RI in the range 1.520–1.530 with a DR of 0.010 and an SG of 2.32. Cabochons can be made from the massive, granular material while the name selenite has been used to describe (mainly) the crystals, which show a moonstone-like glow. Specimens may be mistaken for moonstone of one sort or another but gypsum is far softer and will show signs of wear; should an SG test be thought appropriate moonstones will be found to be denser.

Those collecting the beryllium minerals which can be fashioned into gemstones will not wish to omit *hambergite* which is sometimes cut into small transparent colourless stones in which the high birefringence of 0.072 will show back facets and inclusions doubled. Hambergite's hardness is 7.5 and it has an easy cleavage. The RI is in the range 1.55–1.63 and the SG is 2.35.

Beautiful, very bright, transparent blue crystals of *hauyne* from the Eifel region of Germany are very popular with collectors – the colour is like no other but stones are invariably small. They could be worth imitating perhaps but glass would be the only convincing medium. The hardness is 5.5–6 and the RI 1.496–1.505, SG 2.44–2.50. Hauyne is one of the constituent minerals of lapis-lazuli. A blue GGG reported in 1994 by GIA closely resembled the blue of hauyne but its higher SG and RI would easily make the distinction though they could not readily be tested.

Hematite is a black metallic mineral which is surprisingly often faceted as well as being fashioned as cabochons. Fine crystals come from northern England and while most are sought uncut by collectors some material undoubtedly does appear in jewellery. Hematite is heavy and unsuitable for such items as necklaces but is fine for seal rings. Though appearing black hematite is really a very dark red as its powder proves on a streak plate when a piece is drawn across it (rough material only).

Hematite is imitated by hematine which is a mixture of stainless steel and chromium and nickel sulphides. It is attracted to a magnet where hematite is not. A mixture of granular hematite with magnetite and other materials with an overall iron content of 69 per cent has a bluish-grey colour and gives a dark brown streak and there is some magnetism – it has an SG of 2.33 compared to hematite's 5.2–5.3. Hematite has very high optical properties which cannot be measured by gemmological techniques.

Howlite is best known for acting as a turquoise imitation when dyed. It usually forms colourless nodular masses with a hardness of around 3.5, RI in the range 1.583–1.605 and DR (not often measured) 0.022. Some howlite may fluoresce a brownish-yellow in SWUV and some material from California fluoresces a deep orange in LWUV.

The gemstone *iolite* (mineralogists use the name cordierite) at its finest is a fine dark transparent blue–violet with spectacular pleochroism showing in succession the blue–violet, a less pronounded blue and a pale straw yellow. Cut into cubes iolite makes an attractive object and the mineral is hard enough, at 7–7.5, for most jewellery uses. The RI is in the range 1.522–1.578 with DR 0.005–0.018 and SG 2.57–2.61. Absorption bands may be seen at 645, 593, 585, 535, 492, 456, 436 and 426 nm. Chatoyant material and some near colourless stones have been found in Sri Lanka. The variety 'bloodshot iolite' contains attractive red platelets of hematite.

The pleochroism may suggest tanzanite but the pleochroic colours are not the same and tanzanite has a much higher RI in the range 1.685–1.725.

Kornerupine until the 1970s was known only as a dark brownish-green to greenish-yellow mineral but then some vanadium-bearing crystals were found in Kenya and though specimens soon vanished from the collectors' markets their bright green colour made them especially sought after. Cat's-eyes can also be found in kornerupine. The mineral most commonly has a hardness of 6.5 and RI in the range 1.665–1.682 with DR 0.013. There is strong pleochroism of green, yellow and reddish-brown. Most kornerupines show an absorption band at 503 nm.

Burma has provided some star stones (rare) and some blue specimens are reported from Madagascar: the vanadium kornerupine from Kenya is reported to have RI in the range 1.662–1.675 with DR 0.012–0.013. Some East African material shows strong pleochroism with an intense emerald green in one direction when viewed by polarized light – the body colour of the stones is blue–green. Some Burmese stones are said to fluoresce yellow under both types of UV and a similar effect is reported from the East African material.

Kornerupine has always attracted the attention of collectors and the dark green stones from Sri Lanka can often be recognized among other dark green faceted specimens. It could perhaps be confused with very dark green tourmaline but kornerupine's brown tint is usually evident.

As this text is being written in January 2002 I (MO'D) have just been shown a parcel of dark blue faceted oval *kyanites* reported to have come from India. Most faceted kyanite, which is rare (due to the difficulty of cutting a material with a bladed structure and perfect cleavage in one direction), is blue or green, the latter often showing a stripe of blue. While some of the kyanite that has been around for years does show areas of a fine blue, stones coloured entirely blue are new to the market. Kyanite has a directional hardness of 7.5 and 4, across and parallel to the crystal length respectively. The RI is in the range 1.712–1.734 with a DR of 0.017 and the SG is near 3.68.

Stones are strongly pleochroic showing violet, colourless and cobalt blue according to direction. Some bluish-green material from East Africa contains some chromium and shows an absorption band at 710 nm in the deep red with two others in the deep blue with a dark edge near 600 nm. There may be a faint red fluorescence from these stones. The occasional colour-change stone has been reported with one specimen showing reddish in incandescent light and blue–green in daylight. An emerald-green kyanite from Tanzania had RI 1.714–1.732 and SG 3.68.

This is an interesting gemstone, all the varieties showing unusual properties; it has the same chemical composition as andalusite and sillimanite and is said to be, with them, a polymorph (one of several forms) of aluminium silicate.

Until recently *lazulite* (easily confused with lazurite, a completely different mineral) was known as a slightly obscure blue transparent collectors' stone with a hardness of 5.5, RI in the range 1.608–1.662, DR 0.031–0.038 and SG near 3.08.

Stones show a strong pleochroism of colourless and dark blue. These stones are usually small, Brazil providing attractive material. However, lazulite stepped forward into the ranks of the highly collectable stones when Madagascar began to produce brilliant blue crystals very close in colour to the celebrated Paraíba tourmaline of Brazil, from which lazulite can be distinguished by its lower SG. Transparent crystals of a dark blue lazulite are reported from the Skardu area of Pakistan: crystals from the Nanga Parbat area not too far away have an RI between 1.61 and 1.64, SG 3.12, occurring in a fine deep green as well as yellow to blue green with strong pleochroism – specimens are generally poor in iron or iron-free.

Malachite is easy to recognize from its highly characteristic dark green banding and its frequent admixture with blue azurite. There is little need for it to be tested – the synthetic malachite referred to elsewhere in the book did not flood the market.

Marcasite the mineral is closely related chemically to pyrite but the marcasite of the jeweller *is* pyrite! Pyrite is dealt with later in this chapter.

Odontolite is not a mineral but the fossilized bones or teeth of extinct animals with the organic matter replaced by minerals. It resembles turquoise and may be used to simulate it. Properties vary but are often close to those of apatite which is often the major constituent. Calcite and vivianite are often present, the latter contributing towards the colour. Good commercial quality blue odontolite has RI 1.57–1.63 overlapping with that of turquoise: the SG is near 3.00 (turquoise is just below that figure). Odontolite will not show the copper absorption spectrum of turquoise.

Pectolite is not well known as an ornamental mineral apart from the attractive blue-and-white variety known locally as Larimar which is found in the Dominican Republic. The hardness is 4.5–5 and RI in the range 1.595–1.645 with DR 0.036 though specimens likely to be encountered will be translucent rather than transparent. The best Larimar is a dark blue and translucent. Nothing quite resembles it enough to make testing a serious matter.

Faceted, colourless, transparent *petalite* has a notably bright glassy appearance and is found in many collections. Crystals are large enough to provide quite large faceted stones which have a hardness just over 6. The RI is in the range 1.516–1.523, DR 0.013, SG about 2.4. Some stones but perhaps not the majority have been found to give a weak orange fluorescence, one such specimen giving an absorption band at 454 nm. This stone showed bright orange under X-rays with persistent phosphorescence. Some petalite is pale yellow and a pink chatoyant specimen has been reported from South Africa. This stone had RI 1.51 and SG about 2.34. Petalite is not a difficult species to identify as other transparent colourless stones seem to lack the same clarity.

Phenakite is another transparent colourless material: faceted stones are not too rare and have a hardness of 7.5–8 and RI 1.654–1.670. The DR is 0.016. Some specimens may give a greenish fluorescence under UV: a violet colour under X-rays has been reported. Coloured material does sometimes appear: a greenish-blue faceted stone with RI 1.654–1.670 and SG 3.00 was noted in 1963, the specimen giving a light blue fluorescence under UV and also showed pronounced pleochroism with peacock-blue and violet-red colours. A yellow variety was reported from the former USSR. Phenakite has also been synthesized experimentally and an example is described elsewhere in the book. Phenakite apart from the unusual coloured stones should not be difficult to test.

Phosgenite crystals are sometimes seen at gem shows though faceted stones are rare, partly due to the distinct cleavage. On the other hand the transparent colourless stones show very high RI at 2.114–2.118 and DR 0.028 which makes them very bright with an adamantine lustre though the hardness is only 2–3. Phosgenite shows a strong yellowish fluorescence under UV and X-rays and some of the thicker crystals have been found to show very weak red–green pleochroism. Fine yellow–brown material is known and also pale shades of green, pink or yellow. Phosgenite is especially likely to be seen on the stands of fluorescent mineral specialists.

The zinc phosphate *phosphophyllite* is very rare as a faceted stone and quite rare as cabochons – this is because the beautiful Bolivian crystals are usually better sold as such. Phosphophyllite is a very attractive blue–green with a hardness of 3–3.5, RI in the range 1.595–1.621, DR 0.021–0.033, SG 3.08–3.13. There is a perfect cleavage in one direction which makes fashioning difficult.

Pollucite is transparent and colourless, being remarkable to gemmologists as one of the few colourless gemstones with a single refractive index (1.52) between 1.50 and 1.70 (the other is rhodizite, q.v.). The hardness is 6.5–7.5 and the specific gravity 2.85–2.94. This is a keenly collected gem species despite its unassuming appearance.

Prehnite makes fine cabochons, the most attractive colour being a yellow–brown while dark green, yellow and yellowish colours are also found. It may just be mistaken for jadeite but has lower RI and SG, in the range 1.611–1.665 and 2.88–2.94. A certain oiliness adds to its attraction for many collectors and cat's-eye stones have occasionally been reported. A radiating fibrous structure can usually be detected. Some yellow prehnite from Australia has been faceted.

The mineral *prosopite* may give a blue very like that of turquoise with which it has sometimes been confused. The RI is lower, at around 1.51 (compared to the 1.62 of turquoise), the SG is 2.88 and the hardness 4.5. As the best prosopite comes from Mexico it is likely to be around at American mineral and gem shows along with turquoise.

Pyrite like marcasite (q.v.) is a sulphide of iron, highly collectable (but hard to preserve): it is also the 'marcasite' of the jeweller though marcasite is a valid mineral species in its own right. Pyrite crystals with their familiar brassy-yellow colour and grooved faces are at every mineral show and in every collection. Pyrite has a hardness of 6–6.5 and an SG of 5.0–5.3. The streak is greenish-black. Specimens cut in the rose style with a flat back are often seen – cutting is in fact not too easy as pyrite is both brittle and heat-sensitive.

Rhodizite, like pollucite (q.v.) is one of the only two colourless transparent gemstones with a single refractive index between 1.50 and 1.70: that of rhodizite is 1.694, the specific gravity is 3.44 and the hardness 8.5. Rhodizite may fluoresce a weak yellow in SWUV and a stronger yellowish-green, with phosphorescence, under X-rays. This, like pollucite, is a stone sought by gemmologist collectors for its interesting optical properties (see pollucite, above) but it is also attractive and bright.

While *rutile* is best known to gemmologists from the interesting and useful inclusions it forms in many gem minerals, it can itself be fashioned into small red to dark brown faceted stones with a hardness of 6–6.5 and SG 4.2–5.6. While the very high RI of over 2.90 is too high for gemmological testing, the strong DR of 0.287 shows doubling of inclusions and back facets (if you can see into the sometimes very dark stones). The synthetic form, nearly colourless, shows a very high dispersion too but the effect is masked in the natural material by the strong body colour.

The mineral *scapolite* includes several very attractive and collectable varieties: beautiful pink stones, some with the cat's-eye effect, fine deep violet and purple, citrine-like yellow and an opaque sulphur-bearing variety with a very strong yellow. There is a chance of the pink scapolite being mistaken for rose quartz and the yellow scapolite for citrine, the distinction resting upon the correct interpretation of the behaviour of shadow edges on the refractometer where quartz gives a positive and scapolite a negative reading – this is explained in gemmology textbooks. Scapolite has a hardness of 6 and RI in the overall range (varying with colour) 1.541–1.567. Details of the variations with colour are given on the website.

The DR is in the range 0.005–0.015, again varying with colour, and the SG in the range 2.50–2.74. There is a distinct cleavage in two directions so stones are not very easy to cut.

Pink and violet scapolites may show elements of the chromium absorption spectrum with strong absorption of the yellow and pleochroism is generally well marked with pink and violet stones showing light to dark colours: yellow stones show colourless and yellow.

Scapolite shows a number of fluorescent effects with some Burmese stones showing yellow to orange in LWUV and pink under SW. Some Tanzanian yellow stones fluoresce strong yellow in LW. The opaque sulphur-bearing yellow stones from Ontario, Canada, fluoresce and phosphoresce a very strong yellow in LW: some yellow stones have been found to fluoresce lilac in SW and a strong orange in X-rays. The whole picture of scapolite fluorescence and phosphorescence is perhaps too complicated to be relied upon for identification of unknown specimens which can usually be diagnosed by using standard gemmological tests.

Scheelite is an ore of tungsten and provides bright yellow crystals from which very attractive stones can be cut. The stones show a high DR of 0.016 with RI in the range (unreachable by the gemmological refractometer) 1.918–1.937. The SG is 5.9–6.3 and the hardness 4.5–5. Scheelite may show a rare earth absorption spectrum with groups of fine lines in the yellow and in the green and, most importantly, fluoresces a bright bluish-white under SWUV. When colourless material is faceted the high dispersion gives the gem an extra dimension. Purchasers should beware of synthetic scheelite with the same composition and properties.

Serandite has very great gem potential since it is a magnificent transparent rose to orange–red. Facetable crystals have so far been found only at the celebrated mineral location of Mont St Hilaire, Quebec, Canada, and since that area is a Mecca for mineral collectors gem quality material will be greatly sought after. Serandite has a hardness of 4.5–5 and RI in the range 1.660–1.688 with DR 0.028 and SG 3.32. There is a perfect cleavage in one direction so that cutting so soft a stone raises problems for the lapidary and it is not surprising that few faceted stones are around. The orange–pink colour could be confused with that of other gemstones but gemmological tests should establish the distinction.

The *serpentine* minerals (named from the serpent-like markings in the marble varieties) form part of the kaolinite–serpentine group and in the ornamental context include the williamsite and bowenite varieties of antigorite. *Bowenite* (which is often suggested as a jade simulant) may be yellow, yellow–green, blue–green or dark green and is translucent. It has a hardness of 4–6, RI around 1.56 with DR 0.014 and SG 2.58–2.62. Some of the fine dark green bowenite from New Zealand has an SG of 2.67. There is a good chance that the best dark green translucent material may be mistaken for nephrite which has a higher SG of around 3.00. Absorption bands may be seen at 492 and 464 nm but they are not diagnostic on their own.

Williamsite is a notably translucent apple green and may fluoresce a weak green in LWUV. The hardness is about 4.5, the RI about 1.56 and SG 2.62. The material known as Connemara marble has greenish-white banding with an RI 2.56 and SG 2.48–2.77, the range being due to a mixture of serpentine and carbonate species. The name Styrian jade has unfortunately long been given to an aluminous serpentine with H 3.5, RI 1.57 and SG 2.69, figures well below those for either of the jade minerals. A faceted yellowish-green antigorite from Pakistan showed RI 1.559–1.561 with DR 0.001–0.002.

The iron carbonate *siderite* is common as a mineral but is rarely cut. Faceted stones have a hardness of 3.5–4.5 and RI 1.633–1.873 with DR 0.240. The colour is quite an attractive brown and will appeal to those collectors (including myself) who appreciate this colour.

Sillimanite, with andalusite and kyanite (q.v.) is a polymorph of aluminium silicate and despite its exceptionally easy cleavage gets itself cut into what are rather attractive transparent pale blues to blue–greens and sometimes into pale yellow or pale brown faceted stones. The easy cleavage is no idle story as I (MO'D) once found in the laboratory a stone paper containing part of a faceted light blue sillimanite with many similarly shaped very thin cleavage sections beside it. Such a picture was worth a good deal of text! The hardness is 6.5–7.5 and the RI in the range 1.654–1.683 with DR 0.020 and SG of 3.23–3.27. Pleochroic colours of pale brown, pale yellow to green; greenish or brownish; dark brown or blue have been reported and some dark blue specimens from Burma show a weak reddish fluorescence. In some respects sillimanite could be confused on sight alone with tourmaline of similar appearance but tourmaline's pleochroism is normally two shades of the same colour.

Sinhalite's identification is still remembered by older gemmologists. The attractive transparent green to golden or dark brown stones had been assumed to be olivine but tests in the early 1950s identified it as a borate rather than a silicate. Sinhalite has a hardness of 6.5–7 and RI in the range 1.665–1.712 with DR 0.038 and SG 3.47–3.50. The absorption spectrum shows bands similar to those seen in peridot – 493, 475, 452 nm with the additional one placed at 463 nm – there is general absorption of the violet. The absorption spectrum is enough to distinguish it from varieties of tourmaline or chrysoberyl which can resemble sinhalite.

The zinc carbonate *smithsonite* provides fine green to light blue, sometimes pink ornamental material with a hardness of 4–4.5. As a carbonate it will effervesce with acids and faceted transparent stones show a high birefringence of 0.227 with RI 1.621–1.848. The SG is 4.3–4.5. A particularly fine green variety makes a good simulant of green jadeite and in fact the name Bonamite was coined (in the commercial context) for this particular material. Rare light-coloured transparent stones have a high dispersion of 0.037 but the softness precludes confusion with other highly dispersive species.

The name *soapstone* has traditionally been used for a very soft (hardness 1) material though it is not a valid mineral species but rather an impure talc rock. It can easily be scratched by the finger-nail but this has not prevented it from choice as a carving material. Colours vary from white through grey and green to a brownish-yellow and specimens feel soapy. The RI is in the range 1.539–1.600 though in practice a shadow-edge at about 1.54 would be all that is seen. The SG is in the range 2.20–2.83.

Sodalite while one of the constituent minerals of lapis-lazuli is also sometimes found in a translucent and more attractive blue than that shown by much of the massive material. The best translucent sodalite, from Namibia, has been faceted while the darker blue shades are used for carvings. The hardness is 5.5–6 and the single RI 1.48, SG 2.28 for the massive material. A transparent sulphur-rich sodalite to which the varietal name hackmanite has been given turns a raspberry red from its original colourless when exposed to SWUV, then fades back to colourless. The darker blues may be confused with iolite but sodalite shows no pleochroism.

Sphalerite, often known as zinc blende, shows very high dispersion when cut and this, combined with the bright yellow, green, orange and orange–red possible body colours makes a superb collectors' stone. However, fine large faceted sphalerite is rare since it is soft (3.5–4.5) and has six directions of cleavage. The single RI, between 2.47 and 2.43, is well above the working limits of the gemmological refractometer and the dispersion is 0.156 (above that of diamond, at 0.044). The SG is between 3.9 and 4.1. Absorption bands at 690, 667 and 651 nm have been ascribed to cadmium. Bright orange–red fluorescence under both types of UV enhances the value of sphalerite even more and stones at gem and mineral shows are keenly sought.

While the bright yellow–orange–red stones are not really likely to be confused with diamond, colourless to pale green stones, like those found at Franklin, New Jersey, could cause difficulties when faceted but there will almost certainly be signs of incipient cleavage within the stones.

Deep green sphalerite from Zaire was found to give an SG of 4.18 (another gave 3.98) with the red absorbed to 525 nm and a band in the yellow from 595 to 580 nm. The blue was absorbed from 470 nm.

Sphene (titanite) is very popular at American gem and mineral shows since much of the gem crystals come from nearby Mexico. The colours are yellow through green to brown and the high dispersion of 0.051 gives the otherwise quite soft (hardness 5.5) stones great life. The refractive index is high, in the range 1.843–2.110, with a birefringence of 0.105–0.135 and the specific gravity is 3.32–3.54. The rare chrome sphene is a magnificent emerald green but the high birefringence distinguishes it from emerald: specimens can be distinguished from demantoid garnet in the same way. Some stones may give a rare earth absorption spectrum with groups of fine lines in the yellow and in the green.

The transparent pink kunzite, the fine yellow, attractive green and the very rare emerald-green variety hiddenite make *spodumene* one of the most interesting gem minerals. Kunzite in particular forms very characteristic and collectable crystals. The cleavage in spodumene is very easy to start and kunzite is notably heat-sensitive so that the lapidary needs to take every care in faceting. The hardness is 6.5–7.5 and RI will be in the range 1.653–1.682 with DR 0.014–0.027 and SG near 3.18, more or less identical to that of tourmaline.

In fact pink tourmaline and kunzite can both be found in southern California and it is possible that in local shows some confusion could occasionally arise. The RI can easily distinguish the two when in faceted form. The yellow stones, some fine specimens coming from Afghanistan, have too high an RI for topaz, yellow orthoclase and beryl and too low for yellow sapphire. The normal green spodumene, often misnamed hiddenite (though this name should be reserved for the chromium-bearing variety), looks more like green beryl than any other species – distinction is easy.

True hiddenite is very rare, coming only from sites in North Carolina, USA, and showing a chromium absorption spectrum with a doublet at 690.5 and 686 nm with weaker lines in the orange and a broad absorption with its centre near 620 nm. Spodumene varieties show pronounced pleochroism; kunzite shows purple–violet to colourless, green crystals show green, blue–green and colourless to pale green.

Under LWUV kunzite gives a golden-pink to orange fluorescence with a weaker effect under SW and orange with phosphorescence under X-rays – this type of irradiation may well change the colour of the specimen to a blue–green though this colour may later disappear after exposure to strong light. Yellow–green stones fluoresce orange–yellow in LWUV, the same colour appearing though weaker in SW and stronger again in X-rays, this time with no alteration in body colour. Hiddenite fluoresces orange with phosphorescence under X-rays.

The ornamental mineral *stichtite*, which is made up of flexible purple laminae, is coloured by chromium and attractive examples can be found at gem and mineral shows. A purple ornamental mineral display could include stichtite and sugilite (q.v.) side by side. Stichtite has a hardness of 1.5–2.5, RI in the range 1.518–1.545 with DR 0.027 and SG 2.16–2.22. Pleochroic colours are dark red to light red and a chromium absorption spectrum may be seen.

Sugilite, which has been marketed under the name Royal Azel or Royal Lavulite, is a deep purple massive mineral which occasionally produces material translucent enough for faceting to be attempted – the colour certainly is spectacular in some of the faceted stones. The hardness is 6, the RI in the range 1.607–1.610 with SG 2.74.

The gemstone *taaffeite* just gets into this chapter because of its gemmological history and because it is not quite so rare as appeared at first. Specimens of gem size have been found only in Sri Lanka. Taaffeite is most commonly pale lilac with RI in the range 1.717–1.724 with DR (observation of which the mineral's distinction from spinel was first based) 0.004, easy to overlook under magnification with a hand lens. The SG is 3.60–3.61 and the hardness 8–8.5.

Tugtupite is known to all collectors of fluorescent minerals as it fluoresces orange under LWUV and salmon pink under SW. To the eye in daylight tugtupite (found so far only in Greenland) is a mottled rose red to white and only a few specimens are translucent enough to facet – those we have seen are most attractive. Tugtupite has an RI in the range 1.496–1.503 with DR 0.006. The SG is in the range 2.30–2.57 and the hardness about 6.5. Blue material has been reported. The name reindeer stone has been used.

Popularly known as television stone, specimens of white translucent to opaque *ulexite* may transmit an image along the parallel fibres of which it is formed and in some cases print can be read through the long direction of a crystal a few centimetres in length. The RI is in the range 1.496–1.519 with DR 0.023, SG 1.65–1.95 and hardness 1–2.5. Some specimens have shown a blue–green fluorescence with phosphorescence under SWUV.

Variscite is an attractive ornamental mineral despite its close resemblance to pale green to blue–green turquoise. The RI is usually about 1.56 (turquoise about 1.62) and the SG in the range 2.2–2.57 (turquoise 2.75–2.85). The hardness is 3.5–4.5. Some variscite may show faint green under both types of UV. It will not show the copper absorption spectrum of turquoise.

Verdite is the name given to an ornamental rock composed of green and brown mica minerals with, in material from Zimbabwe, ruby and albite. The hardness may reach 9 if corundum is the principal mineral: the lowest figure is 3 and the SG 2.70–2.87. One or two dark green specimens from the Transvaal have shown a faint chromium absorption specimen. Verdite is especially appropriate for some types of sculpture.

Vesuvianite is usually called idocrase by gemmologists. It is chemically close to grossular and has a notably low birefringence of 0.001 at its lowest. Transparent yellow to green specimens have RI in the range 1.712–1.721 and SG 3.32–3.47. Fine deep green specimens have been found at Asbestos, Quebec, Canada. A massive dark green opaque to translucent vesuvianite long known as californite looks very like jadeite but can be distinguished by its strong absorption band at 461 nm. Lighter transparent specimens may show a weak band at 528.5 nm.

Willemite gives so strong a yellow–green fluorescence and phosphorescence under SWUV light that it is a popular feature of gem and mineral shows. Cabochons and the rare faceted stones may be green or orange with RI

1.691–1.719, SG about 4.10 and hardness 5.5. There is a fairly strong absorption band at 421 nm with other, weaker bands. Willemite occurs with red zincite at Franklin, New Jersey, USA, and specimens in which both minerals occur, with calcite and other species, give a very beautiful variegated fluorescence.

Transparent *wulfenite* is a brilliant orange though faceted stones are almost always small: crystals are more desirable and so are rarely fashioned. The RI is 2.28–2.40 with DR 0.122 and a high dispersion of 0.203. The stones have an adamantine (diamond-like) lustre and are very beautiful. The SG is 6.5–7.0 and the hardness 2.5–3.

Faceted red *zincite* is always small but the red is bright and attractive. The RI is 2.013–2.029 with DR 0.016 and a high dispersion of 0.127. The SG is 5.68 and the hardness 4–4.5. The small size of fashioned specimens could cause confusion with South African or Colorado rhodochrosite.

Chapter 30

Rarely fashioned species

The stones briefly described in this chapter have been reported in fashioned form and are included in the hope that more specimens will enter the market in due course.

Agalmatolite in various pale colours is used in carvings, sometimes under the name figure stone. It is a variety of the mineral pyrophyllite which closely resembles soapstone. Like soapstone, agalmatolite has a low hardness of 1–2 and is sectile (which could assist the sculptor). The material occurs in a range of pale colours.

The name *anyolite* has been used for a bright green massive zoisite in which crystals or ruby are embedded together with black crystals of one of the amphibole group minerals. This material is found in Tanzania.

Green translucent *apophyllite* (mineral names will be fluor-apophyllite or hydroxyapophyllite) has a hardness of 4.5–5 with RI in the range 1.534–1.537. DR may be as low as 0.001, SG 2.3–2.5. Green faceted stones will usually be of Indian origin though a silvery-white variety is also known. While confusion with other stones is just possible, crystals are more desirable and will be found quite often at shows, sometimes under the heading of zeolites though apophyllite is not a zeolite.

White or pale yellow *augelite* has been found sufficiently transparent to facet despite its lack of hardness (4.5–5) and notable brittleness. As most if not all the faceted stones will have come from locations in California, shows held there may well display the odd specimen. The RI is in the range 1.574–1.590, DR 0.014–0.020. The SG is 2.69–2.75.

Gems & Gemology for Fall 2001 illustrated the first known faceted specimen of natural *baddeleyite*. The stone was a very dark greenish-brown with a metallic lustre and gave a high DR (figures not given for this nor for RI). Baddeleyite is the monoclinic form of zirconia, ZrO_2.

Small colourless faceted *boracite* has been obtained from Germany. Its hardness of 7–7.5 and lack of cleavage makes faceting possible and it may well feature in continental European gem shows. The RI is in the range 1.658–1.673 with DR 0.011 and SG 2.95.

Yellow to orange transparent *cancrinite* found in Ontario, Canada, is occasionally of faceting quality and small transparent stones have been cut. The hardness is 5–6, RI in the range 1.495–1.528, DR 0.022, SG 2.42–2.51.

We are not sure of the present availability of *chambersite* since recovery of facetable specimens depends upon divers retrieving crystals from a brine storage well! Nonetheless we have seen some faceted brownish to purple translucent specimens with a hardness of 7, RI 1.732–1.744, DR 0.012, SG 3.49.

Small green pebbles of the fibrous *chlorastrolite* variety of the mineral pumpellyite are occasionally fashioned into cabochons which are usually green and white but may show a fine even green. A pattern resembling that of turtleshell is popular. The hardness is 5–6, RI in the range 1.674–1.722: there is distinct pleochroism of colourless through pale green to brownish-yellow. The finest green may even give rise to thoughts about high quality jadeite.

The name *chloromelanite* is used for jadeite which is opaque dark green to black; surprisingly the colourless transparent mineral *colemanite* is sometimes faceted even through it is brittle, shows perfect cleavage and has a hardness of 4.5. Specimens have RI in the range 1.586–1.614 with DR 0.228 and SG 2.42. There may be some strong yellowish-white or greenish-white fluorescence and phosphorescence under SWUV.

The mineral *durangite* occasionally provides strongly pleochroic fashioned stones (two rays colourless the third orange–yellow, body colour orange) and as material comes from Mexico it is most likely to feature in Californian mineral shows. The hardness is 5 and the RI is in the range 1.662–1.712.

Attractive soft yellow to pink transparent stones are provided by *eosphorite* which has a hardness of 5, RI in the range 1.638–1.671, DR about 0.033, SG 3.05. A strong absorption line at 410 nm accompanied by a weaker one at 490 nm indicates manganese which is the cause of colour.

Herderite is faceted to give colourless, pale yellow or yellow–green, green, pink or violet stones with a hardness of 5–5.5 and RI in the range 1.591–1.624, DR 0.030, SG near 3.0. None of the colours are really strong but some are attractive. May very well be found at American gem and mineral shows as some stones are found in Maine (though most faceted herderites are Brazilian).

At the time of writing there have been reports on *jeremejevite.* Blue–green transparent crystals from Namibia have been faceted and show cornflower-blue to straw or colourless pleochroism. The hardness is 6.5 and the RI 1.640–1.653 with DR 0.007–0.013 and SG 3.28–3.31.

A chromium-bearing variety of the mineral clinochlore has been known as *kammererite* and, with diaspore and fire opal, is one of the gem/ornamental minerals found in Turkey. Crystals appear at gem and mineral shows and the occasional transparent specimen is cut despite the perfect cleavage and the layered mica-like structure: the reason is the very fine purple colour. The hardness is 2–2.5, RI in the range 1.597–1.600 with DR 0.003 and SG 2.64. The pleochroic colours are violet and hyacinth red.

Bright yellow *legrandite* is sometimes transparent and faceted stones exist in large collections. As most specimens come from Mexico examples are sure to be found in Californian gem and mineral shows. The hardness is 4.5, RI in the range 1.675–1.740, DR 0.060, SG 3.98–4.04.

Paperweights and similar ornamental objects have been fashioned from the lithium mica, *lepidolite* which is lilac to purple with a perfect cleavage. The hardness is 2.5–4, RI about 1.55–1.58.

Leucite forms interesting crystals with numerous faces and since they are collectable, colourless and occasionally transparent it is not surprising that some are faceted to make gemstones. The hardness is 5.5–6, RI about 1.50, SG about 2.50.

Apple-green *ludlamite* makes attractive cabochon material with a hardness of 3.5, RI in the range 1.650–1.697, DR around 0.040, SG 3.19.

Marble is not often thought of as an ornamental material but the variety *fire marble* (lumachella) gives an attractive play of colour when viewed at certain angles. It consists mainly of fossil shells: effervescence with acids will distinguish from opal but such a test is not recommended. Opal's play of colour is a diffraction rather than an interference effect and the individual colours are much brighter.

White opaque cabochons with a silky lustre are sometimes cut from the zeolite mineral *mesolite* which has a hardness of 5, RI 1.50 and SG 2.29. Some Indian stones may fluoresce pink: the two directions of perfect cleavage prevent faceting.

Microlite most commonly appears (rarely) at shows as pale yellow to brown cabochons though some green Brazilian crystals have produced faceted stones. The hardness is about 5.5, RI 1.93–1.94 and SG 5.5. Some metamictization may cause anomalous birefringence.

Transparent yellow *milarite* from Tsumeb, Namibia, has been faceted but specimens are rare. The hardness is 5.5–6, RI in the range 1.529–1.551 with a low DR of 0.003 and an SG of 2.46–2.51.

Very richly coloured yellow metallic *millerite* has brassy pyrite-like faces which tarnish to a greenish-grey. The hardness is 3–3.5 and the SG 5.3–5.6. Two directions of perfect cleavage hinder fashioning and the tarnish would distinguish it from pyrite: at least one faceted stone has been reported, as a cloudy yellowish-green specimen from Namibia.

Tsumeb, Namibia, also produces the occasional faceted yellow *mimetite* with hardness 3.5–4, RI 2.147–2.128 with DR 0.019 and SG 7.24. The Namibian material (though not all mimetite) fluoresces orange–red in LWUV. Groups and single mimetite crystals are features of almost all major mineral shows so the likelihood of coming across the odd faceted specimen is quite high.

Musgravite as two greyish-mauve faceted stones each weighing less than 0.50 ct were similar in appearance to taaffeite. The paper describing them, in the October 1993 issue of the *Journal of Gemmology*, gave the RI range as 1.719–1.739, uniaxial negative, SG about 3.65. Specimens show a weak absorption band at 475 nm.

Nepheline may turn up as opaque cabochons in a range of colours: a variety sometimes called elaeolite, with red, green, brown or grey colouration and sometimes nearly chatoyant from the chance placing of inclusions, has a hardness of 5.5–7, RI in the range 1.529–1.546, SG 2.55–2.66. Probably the best-known nepheline comes from the nepheline syenites of Canada but the brick-red stones are not really ornamental.

Associated with benitoite in California, *neptunite* is sometimes fashioned into faceted stones which are a fine red though they appear black. This is one of the gemstones that will attract attention at shows in California; the hardness is 5–6, RI in the range 1.690–1.736, DR 0.029–0.045, SG about 3.20.

Among the metallic ornamental materials *niccolite* can be considered very attractive when it shows a high lustre combined with a peach-red colour. The hardness is 5–5.5 and the SG 7.78. A propensity towards tarnish may be discouraged by the judicious use of nail varnish – not the coloured varieties.

Parisite is known to most gemmologists as an inclusion in Colombian emerald but some small transparent brownish-yellow gemstones have been faceted. The hardness is 4.5, RI 1.676–1.757 with DR 0.081 and SG 4.36. Parisite has a faintly resinous lustre but as it is found in the mineral and gemstone-aware state of Montana, specimens will be found at gem shows occasionally.

The colour of translucent *purpurite* may be fine and deep, specimens having a hardness of 4–4.5, RI 1.85–1.92, SG 3.69. A strong pleochroism of grey to purple can be seen in the more translucent specimens.

Metallic dark red (black-appearing) *pyrargyrite* has been faceted. The hardness is 2.5, RI 2.88–3.08, DR 0.200, SG 5.85, with a purplish-red streak. The red stones (from Bolivia) can look very fine.

Pyroxmangite can be found as a very bright transparent red faceted stone, hardness 5.5–6, RI in the range 1.726–1.764 with DR about 0.018 and pleochroism in shades of red and pink. There is a close resemblance to the rare faceted rhodonite but pyroxmangite has no manganese absorption spectrum.

Sapphirine despite the name has no connection with sapfsphire though most specimens are blue. The blue is rather dark and reminiscent of iolite with a similarly strong pleochroism: in recent years an orange–red variety has been found in Tanzania. The hardness is 7.5, RI in the range 1.714–1.723, DR 0.006, SG 3.4–3.5, both RI and SG well above those of iolite.

A small dark green stone with RI 1.697 and 1.704 showed an absorption line at 475 nm and was identified by GIA as *serendibite* in 1997. From the literature the hardness is 6.5–7 and the SG 3.47.

A new mineral, *shomiokite-(Y)*, was first reported in 1997 in a rose-coloured faceted transparent form. The 0.61 ct stone reported in *Gemmologie* had RI 1.530–1.539 with DR 0.009 and SG 2.64.

Transparent bright orange–yellow *simpsonite* may reach Australian or only Western Australian gem shows in faceted form. The hardness is 7–7.5, RI in the range 1.986–2.034, DR 0.058, SG about 6.0. Under SWUV there may be a bright blue–white, pale yellow or light blue fluorescence.

Staurolite is famous for its cross-like crystals but small brownish-red faceted stones do turn up in the specialist market and at gem shows. They have a hardness of 7–7.5, RI in the range 1.739–1.761 with DR about 0.013 and SG 3.65–3.83. A *stibiotantalite* in transparent pale brownish-yellow faceted stones can look attractive but is very rare in this form. The hardness is 5.5, RI in the range 2.37–2.46, DR 0.090, SG about 7.46 for the Californian gem-quality material. The pale colour is helped by the high dispersion of 0.090 and the near adamantine lustre.

Thomsonite is a very popular translucent to opaque material at gem and mineral shows in the United States with its individual markings showing eye-like forms in reddish, yellow, brown or whitish colours. Nothing else really resembles it. The hardness is 5–5.5, RI in the range 1.490–1.544, SG 2.25–2.40.

Vivianite is interesting in that the colour darkens on exposure to light and for this reason green stones have occasionally been cut despite the perfect cleavage and sectility of the crystals. The hardness is 1.5–2, RI in the range 1.579–1.675 with DR up to 0.059. The pleochroism is noticeably strong with different shades of yellow, green and blue.

Vlasovite is found in Russia and Canada and some transparent colourless specimens are known. The hardness is 6, RI in the range 1.603–1.625, DR 0.020, SG 2.92–2.97.

White crystals of *wardite* when mixed with green variscite make attractive cabochons with a hardness of 5, RI in the range 1.586–1.604, DR 0.009, SG 2.81–2.87.

Cabochons of *wollastonite* with a reddish colour and with a fibrous structure sometimes producing chatoyant effects have a hardness of 4.5–5, RI in the range 1.616–1.653, DR 0.015, SG 2.8–3.09. There is a blue fluorescence under SWUV with a yellow phosphorescence.

Yugawaralite has been found in very fine transparent colourless crystals from a quarry in the Bombay area of India and some faceted stones turn up occasionally. They have a hardness of 4.5, RI in the range 1.495–1.504, DR 0.009, SG 2.24.

Very rare, colourless, transparent *zektzerite* is found in the state of Washington, USA, and I have seen faceted examples. The hardness is 6, RI in the range 1.582–1.584 with DR 0.002 and SG 2.79. Specimens show a light yellow fluorescence in SWUV.

Chapter 31

Gemstones in their rough state

The collector of gemstones will want to know what they are whether they are fashioned or in their rough state. While it is true to say that rough specimens of the major gem species will not generally be obtainable by the amateur, the better gem and mineral shows display material that can be of very high quality so that considerable financial loss could be incurred if misidentifications occur – and they do, though not very often at the better shows and rarely through fraud.

The only way to get to many important gem deposits is via some sort of attachment to a geological/mineralogical institution or government department which may send you to areas on which they expect you to provide a professional report. This will mean that you will have to have a good deal of professional experience.

For the purpose of this chapter we are assuming that most readers will be interested in the features that such a reporter may want to mention and which the lapidary may consider before fashioning a specimen. They can be looked for at gem shows quite easily but we would warn against straying into the mine-strewn field of values!

Despite being told quite often that there are price lists for all kinds of gemstones – and there *are* some – we are also told, sometimes with some heat, that dealers get together to arrange prices for their own convenience and with the aim of forcing the customer to pay higher than they need. In our experience this is not the practice among the major dealers – how could they retain business if it was? Dealers in gemstones are also fiercely independent!

Alphabetical listing

It will probably be easier for the reader if we look at the stones concerned in alphabetical order of the names by which they are best known, thus *aquamarine* rather than *beryl, aquamarine* and *alexandrite* rather than *chrysoberyl, alexandrite.* We shall try to give criteria for high quality though naturally there are many version of 'high quality'. In general the deeper the colour the better the stone but very dark red garnets, for example, will appear black once cut (into cabochons) and the same may be true of very dark blue sapphires. In both cases merely examining the nature of the rough will already make this outcome appear likely.

Alexandrite crystals are often seen as apparent hexagons known as trillings though they are really three crystals grouped together with re-entrant angles between their ends. In many cases these angles are filled up so that no re-entrant angles appear on the edge and the whole specimen looks quite solid. The same form can be seen in yellow to yellow–green chrysoberyl. Alexandrite crystals of this kind are usually a very dark green and may show red only on the thinner edges – cutting of course allows the finished stone to be transparent. The best trillings have specimen values to the limit of their suitability for cutting which usually takes precedence.

Faced with an apparent alexandrite trilling the fibre-optic light source will be helpful as any red will show up. Both alexandrite and yellow–green chrysoberyl are also found as water-worn material.

Almandine crystals may be found as slivers or may show very attractive rounded forms with numerous faces – these have some specimen value. The most suitable red garnets for faceting (the darker the red the more appropriate the hollow cabochon (carbuncle) form). This will allow some light to pass through the base to the surface and the stone will not appear black by reflected light. It will usually be the pyrope–almandine or rhodolite crystals which will be a lighter red and be more often found as slivers rather than as crystals with faces. In recent years Tanzania has produced fine large rhodolite rough, some of which approaches amethyst colour. Smaller pyropes occur as little more than grains in fairly soft schists or alluvially.

We have seen very fine *star almandine* from the emerald Creek deposit in Idaho, USA. This material (or at least the specimens we saw) had been tumbled after mining to give matrix-free surfaces on which the rays of the star were clear to see. In general the signs of chatoyancy or asterism when suspected can be brought to the attention of the lapidary when a drop of liquid is placed on that part of the surface where the eye or the rays of a star appear to be lurking: water will do as a drop will act as a condensing lens just as well as any other clear liquid.

Amethyst crystals usually occur in groups but even in isolated individuals the colour is always concentrated within the pointed tips. It is quite rare for the colour in this area to be even and not to show whitish streaks. The best colour is, naturally, unalloyed purple: traditionally the very best colour should change from blue–purple in daylight to red–purple under incandescent light. The names Siberian, Uruguay and Bahia are still used, in descending order of desirability (specimens do not have to have originated in these places). Smokiness and heavy inclusions are a drawback as in most coloured stones. The synthetic corundum coloured by vanadium can look quite like amethyst but will show the strong absorption line at 475 nm where amethyst shows none.

The pleochroism of *andalusite* (greenish- to brownish-red) will be immediately apparent in crystals which generally turn up as water-worn pebbles in alluvial deposits. There is always the possibility of confusion (genuine or assumed) with the colour change of alexandrite but the colours are different and, in andalusite, do not depend on changes in the incident light but on direction of viewing.

For the collector the different colours of *apatite* (yellow, blue, green, purple) have strong appeal and the transparent hexagonal crystals are often very well formed. There are some cat's-eyes. Crystals are soft and need careful handling, the rarest purple ones coming from such sources as the Pulsifer Quarry, Androscoggin County, Maine, USA, and blue crystals from Burma. By far the commonest colour of apatite is yellow as found in crystals from Durango, Mexico, and these are common visitors to gem shows (and to the practical sections of gemmological examinations)!

Aquamarine crystals, growing unlike their beryl family relative, emerald, into large sizes, fetch the highest prices when their colour is the finest blue with no hint of green. Many greenish-blue beryls are heated to give the aquamarine colour. The name Fortaleza has been used to denote the finest blue, the name arising from an aquamarine location in Brazil. Crystals are often found clear enough for large stones to be fashioned from them and their shape is always easy to distinguish from that of topaz. Aquamarine has a hexagonal and topaz a rhombic (playing-card diamond) cross-section. Aquamarine crystals will not show the rainbow-like bands of interference colours, denoting cleavage, that are usually present somewhere in topaz. As soon as green begins to take over from blue the value of the crystal drops, though many connoisseurs, including the present writers, prefer the 'sea-green', pre-heating beryl.

Axinite crystals can be faceted into rare and attractive gems but as so often happens the characteristic wedge-shaped crystals also command quite good prices. The colour is hard to mistake once seen with a cinnamon-brown (clove-brown) predominating among the pleochroic colours which include purple, seen especially in crystals from Baja California, Mexico.

Benitoite is highly collectable both in faceted form and for its crystals which are unique among minerals in showing a three-fold axis of symmetry with a plane of symmetry at right angles to it (see *Gemmology* for a simple explanation of these terms). Neither the crystals nor the stones fashioned from them are large: colours are blue and much more rarely pink or colourless. As they have been found so far at one location only on Earth they will continue to be sought after.

Beryl (other than emerald and aquamarine which are described under these names) may be yellow, colourless, pink (morganite), red or green (non-emerald). The latter may be coloured by vanadium and such crystals are a beautiful soft grass green with none of the bluish overtones sometimes seen in emerald. Like all the beryls they show a hexagonal cross-section: those green beryls which are not coloured by vanadium (or chromium, which would make them emerald) may be chartreuse colour (difficult to define) while some crystals from the Ukraine are a recognizable light leaf green and can reach quite large sizes. Some brownish-green crystals may turn blue on heating.

Red beryl occurs in small, often damaged crystals which are unlike those of any other mineral. They show the normal beryl hexagonal cross-section.

The pink beryl *morganite* customarily forms large crystals which may begin life with an orange or apricot colour which fades to pink (to which colour it can also be heated). In general the pink is not quite strong enough to catch the eye of the general gem-buying public but fine large crystals are collected for their own sake.

Some *star beryls* appear on the market from time to time, the rays arising from platy crystals of brown hematite. This brownish material usually comes from Brazil.

Golden beryl is still called heliodor in some circles so naturally all yellow beryl is heliodor. As much yellow beryl is quite similar to citrine, which is easier to obtain, yellow beryl may not be receiving the attention it deserves. Transparent yellow crystals are sought by collectors but deep golden specimens get faceted as soon as possible after recovery.

The sodium beryllium phosphate *brazilianite* may occur in large greenish to pale yellow crystals not unlike yellow beryl or citrine: it takes a different form from either and is far rarer.

While the material known as wood tin is a form of the tin oxide *cassiterite* and turns up at mineral shows, this mineral may also be found in colourless outer zones of brownish crystals, usually from the tin deposits of Bolivia. This material is distinctive and unlikely to be confused with any other. Wood tin (usually from Mexico) is brown while the lustre of the colourless pieces is near adamantine though the colourless parts of the crystals are often streaked with brown.

Chrysoberyl other than alexandrite, which has been described separately, may occur in the trillings already described or as pebbles in alluvial deposits. Chrysoberyl gives the finest cat's-eyes, the lapidary looking for parallel lines or a silkiness on the surface of the rough. Such material is not likely to be offered at gem and mineral shows. The beautiful yellow–green and chartreuse-coloured transparent chrysoberyls are priced higher than those which are merely yellow because citrine of similar colour is cheaper. This is not the case with the chartreuse-coloured material.

All *corundum* varieties are entered under *ruby* or *sapphire*.

Crystals of *danburite* (and fashioned stones) may be mistaken for topaz and do appear at gem and mineral shows. As in topaz, danburite crystals have a rhombic cross-section but in general are more wedge-shaped and less blocky than topaz: most are colourless but attractive pale yellows are found in Madagascar while crystals of a pronounced orange have been found in the former USSR – some may have gained their colour from irradiation. Danburite crystals have no distinct cleavage as topaz crystals do so that they will not show the bands of rainbow colours where there are internal cleavages. Danburite also gives a sky-blue fluorescence under LWUV where topaz is inert.

Diamond crystals are fairly easy to find at shows, the perfect octahedra being the most highly priced. The unmistakable adamantine lustre distinguishes them from crystals of colourless synthetic spinel or synthetic corundum both of which show triangular markings on the octahedron faces: the edges of such trigons in diamond do not follow the octahedron edges but are reversed: in the synthetic spinel and corundum the trigon edges do follow those of the face. While diamond crystals may fluoresce sky blue in LWUV colourless synthetic spinel shows a similar effect only under SW light and corundum is inert. The diamond crystals in which the fluorescence is observed belong to the slightly yellow Cape series.

Diamond crystals of fancy colours are virtually unobtainable outside the diamond trade, specimens purporting to be such need very careful examination as synthetic materials such as CZ are fashioned into octahedra.

Diopside crystals are usually a very dark green and appear stubby and well developed. There is a perfect cleavage in one direction. In general they will not feature prominently at gem shows and the name may be confused with that of *dioptase* whose crystals are always among the most spectacular at these venues as well as in museums. Dioptase crystals are a bright emerald green but are too small and brittle (in most cases) for faceting. Crystal groups from Zaire, for example, are often of museum display quality.

Emerald is more plentiful than the other classic gem minerals and crystals more easily obtainable. Do not think of visiting Colombia to buy them – such an enterprise could be very dangerous. You will see from crystals in museum displays that, like other members of the beryls, emerald crystals have a hexagonal cross-section. They do not usually occur in clusters so any cluster of well-coloured crystals is likely to be synthetic (and highly priced).

If you are still not sure and no testing instruments are at hand the end faces (pinacoids) of this type of synthetic emerald may show curved structures of a type not seen in the natural material.

The finest colour, a deep bluish-green, has traditionally been called Muzo after the major mines in Colombia though not all 'Muzo' emeralds may have been found there. Sandawana emeralds from Zimbabwe are a very bright yellow–green. Crystals of this material virtually never appear on the market.

Crystals of *euclase* are not of great interest (though some can show an array of faces), save in one particular case – the spectacular sapphire-blue specimens from the Miami mine in Zimbabwe. These crystals occur in groups, sometimes with a rock Matrix, and despite their very easy cleavage they sell at high Prices – not usually to lapidaries. Euclase crystals of pale yellow, green and blue come mostly from Brazil.

Crystals of *fluorite* show a great range of colours, may occur in large sizes and in groups of interpenetrating twins and have a very easy octahedral cleavage. With few exceptions the fluorite octahedra seen at gem and mineral shows have begun life as cubes whose corners have later been knocked off. This does not matter but the process if too clumsily done may have started cleavages elsewhere in the specimen and these should be indicated by the presence of rainbow-like interference colours.

The massive Blue John variety usually shows chevron (V-shaped) markings and shows no fluorescence under LWUV: most other fluorites glow purple. Blue John is easily mistaken for amethyst in certain carvings but will often show yellow or colourless markings and in any case occurs only in masses rather than individual crystals.

Fluorite crystals from Illinois may be large, multi-coloured and show prominent vicinal faces on the cube faces, this giving a striated effect.

The *garnet* group of minerals includes the species and varieties almandine, andradite (variety demantoid) grossular, hydrogrossular, pyrope, spessartine and uvarovite. We have already dealt with almandine and pyrope together (under almandine) as chemically pyrope shades into almandine with increasing iron content to form an isomorphous series. There is also a series between spessartine and almandine. Because of this dark red crystals of spessartine (rich in iron) are not distinguishable by eye from dark red almandine.

Spessartine is the manganese-bearing member of the garnet group and when well away from almandine in the series provides very fine and highly collectable orange crystals which may be heavily included by fluids resembling flags. Crystals show the usual forms for garnet.

Demantoid is the chromium-bearing variety of andradite and though magnificent but small faceted stones come onto the market and fetch high prices the crystals seem never to appear at shows, at least not gem-quality ones. In any case they are small: some demantoid with a distinctly yellow colour is shown from time to time. This is from the Italian Alps. No Russian demantoid crystal can be expected to make it to the shows if there is any chance of it being faceted.

The orange–yellow–brown *hessonite*, a variety of grossular, may be confused with spessartine in the rough as colours can appear similar. Hessonite is most commonly shapeless and it would probably be wise to obtain the absorption spectrum (from spessartine) if difficulties arise over distinguishing the two materials: hessonite shows no significant absorption in the visible while spessartine gives a combined manganese–iron spectrum whose relative absorptions depend on the depth of the colour.

The green chromium- or vanadium-bearing *tsavolite* variety of grossular usually shows no outward crystal form: crystals occur in a soft rock and sometimes in potato-like formations which split open easily and release their crystal contents which are usually small with no perceptible form. Though the resemblance to emerald is strong as far as colour goes there is no resemblance at all in the aspect of the crystals. Tsavolite crystals do not show the hexagonal symmetry of the beryl minerals.

Hydrogrossular occurs as green masses which can resemble either of the jade minerals but there are no crystal forms visible.

The apparently black *hematite* (really red as a streak test will show) is commonly sold by mineral dealers in the form of 'kidney ore', masses whose notable metallic lustre separates hematite from any contenders.

Mineralogists call *cordierite* the gemstone we know as *iolite*. This is distinguished by its pleochroism (different colours seen in different directions), used by the lapidary to place the best (blue) colour on the table facet. The other colours are a pale straw yellow and a lighter blue. Crystals show these colours and are not heated to enhance or alter them.

Iolite can be confused only with one other gem mineral but this is an important one, tanzanite, which also shows pleochroism. In tanzanite, whose crystals need to be heated from the reddish-brown colour they generally possess when mined to the blue seen in jewellery, the pleochroic colours differ from those of iolite and are more spectacular. We have seen heated tanzanite crystals which show not only the 'final' blue but also ruby red and emerald green in different directions.

Iolite crystals have a rectangular cross-section and a distinct cleavage in one direction. The variety bloodshot iolite contains profuse reddish hematite inclusions and these may be visible in the rough forms.

Jadeite and *nephrite*, the only two minerals to which the name 'jade' can be given, both occur as boulders or pebbles rather than individual crystals. Boulders of jadeite usually have a brown skin which causes the specimen to resemble a potato. Windows are cut to ascertain quality. The finest jadeite from Burma is sold at state-controlled auctions and there is no chance for the individual who is not a dealer to take part. Jadeite from central America and elsewhere is also massive but central control of sales is not established other than in Burma. The quieter colours of nephrite are not sold centrally in any country as prices for the finest material are much less high: the best green nephrite comes from British Columbia, Wyoming, Siberia (one of the traditional sources) and Alaska. The only control over the export of nephrite is exercised in New Zealand and it is likely, according to Miller and Sinkankas *Standard Catalog of Germ Values* (1994) that material advertised as originating in China, Taiwan or Korea has in fact come from British Columbia.

Rough jadeite may be tested by searching for the 437 nm absorption band or by specific gravity but in practice by the time the jade has been worked its identity will have been established. Only somewhere along the retail chain will substitution normally take place.

Rough nephrite cannot be tested with the hand spectroscope so that specific gravity tests need to be used, at least in the first instance. This may entail the use of large containers, the process being fully described in Peter Read's *Gemmology*.

Kunzite is the pink transparent variety of the mineral spodumene and despite its very easy cleavage, heat sensitivity and brittleness it is often encountered in jewellery. Rough kunzite is unmistakable, occurring as flattened crystals with ragged ends through which the best pink colour is seen. Kunzite is one of the gem minerals that are recognizable on sight.

Crystals of *kyanite* are also very easy to recognize. Well known to the gemmology student and the lapidary for its pronounced directional hardness (harder across than along the crystal) kyanite has a very easy cleavage but a beautiful blue or green colour, the latter often showing a straight blue stripe. As this book was being written I (MO'D) was shown very fine transparent dark blue faceted kyanites from India. Cutting the flattened bladed crystals is like producing a faceted stone from the pages of a book.

Labradorite is one of the constituent minerals of the feldspar group and one variety is perhaps best known for its almost opal-like appearance though the colours seen against a pale to dark grey body colour are less bright and less varied, blue being the commonest. This is because they are caused by interference rather than by diffraction as in opal. This variety of labradorite occurs in masses (material from Finland is known as spectrolite) but there are other varieties: Miller and Sinkankas (1994) suggest that transparent grains of straw yellow and weathered from lava, offered as labradorite, are in fact sanidine (q.v.).

A variety of labradorite is known as sunstone from the spangled effect given by profuse inclusions of hematite or goethite, both iron oxides. As with moonstone (q.v.) sunstone needs only to be a variety of one of the feldspar group minerals and to look like sunstone. The golden to reddish inclusions are seen in a transparent near-colourless host. Crystals show no easily recognizable form but the inclusions are easy to see.

Another variety of labradorite very closely resembles rich red garnet though the colour arises from inclusions rather than from the presence of impurity colouring elements. The inclusions in this case are of copper and the finest examples come from Oregon, USA. This is a very highly priced material.

Though the red labradorite shows no very clear crystal forms, one of the potassium feldspars, orthoclase, does provide very attractive crystals. Transparent crystals of yellow orthoclase may reach lengths of several centimetres, Madagascan pegmatites providing excellent material. Some of the smaller ones do reach gem shows; they can be distinguished from yellow beryl, whose colour comes very close, by the presence of absorption bands in the blue (yellow beryl shows no significant absorption in the visible). It is always worth taking your hand spectroscope to a mineral show!

Lapis-lazuli is a rock composed of a number of minerals. The finest material is probably now found in the Peshawar area of Pakistan, having been brought over the border from neighbouring Afghanistan where major sources occur. The finest grades show no signs of white spotting (by calcite) and do not need to be kept wet when showing to customers (though this helps). Lapis from Chile may be heavily streaked with calcite. Bright brassy-yellow pyrite crystals if attractively distributed do not necessarily detract seriously from value.

The green banded *malachite* is easily recognized: an experimental synthetic malachite has been produced in Russia but the trade has not been affected by it in any way. Malachite is often mixed with the blue banded azurmalachite. While there are no recognition problems Miller and Sinkankas (1994) remind readers that excessive handling may cause a black surface coating to develop. Russian sources are reported to be near exhaustion but most malachite on the market today comes from Africa.

Microcline, often referred to as amazonite or amazon stone, is one of the more easily recognized feldspars, forming opaque green crystals streaked with white. No other mineral crystals look anything like them and the gemmologist's best tool, familiarity, is all that is needed.

Moonstone, like sunstone (though more important), needs only to be a variety of one of the feldspar group minerals, and to show the moonstone effect (adularescence). Crystals show no outward form but the surface may give a clue to what is possible. Sometimes the adularescence presents on a thin edge causing problems for the lapidary. Not all moonstone is colourless, pale shades of green, blue or yellow occurring quite frequently. The finest material is still found in Burma and Sri Lanka though less frequently. Moonstone rough is not easily confused with anything else.

Obsidian and the other *natural glasses* cannot form crystals but do turn up in great quantities at gem and mineral shows. The patterning varies and a large number of names are used; Apache tears is perhaps the best known with white flecks against a black or dark background. *Libyan desert glass* occurs in quite large transparent lemon-yellow lumps.

The finest rough *opal* is sold at the mine and will scarcely ever reach a gem and mineral show though there are some specialist dealers in rough. Black opal is far more costly than white opal with water opal and fire opal also commanding good prices when at their best. Rough opal cannot easily be confused with any other gem material. The highest prices will be asked for specimens in which the widest range of colour appears (prominent red patches are very desirable). Opal frequently occurs in very thin seams on the host sandstone rock and buyers should bear in mind that such material may not be suitable for single stones but may be appropriate for doublets and triplets.

Peridot, the gem variety of the mineral olivine, does occur as very fine well-formed crystals but such specimens are exceptionally rare on the general market – most of the finest specimens from St John's island are now in museums. The bright peridot from Arizona usually occurs as rounded pebbles.

The gem varieties of *quartz* are easy to find in the rough state as the mineral is by and large plentiful, although fine specimens of rose quartz, for example, do not occur very frequently. We have already looked at amethyst, the most important quartz variety. The yellow–orange–brown *citrine* may have begun life as amethyst which is then heated but whether a particular specimen is or is not heated is immaterial in this species, disclosure being impossible as the outcome of testing for evidence of heating is very uncertain. The finest colour of citrine (colour names are quite likely to feature in shows) is probably obtained from heating amethyst and the names Rio Grande, Madeira and Palmyra are used as location adjectives denoting increasingly deeper colours. Miller and Sinkankas (1994) give the names ox-blood and sang de boeuf for natural citrine with a dark reddish colour.

Citrine, like amethyst, occurs in well-formed easily recognizable crystals and virtually all specimens on the market come from Brazil. There should be no possibility of confusion with any other of the similarly coloured gem materials.

Rock crystal occurs in so many fashioned forms and is so plentiful that there should be little trouble in identifying the rough since the only form worth offering for sale will have well-developed crystal faces. Such specimens can be quite keenly sought when occurring in large sizes and these are a feature of European shows, rock crystal from the Alps being collected by *Strahlers* (*cristalliers*). Smaller specimens of rock crystal with fine displays of faces, specially the 'Herkimer diamonds' of New York state, can be quite expensive.

Rose quartz rough is seen all over the place and is almost invariably formless in this state. Most of this low-priced material is cabochon quality at best. Nevertheless it is possible sometimes to obtain fine crystals. Rose quartz may show the star effect. Only pink scapolite really resembles rose quartz and can be distinguished from it by its generally well-formed tetragonal crystals.

Smoky quartz is easy to find in its rough state and nothing really common is likely to be confused with it. The name Cairngorm usually denotes a rich brown colour and darker shades have traditionally been called morion. The name smoky topaz comes up quite often in this context but there is no such material. Heat treatment may be used to remove the smokiness when the original crystals combine citrine and smoky colours. Crystals usually take the characteristic quartz form with pointed terminations.

The non-transparent quartz varieties do not in general present any problems in their rough state nor would it matter very much if they did. It is possible that fire agate may be mistaken for opal but the colours do not show in small patches as in opal but are seen as successive bands of a particular colour.

Rhodochrosite and *rhodonite* may resemble one another though rhodochrosite is likely to include white calcite streaks and rhodonite black veining of manganese oxides. Both minerals may be faceted: transparent rhodonite is very rare but some South African rhodochrosite occurs in very fine transparent crystals of deep orange–pink colour. Fine Argentine rhodochrosite is more likely to be massive but fine crystals occur in Colorado, especially at the Sweet Home mine.

While many texts may appear to state that all rubies occur as tabular (flat) crystals and all sapphires are bipyramids (two pointed pyramids joined base to base) this is not a rigorous distinction. Many examples of both varieties occur as formless water-worn pebbles: exceptionally fine ruby crystals with many faces and considerable transparency may just turn out to be productions of Professor P. O. Knischka (see Chapter 3) which may show faces not found so far in many, if any, natural rubies.

The best facet-grade *ruby* never gets near the collector as crystals are sold to dealers encamped at the mines or through state-controlled gem auctions. It may surprise readers to learn that some of the very finest crystals seem to avoid either route to the lapidary and setter! At gem shows there are, however, plenty of opaque crystals, often showing attractive and characteristic forms. While the finest colours come from Burma, Thailand and Sri Lanka also provide expensive specimens and since the geology of northern Vietnam is very similar to that of the ruby-producing areas of Burma it is not surprising that fine rubies are now coming from there. In recent years ruby from Mong Hsu, Burma, have come onto the market: crystals can often be distinguished by their purple cores but again, the finest ones will not be easily available. Rubies from East Africa often show good crystal form and may be found from time to time but good forms and transparency do not coincide in crystals available to the general collecting public. Most of the opaque to translucent available ruby crystals come from India.

Sanidine, a variety of orthoclase, when fashioned can very easily be confused with smoky quartz but crystals show no outward forms while they would be expected from quartz.

The finest *blue sapphires* come from Burma and Kashmir and, like the finest rubies, will not be available in their rough state to the general buyer. There are usually plenty of rough opaque crystals with fair crystal shape around for the collector at shows. Sapphires in Montana, USA, for example, which occur in old river gravels, can be dug on a fee basis: the visitor buys a bucket of gravel or digs for it himself in the soft material. The contents are washed and the crystals extracted and kept. Quite often the visitor turns up a gem-quality specimen which, in the case of these particular Montana sapphires, can then be heat treated to give a very good and often distinctive colour. Montana produces sapphires in all colours: the Yogo blue sapphires, however, mined from a hard rock deposit are not available for digging as they are too finely coloured, scarce and in demand for fine jewellery. If some do turn up as crystals they will be distinctively flattened rather than showing the bipyramid or tabular (flat) form.

Sapphires of different colours from Australia and elsewhere, often so dark a blue as to appear black and sometimes a dark green, are available fairly easily.

Sphalerite or zinc blende is attractive enough to be found at gem shows, the yellow or orange crystals showing signs of at least one of the six possible directions of cleavage. Green sphalerite is also found and may be pale enough to arouse thoughts of green diamond. The cleavage signs and the very high specific gravity of about 4.0 make distinction easy and it would certainly be a rare occurrence to find green diamond crystals of polishing size available.

Sphalerite has a hardness of only 3.5 but the dispersion is high enough to make faceted specimens very fine to look at. Most faceting material will have come from Spain or Mexico and a variety of crystal shapes is possible.

Sphene does not show notable crystal forms but the crystals will at least hint at the high refractive index, birefringence and dispersion possible in a stone fashioned from them. Crystals come mostly from Mexico and will most commonly be a yellowish-green: the occasional chrome-green crystal, also from Mexico will be too rare for anything but a museum collection or faceting. Material not of gem quality may, however, be sold as characteristic wedge-shaped and twinned crystals.

Spinel forms desirable crystals which are often perfect octahedra: red specimens are especially sought and quite often found though good ones are never common. The finest reds come from Burma and specimens from that area are always hard to find. Blue spinel octahedra are also attractive. Sri Lanka produces both reds and blues. Some imitations of red spinel crystals exist. Colourless octahedra of spinel are, up to now, always synthetic.

Spodumene of which the pink variety *kunzite* has already been described may also be found in yellow or green colours, almost invariably from Brazil though some fine transparent yellow specimens have come from Afghanistan. As with kunzite the crystals are typically flattened and show a very easy cleavage, signs of which may be apparent from the presence of bands of interference colours. Though the green Brazilian spodumene has often been given the name *hiddenite*, this name should be reserved for the very rare chrome-green spodumene found so far only in North Carolina.

Topaz is easy to find in almost any quality as the mineral is not rare and specimens attractive and collectable. They show the characteristic rhombic cross-section and termination of pyramid and dome forms (the dome resembles the pitch of a house roof). The crystal base is a cleavage direction and will appear notably flat with a typical slightly pearly lustre. Topaz colours range from colourless through shades of orange and yellow to blue: the blue in naturally coloured topaz is pale, the stronger blues seen in commerce today being the result of irradiation.

The varied colours and relative abundance of *tourmaline* together with its hardness and lack of easy cleavage make it one of the most desirable gemstones. Crystals are easy to find at gem shows and always display prominent striations (grooves) parallel to the long direction.

This direction also absorbs light most strongly so that stones faceted with their table at right angles to it will appear as dark as possible for the colour concerned: thus, if the crystal is green, this will be the darkest green direction.

As far as the lapidary is concerned the best colour should be that seen through the table facet and this facet can be placed at any desired angle so that too light or too dark shades of the crystal's colour may be avoided.

Red as usual is the colour most in demand but fine chrome greens are also desirable. Tourmaline cat's-eye is less sharp than that of chrysoberyl. The well-known water-melon tourmaline may be obtained when suitable crystals are sliced across their length and slices are usually on offer at gem and mineral shows. Tourmaline is very distinctive, the pleochroism being light and dark shades of the same colour and very pronounced.

Turquoise of gem or ornamental quality occurs only in masses which are not hard to distinguish from imitations. Turquoise found in seams is of lower quality than that found as nodules according to Miller and Sinkankas (1994). Turquoise is very often enhanced by waxing but it is not necessary at shows to try to find out whether or not a particular specimen has been so treated.

Zircon crystals of gem quality are not too common at shows as the reddish-brown material found in Thailand and neighbouring countries is heat treated to give the colourless golden yellow and blue colours seen in jewellery. Green zircon may be metamict and show no crystal forms.

Zoisite is unfamiliar as a mineral species: the name tanzanite, however, is much more familiar as the transparent bright blue gemstone introduced to the markets in the 1960s. Crystals are heated for the blue to be useful in faceted gemstones and show very strong pleochroism which is unlike that of iolite (q.v.). In any case the blue of tanzanite is much brighter and of a different quality. It is just possible that the less electric blue of the Paraíba tourmalines could be confused with tanzanite but this would presume that the tanzanite had been heated.

Chapter 32

How crystals are grown

An understanding of the ways in which gem-quality crystals can be synthesized is of great importance to the gemmologist but almost all of the literature on crystal growth (which is very large) is inaccessible to the average gemmologist who will not have access (without special permission) to the libraries in which most of it is held. It is for any serious purpose all journal based, monographs in this field as in so many others being so quickly rendered out of date and only the very largest university or national libraries can afford the very high subscriptions. A guide to some of it can be found in O'Donoghue, *Crystal Growth: A Guide to the Literature*, The British Library, 1988. This guide is particularly useful as a route to the monograph literature from the late 1960s to the early 1980s since these were years during which many transparent crystals with gem potential were being tried out (some just once) though the research aim was not ornament. These monographs went quickly out of print and are now virtually unobtainable since hardly any of them reached a second edition and the print run of the first will have been small. In a sense this was the golden age of crystal growth literature as far as the academic gemmologist was concerned.

Though readers who have been gemmology students will be familiar with some aspects of the following this will not be the case with those new to the subject. Some terms or phrases can be explained here, others in the text.

Crystals are solids but may have to grow from other states of matter as well as from other solids. Gem-quality crystals have to be large enough to see, be hard enough to withstand wear in jewellery and resistant to the type of chemical attack that may occur during normal household life (this may happen when the crystalline substance is porous). A successful gem crystal also needs to look attractive and have a memorable name: lastly, it must not, however beautiful, be a one-off since nobody will be able to sell it; repeat orders are essential to the gem dealer.

Flux growth

Most synthetic gem crystals grow from a molten *starting material* which is sometimes a *feed powder*. The melting-point of the desired substance is critical to whether or not a crucible (heat-resistant container) is used and research into methods of growing crystals at lower temperatures than their melting points has had results which closely concern gemmologists. The problem is surmounted in the case of ruby and emerald (as well as some other gem crystals) by first dissolving the starting material (with the composition of the crystal desired) in a molten *flux* compound(s). This is where the term flux-melt comes in. Growth of ruby (as it might be) then takes place at temperatures below the 2037°C melting point of ruby.

This method uses a precious metal crucible so that unwanted compounds from the crucible will not form with the substance being grown: the cost of the platinum or iridium crucible is one reason why flux-grown ruby and emerald are so surprisingly expensive. The other reason is that growth of crystals large enough to be used as gemstones takes up to one year, most of that time being taken up by computer-controlled cooling of the melt which, with an open crucible, can be replenished as growth proceeds.

Normally gem crystals grown by this method will be encouraged to take particularly desired forms (crystallographese for shape) so that very thin crystals, for example, do not form when the grower wants to sell gem-sized material to the cutter. The secret is to allow growth to take place on 'seeds' of the wanted substance so that the crystal knows which faces need to be developed at the expense of other ones.

All this may produce a very fine crystal at the end of the run but the growth process leaves clues behind for the investigator. Some of the flux may be left behind to form highly characteristic twisted veils which at first sight may resemble the flat veils ('fingerprints' or 'feathers') seen in many natural gem minerals, especially ruby and sapphire. The high temperatures involved in growth may cause metallic fragments to break off from the crucible wall and become incorporated in the grown crystal as recognizable angular shapes.

Most of all, though, the grown crystals will contain no *natural* solid inclusions (although flux-grown emeralds may contain crystals of the mineral phenakite which occasionally forms when growth temperatures slip out of control for a time). Synthetic gemstones usually look suspiciously inclusion-free, as of course they are but to be able to evaluate gemstone interiors needs quite a lot of experience.

While crystals grown by this method usually finish up as faceted stones some of the better-looking crystals or crystal groups are sold unfashioned and can often deceive.

Hydrothermal growth

Some emeralds are grown in a closed crucible (autoclave) under pressure: growth takes place on seeds as in the flux-growth method but there are no twisted veils of flux because no flux is involved. Fortunately these emeralds are much less common than the flux-grown ones since identification is more difficult, but there are no natural solid inclusions so that the tester should suspect artificial origin from the first. This *hydrothermal* method of growth is used above all for the production of quartz for electronic devices and does produce synthetic rock crystal, amethyst, citrine and a greenish transparent quartz.

Crystal pulling

The invention of the laser made it necessary for crystals of very high purity and with laser potential to be available. Early (and successful) crystals for this purpose included ruby so that a new growth method was needed. This was because laser action could not take place in the presence of unwanted impurities, these including flux or crucible material.

The solution was to lower a prepared seed crystal to the surface of a melt of the same desired substance and then to raise it slowly. If all went well the seed would take the melt up with it to form a cylinder or rod which would then be suitable for laser (or gem) use. Such *pulled* crystals are virtually inclusion-free and because of this should be regarded with suspicion.

All crystals so far described would be colourless: the red of ruby and the green of emerald being achieved by doping – adding small quantities of elements not part of the normal composition of the growing crystal to achieve some definite end, in this case, colour. Doped crystals and the gemstones fashioned from them usually show a more even colour than their natural counterparts in which irregularities of colour distribution are almost the rule.

While synthetic ruby and emerald, with some quartz varieties, are the main counterparts of natural gem minerals, some of the synthetic gemstones coming onto the market in the 1960s and later have no natural counterpart. These are oxides (oxides and silicates make up the bulk of the major gem mineral species) which have been grown either by both the flux method or by pulling.

Synthetic garnets

The chief among these substances are the synthetic garnets which have the garnet-type structure but are oxides rather than silicates. The well-known YAG and GGG are members of this group and are doped to give a wide range of colours. None will show natural mineral inclusions but flux traces usually can be found in some part of the finished stones.

Skull melting and cubic zirconia

The best-known simulant of diamond, cubic zirconia (CZ), has so high a melting-point (2730°C) that no crucible can be used. Instead the crystals grow inside a block of their own powder – the skull-melting technique. Here again there are neither mineral nor flux inclusions and the clarity of the product immediately catches the eye.

Verneuil (flame-fusion) growth

We have looked at the more important examples of gemstone synthesis before the method which represents by far the highest quantities of stones produced. Growth from a feed powder by melting it in an oxy-hydrogen flame is commonly known as the Verneuil process and ruby, colourless and the different colours of sapphire, colourless and coloured spinels, rutile and strontium titanate are grown with dopants added to the feed powder where appropriate.

Coloured products of this method usually show curved growth bands and colour zoning which are harder to see than might be expected from some texts. Continual adjustment of specimen and lighting may be needed although large well-formed gas bubbles can usually be seen. No natural mineral inclusions can be present.

While the refractive index and specific gravity of flame-fusion corundum echo those of the natural material, those of the spinel are higher due to a necessary increase in the amount of alumina needed for satisfactory crystal growth to take place.

Further details of testing methods can be found in the appropriate species chapters.

Other artificially made products, including opal, coral, turquoise and lapis-lazuli, are not true synthetics but imitations and a detailed account of how they are made is not needed (even if details had been published).

The growth of diamond

The complex question of how synthetic diamond is grown has not been covered in this book since the methods used are in general not published and the rate of development is such that the reader is advised to consult the journal literature.

Chapter 33

Synthetic crystals for the collector

We have already looked at the synthetic versions of the major gem species (when they exist). Though fine details of growth methods are out of the scope of this book they are discussed in a general fashion elsewhere in this text.

Crystal growth needs constant experimental advances to ensure that the grower achieves satisfactory production of a material with all the properties that the customer wants. The experimental process ensures that many crystals of a very wide range of species are grown and later rejected for one reason or another. Many have been accepted for their previously unsuspected ornamental use and sometimes value.

The synthetic garnets and cubic zirconia are discussed elsewhere as well as the longer-established corundum varieties, emerald, spinel and alexandrite, with imitations such as lapis-lazuli, turquoise, opal and coral. Nonetheless there remains a number of less well-known productions which have appeared for a while and then disappeared, often so completely that specimens are very hard indeed to obtain – in fact the development of a synthetics collection involves just as much, sometimes even more, ingenuity as the building-up of a collection of natural gemstones – all you need for the latter is money.

Among the rarer synthetic products, many of them named for the first time with references in *Synthetic gem materials* (1976), are as follows.

Lithium niobate, colourless or doped to give a range of colours, one of the more startling ones being a bright violet: hardness over 5, RI 2.21–2.30 with DR 0.090 and SG 4.64–4.66. The colourless version at least was offered (not widely, in all probability) as an imitation of diamond and since the dispersion is 0.130 (three times that of diamond) the stones are worth examining if you can find a specimen. The low hardness and high birefringence rule out diamond when specimens are tested. Linobate was one of the trade names used.

Lithium tantalate is colourless with a hardness of 5.5, RI 2.175–2.22, SG 7.3–7.5, dispersion 0.087 (twice that of diamond).

Yttrium aluminate is colourless but doped to give a wide range of colours. Much less common than the synthetic garnets with a hardness of 8.5, single RI 1.94–1.97, SG 5.35, dispersion 0.033. Some doped specimens show a rare earth absorption spectrum.

Another singly refractive material also containing yttrium was given the trade name *Yttralox*. It has a hardness of 6.5–7, RI 1.92, SG 4.84, dispersion 0.050. When grown this material is colourless but reports suggest that some specimens at least turn yellow over time when the composition varies from the ideal. This would explain its scarcity as alteration would probably not commend itself to the original research purposes.

The hard, highly dispersive materials may well be taken for diamond if only a single refractive index exists: one or other of the diamond reflectivity meters or testers should make the distinction easy.

While cubic zirconia (CZ) is well established as today's most successful diamond simulant, the analogous material *hafnia* (hafnium oxide) has also been grown in colourless transparent form. This has not come to be used ornamentally perhaps because it is more expensive to produce on a sufficiently large scale. There is no trade name.

The fact that germanium and silicon have some properties in common has led to the growth of germanates, in particular *bismuth germanate*. There is more than one possible composition: when transparent and faceted stones are spectacular, with high dispersion. Most are soft, however, with a hardness of about 4.5. The single RI is 2.07 and the SG 7.12. The body colour is a bright golden orange. As usual the reflectivity meter will show that specimens are not diamond and in any case the very high SG will make faceted stones notably heavy. We have seen only large faceted specimens. A *bismuth silicate* has been grown in colourless and in orange to brown forms. No trade names are reported.

Bromellite, it is hard to believe, has been faceted. The note of surprise is introduced because the dust from this beryllium oxide is highly toxic. Stones are colourless with a hardness of 8–9 with RI 1.720–1.735 with DR 0.015 and SG 3.0–3.02. There may be a weak orange fluorescence under LW.

Phenakite, the beryllium silicate, has been grown for experimental purposes. Doping with vanadium gives an attractive light blue and crystals are slender, making them collectable if they could be found (see the third paragraph of this chapter). Some of the specimens we have seen have small well-shaped crystals growing from some of the larger faces

We have seen that natural *zincite* is a fine red: zincite has been grown by the hydrothermal method to produce colourless, orange, yellow and pale green varieties. At the time that *Synthetic, Imitation and Treated Gemstones* was published (Butterworth-Heinemann, 1997) some large, clean transparent faceted zincites in a red–orange and a green form became available. The stones were claimed to be natural but it was perhaps coincidental that a few stones came onto the market at a time when the synthesis was proceeding. The supply of stones appeared to dry up after a short time so there has been no certain decision on their true origin.

Scheelite has been grown on a fairly large scale for industrial purposes and some crystals have been cut. They fluoresce a strong sky blue under SWUV as do their natural counterparts: some crystals have been doped to give different colours but the strong birefringence rules out a possible diamond imitation. Neodymium doping has produced a purple specimen and this shows the usual rare earth spectrum with two groups of fine lines in the yellow and in the green.

Despite its low hardness (4) and easy octahedral cleavage, crystals of *fluorite* have been grown for purposes other than ornament and some have found themselves faceted. The properties are the same as for the natural mineral but doping has produced some rare colours: one is a red in which uranium is reported to be the dopant. One red uranium-doped fluorite crystal showed a fine line absorption spectrum in which the strongest line was at 365 nm (outside the visible region). A brilliant green fluorite gave an exceptionally long phosphorescence after X-ray irradiation – the dopant was reported to be indium. Another red fluorite was found to contain a number of cavities, straight growth planes and crystallites, the combination giving a deceptively natural appearance.

Periclase in colourless faceted form has been marketed under the name Lavernite. Specimens have a single RI of 1.73 and SG 3.5–3.6. Some stones show a whitish glow under UV. Irradiation has turned some specimens blue, dark blue or green. The hardness is 5–6. Periclase is magnesium oxide, MgO.

Greenockite, the transparent orange cadmium sulphide, is grown by the vapour-phase method to give quite large crystals and faceted stones which have a hardness of 3–4, RI 2.50–2.52, SG 4.7–4.9. The faceted orange stones have too high a refractive index to look like the orange garnets, still less fire opal, but they are attractive. This is quite a rare material.

Strontium titanate (the trade name is Fabulite though we have not seen or heard it for some years, presumably since the arrival of the synthetic garnets and then of cubic zirconia) is grown by the flame-fusion method so faceted stones are cut from the boule. Completely colourless stones can be obtained: the dispersion, at 0.190 (about four times that of diamond) and the clarity make stones very dangerous in small sizes and when they are used as the base in composites, when the low hardness of 5–6 is offset by the use of a colourless synthetic corundum or synthetic spinel top. There is quite a good resemblance to diamond if you are not familiar with diamond's almost inevitable mineral inclusions. The single RI of 2.40–2.41 is very close to diamond's 2.42 and the SG, at 5.13, is immediately felt to be greater than diamond's 3.52.

Strontium titanate can be doped to give interesting colours which certainly could be mistaken for those of coloured diamonds with which customers may not be familiar and which in any case vary widely. On the whole the range of possible colours achieved from doping runs from deep red or reddish-brown through orange, yellow to blue, purple and black. The lighter colours are the most likely to be mistaken for those of diamond. Gas bubbles can be seen in specimens as with all flame-fusion stones.

Also grown by the flame-fusion process is *rutile* but stones are so strongly birefringent (DR 0.287) that diamond could only be considered by those quite unfamiliar with diamond. Stones are never completely white and while this may suggest Cape diamonds the colour is not quite the same and in any case no absorption can be seen at 415.5 nm. In addition the dispersion is so high, at 0.28–0.30, about six times that of diamond, that rather than showing diamond's flashes of colour as the stone is moved under a spotlight, rutile's dispersion is more like the play of colour in opal. The hardness is 6.7 and the RI 2.61–2.90, unmeasurable on the gemmological refractometer: the SG is 4.25. An absorption band at 425 nm acts as a cut-off to the spectrum in the violet.

Light blue colours are achieved by oxidation after growth. The addition of cobalt or nickel, without oxidation, produces red, amber to yellow. Red can be achieved by the addition of vanadium and chromium: beryllium gives a bluish-white. A less yellow cast can be obtained by the addition of aluminium to the feed powder. Star stones are made by adding approximately 0.5 per cent magnesium oxide to the feed powder and annealing the boule (reheating after completion of growth) in oxygen.

The various diamond testers would reject rutile but in the absence of the appropriate instruments (including the lens) the amateur could be deceived, since rutile can be most spectacular.

The calcium molybdate *powellite* has been doped with rare earths to give pink, perhaps from holmium (this colour has been reported but no doubt other colours can be made). A variety of fluorescent effects are achievable so despite the low hardness of 3.5 stones do get cut from crystals intended for applications other than ornament. The RI is 1.924–1.984 and the SG 4.34.

Wulfenite is the lead analogue of powellite as it is lead molybdate. Colourless synthetic crystals have been grown but while they do not show the magnificent orange of natural wulfenite (q.v.) they do possess a high dispersion.

Proustite is a magnificent transparent deep red but soft and liable to surface alteration if exposed to light for prolonged periods. Natural proustite could well have been mentioned among highly collectable natural gems but the properties of the synthetic material are the same as for the natural. Here the hardness is 2.5, the RI 2.79–3.08, giving a birefringence of 0.296, and the SG is 5.57–5.64. The faceted synthetic proustite specimens we have seen are quite large compared to almost all faceted natural specimens. However, should a faceted proustite turn up at a show there will be a great deal of hype surrounding it (this is always to be distrusted, at least at the beginning) and the true nature of the specimen will not be easy to establish.

In the late 1960s and early 1970s minerals of the *spinel* group were being grown for a variety of research purposes. Many were grown by the flux-melt method and some crystals, though not in great numbers, must have reached the mineral market. A variety of compositions and colours were grown and small crystal groups achieved. These groups will hardly ever get near the market and there may be very few of them in any case. In a collection made by one university some vanadium-doped crystals were blue, some crystals doped by copper were an attractive green, some doped by manganese were red. All were well-shaped octahedra and grown by the flux-melt process. Details of Russian synthetic spinels aimed at the gemstone market can be found in Chapter 12.

The mineral *gahnite*, also a member of the spinel group but containing zinc instead of magnesium, has been grown to give blue octahedra with RI 1.805 and SG 4.40. The hardness is 7.5–8. It is possible that some crystals may have been faceted.

A synthetic blue *forsterite* coloured by cobalt gave RI 1.635–1.670, DR 0.010, SG 3.26. Very small gas bubbles could be seen and there was a strong pleochroism giving violet, blue and purple. The high birefringence distinguished the stone from tanzanite which it was presumably intended to imitate.

Synthetic forsterite grown by crystal pulling can be doped to give a variety of colours but the green is not very like that of peridot though the properties overlap. The SG is lower than that of natural peridot.

A zinc sulphide with the same composition as *sphalerite* has been grown with a single RI of 2.30 and SG 4.06. The hardness is 3.5–4.0.

A borosilicate glass with the name *Laserblue* owes its colour to copper as a dopant. The colour is an intense medium dark blue and specimens give an RI of 1.52. The hardness is 6.5 but the material is reported as heat-sensitive and thus difficult to cut. Another 'named' glass is *alexandrium*, a lithium aluminium silicate glass doped with the rare earth neodymium to give a light blue to lavender range of colours. This glass, with a hardness of 6.5 is heat-sensitive and has an RI of 1.58. The neodymium will give the usual fine line absorption spectrum, a test which will immediately rule out any natural material.

The name *Bananas* (derived from barium sodium niobate) has been given to a yellowish, highly dispersive material with a single RI of 2.31.

Perovskite is calcium titanate and has been synthesized to give a number of colours of which we have so far seen only pink. The hardness is 5–6, mean RI 2.40 and SG 4.05. Colours are achieved by doping and no doubt some will show rare earth absorption spectra.

The tin oxide *cassiterite* is reported to have been grown but we have seen no specimen. The crystals are said to be colourless to slightly yellow, with RI 1.997–2.093, DR 0.098, dispersion 0.071 and SG 6.8–7.1. The hardness is 6–7.

Synthetic *fresnoite* was described in *Australian Gemmologist* for April 2000. The specimen in question was a 6.11 ct faceted stone of yellow colour. The RI was measured at 1.765–1.770 with DR 0.019, SG 4.45, hardness 3–4. Specimens show rounded gas bubbles. The material is the synthetic version of a barium titanium silicate found in California but in sizes too small for ornamental use.

The name *langasite* has been given to a synthetic lanthanum gallium silicate which was reported in transparent orange to yellow transparent faceted form in the Fall 2001 issue of *Gems & Gemology*. The RI is in the range 1.909–1.921 with DR 0.014, dispersion 0.035 and SG 4.65. The hardness is about 6.5: there is no significant absorption in the visible region. The cause of the colour was not determined in the report. There is some resemblance to zircon. Two-phase inclusions in the form of parallel needles were noted in one specimen.

A similar material with the composition *strontium lanthanum gallium oxide* was reported in a transparent light pink seen in daylight, altering to light yellow in incandescent light: the change is not instantaneous as in alexandrite. The RI is near 1.82 and the SG near 5.25. The absorption spectrum showed a doublet in the green near 525 nm and a line in the blue near 485 nm. The specimen gave intense green fluorescence under LWUV with a weaker green under SW. Whitish breadcrumb-like material and short, sometimes interrupted cylindrical tubes were seen under magnification.

Chapter 34

The enhancement of gemstones

A summary of enhancements shows that relatively few treatments can be said to be recent and widespread. This implies that the trade and customers have accepted (or never considered) the long-established practices of making green beryl into aquamarine, the heat treatment of zircon to obtain colourless, golden yellow and blue stones, the heating of zoisite to obtain blue tanzanite: in more recent years the production of dark blue from pale blue topaz seems to have aroused little comment and the reported treatment of tourmaline to give a more 'Paraíba-like' blue has reached only the gemmological press.

Assuming that a certain proportion of the people who sell coloured gemstones rather than diamonds don't even realize that treatment *can* take place, much less than it does take place, it is quite likely that the general public do not in fact know that stones get treated – they know that there are 'fakes' or 'the jeweller swapped my stone' (very unlikely – dishonesty in the jewellery and gemstone trade is rare) and it is only the few who may read auction house catalogues, buy high quality gemstones from major dealers or take classes in gemmology who will keep up with all the treatments reported today – many of them are one-offs.

So perhaps, *with the exception of diamond*, we can review and perhaps amplify the remarks on treatment made in the course of the chapters on the individual gem species.

Amber

The treatment of amber depends, like all other treatments, upon what the dealer thinks will help to sell the specimen. The piece may need to be darkened, lightened or clarified (not all included matter is detrimental to the specimen's appearance and attractive insect or plant material can be positively desirable). Some pale amber is heated to darken it but specimens can be coated to achieve a similar result. Dyeing amber has been known since Classical times and the use of heated rapeseed oil (among other substances) to remove myriads of minute gas bubbles was also known 2000 years ago. While heating to remove the bubbles may crack the piece the famous 'nasturtium leaves, sun spangles' can also develop and in fact improve its appearance.

Reconstitution of relatively large amber pieces by consolidating and combining fragments usually leads to the formation of elongated bubbles and flow marks. As at other stages in amber treatment, dyes can be added during the reconstitution process – sometimes they may fade.

Inclusions can be added in quite ingenious ways: insects can be set in plastic which can then be placed in amber or in contemporary resins.

Many of the practices described can be identified with careful observation and familiarity. We should remember that amber chips where plastics peel and that heat will release an aromatic smell from amber and a pungent one from plastics. Very brilliant fluorescence may suggest pressed amber while lack of any fluorescence may suggest a coated specimen. Ether affects contemporary resins more than amber.

Ammolite

Ammolite or korite is fossilized ammonite shell, usually forming part of a composite, the convex cover of colourless synthetic spinel enhancing the interference colours beneath. The occasionally fragile surface of the shell can be plastic coated. It is unlikely that uncoated or shell pieces on their own will be used in jewellery.

Apatite

Some gem minerals of the *apatite* group may have their colour lightened. Dark green Mexican stones have been heated to give a lighter green or even a blue–green but the practice cannot be common.

Azurite and malachite

Azurite and *malachite*, which very often occur together, may be made into *stabilized azurite–malachite block* by compression and impregnation with plastic, according to a report in *Gems & Gemology,* 1989.

Beryl

In the beryl gemstones blue *aquamarine* may have started as green beryl, heating accomplishing the change from green to blue, or it may have started as blue. In either case the colour is stable to light (irradiation might reverse it but there would be no point in returning the stone to a colour considered less desirable today). No gemmological test will identify the colour of the original.

Very *dark blue beryl* of the *Maxixe* (colour from natural irradiation) or *Maxixe-type* (colour from man-made irradiation) will show a quite different absorption spectrum (with bands in the red and the yellow) from that of aquamarine which is never as dark. The colour of the Maxixe and Maxixe-type beryls will fade but not quite so quickly (unless heated) as was thought at first.

Yellow and golden beryls are neither heated nor irradiated and their colour is stable.

A beryl containing iron and manganese may be heated to give the pink of *morganite* from an orange colour but again there are no tests available or necessary.

The most important of the beryls is of course *emerald* and specimens are neither heated nor irradiated. They may, however, have the scattering of light from profuse inclusions minimized by oiling or filling via surface-reaching fractures. Traces of oil may be felt or smelt and may stain the parcel in which stones are kept. Filling with glassy or polymeric material will cause a flash of an alien colour to be seen inside the stone – usually orange or purple. Examination of the suspect area may reveal bubbles if the filling is a glass but since the RI of filler and host may be close detection of the filled area may be difficult if no colour flash is seen.

However, some fillings do deteriorate over time and it is worth checking a specimen after cleaning to see if there have been any obvious changes: ultrasonic cleaning can play havoc with oil in a stone and may also damage polymer fillings. Heating the stone during jewellery setting may cause it to fracture (emerald is notably brittle). Some oil may be coloured and stones need only to be immersed for the colours to be seen. The microscope may show the edges of fractures but in practice they are by no means easy to see and the gemmologist needs to be prepared for constant adjustment of the lighting used.

Of the enhanced beryls only the Maxixe and Maxixe-types can be identified by straightforward gemmological testing.

Chalcedony

The enhancement of chalcedony by *dyeing* is described in Chapter 6 on quartz and it needs only to be said here that dyeing has always been perfectly acceptable and needs no disclosure. Colours are always permanent and may be said in some cases to improve the appearance of the stone.

Coral

Coral when black may be bleached to give fine golden colour skin shades of red and pink. Dye can often be detected with acetone or nail polish remover.

Corundum (ruby and sapphire)

Corundum (with diamond) is the mineral in which enhancement techniques find their fullest development. As far as *ruby* is concerned the ideal is to get a stone up to the colour of a Burma ruby of good quality and this has meant the elimination of iron as far as possible. In practice the Siam (Thai) rubies with a slight brownish tint are found only in older jewellery since the routine now is to heat them, this 'improving' the red by removing the brown. This practice need not be too great a problem for the gemmologist providing the experience and confidence are there. In rubies the inclusion pattern by which country of origin is determined can become familiar as solids are involved to a considerable degree. Details are given in Chapter 3 on ruby. On the other hand where locality information is requested the stone is usually an important one and the laboratory needs to get it right so care is needed.

It is possible for the star effect to be enhanced or diminished by heating (why diminish it? – too strong a star suggests a synthetic product) but the practice cannot take place too often since there seem to be few reports on it. In any case star stones are not always reverenced equally in different cultures.

In the same way the padparadschah pink–orange colour is particularly important to the gem dealers of Sri Lanka and heating may be used to alter suitable stones to a colour as close to this as possible. The only way to tell whether or not this has been done is to look for surface signs of heating (fractured facet edges) or internal ones (exploded inclusions – they may show disc-shaped fractures surrounding them). Abraded or multiple girdles are frequent. These may be seen on any heat-treated variety of corundum.

When we turn to sapphire the real trouble is centred around the yellow variety – or it would be were this colour more popular with customers and dealers generally. There are seven types of yellow sapphire and the colour is not always stable (see Chapter 3 on corundum). There are no gemmological tests which can be used on a yellow sapphire to detect whether or not the colour has been enhanced though natural stones can be distinguished from synthetics.

Some yellow sapphires started life as yellow, green, sometimes colourless sapphires which were then heated to produce a more acceptable yellow colour. The presence of iron is essential to the colour enhancement process.

Many yellow sapphires were treated in about the early 1970s, the stones appearing a much darker golden yellow than usual. Experience showed that some of them at least had been irradiated. Though it has sometimes been reported that heat-treated yellow sapphires fade, it is much more likely that irradiation would be responsible since this produces a colour-centre which when subjected to different energies (in this case those of visible light or heat) allows the coloration process to be reversed.

Irradiation and heating of a sapphire containing chromium may produce the pink–orange padparadschah colour which is stable.

Both iron and titanium have to be present when blue sapphire is needed and although the geuda material of Sri Lanka is whitish or pale blue with titanium dioxide (rutile) to give 'silk' a much deeper blue is obtained by heating (no irradiation takes place): details can be found in Nassau (1994). The colour is stable once formed, but no gemmological tests can establish the nature of the original geuda material, though the high temperatures involved (perhaps up to 1900°C) may well lead to fractured facet edges.

Heating may lighten the dark blue sapphires found in Australia and elsewhere and remove unwanted blue spots from yellow sapphires. Stones from the river gravels of Montana are always heated, the process producing a wide range of fine colours. Here again small spots may be developed into colour which fills the entire stone – this is especially effective with the orange and yellow sapphires whose colour is stable.

Heating can be carried out on flame-fusion synthetic rubies and sapphires, the results including a weakening of the effect of the curved growth lines. This is achieved by extended high temperature heating which gives a specimen in which the concentrations of impurities are diffused around the stone and, as a result, the Plato test between crossed polars is less positive.

Natural-looking inclusions may be introduced into the cheaper synthetic rubies and blue sapphires by quench-crackling – a heated stone is dropped into water or liquid nitrogen (Nassau, 1994, points out that the latter is a gentler process). Cracks thus produced can be filled with coloured substances. They could also be sealed by further heat treatment or by a synthetic overgrowth though this would not be common.

Diffusion-treated blue sapphires have been remarkably popular bearing in mind that the coloured layer extends only about 0.4 mm into the stone even when 'deep diffusion' processes are used. In the absence of both iron and titanium in the (colourless) sapphire intended to be treated, both elements have to be diffused in. Diffusion may be used to produce asterism (titanium needs to be added) but the lack of colour depth should still be apparent. It is surprising how extensive the practice of diffusion has become with even the smallest faceted stones and crystals being treated. Should the stone later need to be recut, as might happen in a repair, the colour would quite likely be lost.

Readers may well be wondering why it seems only to be blue which is diffused into sapphires. Surely it would pay better if red could be obtained by so simple a method? Nassau (1994) explains that it is much harder to achieve: a satisfactory blue needs only about one hundredth of 1 per cent of Fe and Ti impurities whereas to obtain a satisfactory red about 1 per cent of chromium needs to be evenly distributed round the specimen. Patchy colour, high RI, abnormal pleochroism and a patchy white fluorescence under SWUV have all been observed in experimentally grown chromium-diffused rubies.

Diffusion treatment may not only show on immersion but the colour may bleed round fractures or fissures and at open pits on the surface.

Fracture filling as in emerald is also routinely carried out on rubies. The treatment produces localization of colour, best seen when the stone is immersed, as well as gas bubbles in glassy fillings. Oiling produces the unmistakable smell from a parcel of stones. The nature of the polymer filling cannot be established by gemmological testing. Magnification may show the outlines of filled cracks.

Most texts like to mention that heated corundum 'plonks' on a hard surface while unheated corundum 'plinks'.

Diamond

See Chapter 2.

Garnet

Not much is done to *garnets* in the way of treatment but Nassau (1994) reports that an almandine which was heated in air developed a silvery lustre which was attributed to the formation of specular hematite (having a mirror-like surface) (report in *Gems & Gemology*, 1988).

Lapis-lazuli

The only treatment carried out on *lapis-lazuli* is with the aim of presenting as bright a blue as possible. For this reason specimens are often shown under water or will have been frequently sprayed if in a hot, dusty exhibition hall. The colour may also be enhanced by waxing or dyeing. Dye can be detected by acetone or by a nail-polish removing fluid. If the final coating is with a colourless wax the acetone test may not work (Nassau, 1994).

Opal

Opal may easily be dyed as it has a porous structure and treated opal matrix (described in Chapter 5 on opal) can be identified by the profuse black dots found throughout the whole specimen. They are not unattractive and aid the gemmologist.

Colourless oils and waxes are routinely used to improve the surface of opal, their presence being detectable by response to the thermal reaction tester when it is brought near.

Quartz

The single-crystal *quartz* varieties can have their colour enhanced or at least altered by irradiation and/or heating. Rock crystal can be turned a stable smoky colour by irradiation but there is no way of distinguishing between stones which have a naturally or artificially induced smoky colour. Natural or artificially irradiated smoky quartz may be heated to render it colourless: before this stage is reached the specimens become greenish-yellow, a stable colour which is similar to that shown by some citrine. Gemmological tests are not needed.

Amethyst colour can be produced in quartz containing iron if specimens are irradiated. Some amethyst fades in strong light and heating can lighten dark colours but in general amethyst's colour is reliable.

Citrine can easily be obtained by heating amethyst to around 450°C. The resulting colour is stable. Some inclusions may alter during this process. Nassau (1994) quotes a paper in *Australian Gemmologist*, 1987, that on heating some goethite may alter to hematite.

The material known as 'ametrine' combines a pale yellow citrine with amethyst and while natural specimens are mined in Bolivia the colour combination can be created by heating amethyst to about 400–450°C. There are no real means of distinguishing natural from heated ametrine.

A transparent green quartz of a rather quiet though pleasing colour, to which the name prasiolite has been given, may be obtained by heating amethyst. Again there are no gemmological tests.

Rose quartz which is often pale can sometimes be darkened by irradiation. Testing is not necessary. A fine transparent blue colour arises from minute crystals of rutile or other minerals or fine cracks (Nassau, 1994), scattering being the colouring mechanism.

Rainbow quartz is made by heating and quenching rock crystal in water, the resulting cracks causing rainbow-like interference colours. Dark tiger's-eye can be lightened by bleaching with chlorine bleaches while red colours can also be obtained by heating.

Fractures in transparent quartz have been filled with Opticon to hide cracks.

Spinel

Red spinel has been improved in colour by a heat treatment which removed a brown component to leave a pure red (Nassau, 1994). Nassau also reports a colour-change spinel found in a parcel of grossular garnets, the spinel showing a milky pink in artificial light and purplish-brown in daylight. The presence of tension cracks inside the stone suggested heat treatment.

Spinel triplets may show various colours: some reported examples have a central layer consisting of uranium glass in yellow to green colours. These stones are radioactive.

Spodumene

The pink spodumene kunzite has been irradiated to give a rapidly fading green colour. It may be offered as hiddenite as the naturally coloured green spodumene often is but the fading will be apparent.

Tanzanite

Tanzanite obtains its blue colour from the heat treatment of brownish, strongly pleochroic rough. Disclosure is not essential.

Topaz

All colours of topaz (except the pink which is coloured by chromium) are coloured by the operation of colour centres so that the colour may be altered quite easily. The colour of the blue topaz on the market today is very largely acquired by irradiation. Most topaz when irradiated turns brown but if the irradiation energy is stepped up the specimen will turn a greenish-brown, then to green and finally to blue. Some green topaz is sold as it is. Heating after irradiation can remove the yellow or the brown, leaving the blue which is stable to light.

One of the (at least) four processes used renders the stones radioactive at first but specimens on the market are not radioactive. On the other hand it is not easy or even possible to say whether or not a particular blue stone has been treated though if it is a dark 'London blue' it is virtually certain that it will have been. Disclosure will not be needed.

As with quartz, topaz crystals have been coated with thin films of gold to give the Aqua aura effect.

By taking the refractive indexes of specimens it might be possible to determine whether or not a particular topaz started life as colourless as, in general, they have a lower refractive index and specific gravity than the natural browns. Thermoluminescence has been advocated as a test for irradiated blue topaz but seems hardly necessary in the commercial context: details can be found in Nassau (1994). Only a fade test can separate the brown stable from the brown fading topaz.

Tourmaline

Tourmaline is treated only to lighten colours found too dark to cut. If the material has been irradiated heating may reverse the effect achieved but irradiation is not a common practice with tourmaline.

The electric blue tourmaline known as Paraíba tourmaline whose colour is caused by copper occurs in violet–blue to grey–blue and some pink colours (Nassau, 1994). Heating to 550°C removes the pink which is caused by manganese and produces the startlingly bright blue.

Tourmalines are or were frequently surface treated to hide cracks and light stones have been foiled.

Turquoise

Turquoise which is so often found in pale colours has a porous structure making impregnation easy. The colour may be affected by cosmetics or perspiration or filling by oil, wax, or plastics. Paraffin wax impregnation is common though today polymers are more often used as they are less likely to degrade. Sodium silicate solution (waterglass) has been used with concentrated hydrochloric acid, the ingredients forming a silica gel inside the pores: colouring agents are used in addition. Turquoise is not often dyed, blue waxes or plastics being preferred: Nassau (1994) quotes old recipes. He also mentions that green turquoise was preferred to blue by the Egyptians and that and oil–vinegar mixture was used by them to turn blue turquoise to green. The practice is reported to be still undertaken in Egypt.

Turquoise can be coated with colourless material with the aim of protecting impregnation: the surface can be treated with acid and a blue epoxy resin used as a coating. Nassau (1994) describes turquoise beads which were first painted blue, a 'matrix' painted on with black paint and the bead finally coated with clear lacquer. It seemed surprising that the turquoise was found to be necessary!

The porous structure of turquoise means that specimens will inevitably become dirty if not carefully handled and stored. While a number of solvents (ether, petrol, ammonia, hydrogen peroxide) have been advocated at various times none are really efficacious and some are dangerous.

The microscope is really the only useful tool in the testing of turquoise. Treated turquoise tends to show a stronger absorption spectrum (two bands deep in the blue) than untreated material but the spectrum is hard to see in any case. The thermal reaction tester may melt wax and plastic coatings, producing characteristic smells: dye can be removed with an ammonia-soaked swab. Colour may be seen concentrated in surface-reaching cracks.

Zircon

There is no need to disclose the enhancement of zircon as the colours of blue, golden and colourless specimens have long been recognized as the result of heat treatment. The question of colour stability is more important as blue and colourless zircons may discolour (rather than fade) on exposure to strong sunlight over time. Apart from fade tests there is no way of ascertaining whether or not a particular stone will discolour.

Chapter 35

Locality information

There is an increasing tendency for buyers of large or otherwise important gemstones to ask for locality information as the right kind of place of origin can add a good deal to the price of the item. While the provision of locality information has always been sought it is now more likely to be demanded: this means that laboratories issuing certificates will have to investigate at least the more important stones passing through their hands.

The 'right kind' of locality information would include 'Burma' for ruby and blue sapphire, 'Colombia' for emerald and 'Sri Lanka' for the lighter, brighter blue sapphires. Other species and places are less likely to need placing in a particular area but for the collectors' market the whole thing begins all over again but starting with lower prices.

The best place to find out about locality information requirements in the gemstone and jewellery trade is to look in the jewellery sales catalogues of the major auction houses. The main ones are Christie's and Sotheby's both of which hold regular sales during the year not only in London and New York but also in Hong Kong (specializing in but not confined to jadeite jewellery – both firms) and in Switzerland. To subscribe to the whole list of catalogues for both firms together will cost around £1200 a year but they are easy to consult by calling at the salerooms.

It is a good thing (?) that it is not so far possible (for the gemmologist or jeweller) to be able to assign a place of origin for a polished diamond: though research work may one day make this possible it is at present hard to see how the techniques needed could be made sufficiently inexpensive. It is sometimes possible for sorters to tell where a crystal has come from but diamond crystals do not appear very often in the gem and mineral market: small-sized ones seem to be the exception and we have seen many small but beautiful examples.

The locality most in demand is Burma for ruby and blue sapphire and many gem testing laboratories now put their signature on certificates. To establish place of origin the gemmologist needs to be able to examine the interior of a fashioned stone and try to establish the nature of the solid and liquid inclusions inside (any stand-alone gaseous inclusions will indicate glass or one of the commoner synthetic stones so that locality information will not be needed).

While details of the type of inclusions to be sought are incorporated in the information given in the descriptive section as part of the species notes, it seems worthwhile giving a conspectus of the main points here so that they can be picked out more easily.

The *Burma ruby*'s most characteristic mineral inclusion is apatite which forms six-sided crystals: this is followed by calcite which forms white or light-coloured rhombs. If these are accompanied by a swirliness of colour (to which the indefinable name of 'roiling' has been given) we are close to a locality diagnosis. Apatite crystals are particularly characteristic.

Naturally there are other inclusions to back up these – they include crystals of corundum itself, occurring as dark flat crystals with a perceptible hexagonal shape and the occasional octahedron of spinel. While needles of rutile (golden to brown) and other minerals are often present they may be found in rubies from other places and are not therefore a sure sign of Burmese origin. In Burma ruby the rutile needles are arranged crystallographically as they have come out of solution (exsolved). The features just described are typical of rubies from the best-known ruby-mining area of Burma, the Mogok Stone Tract.

It is worth noting that rubies from the Mong Hsu mining area of Burma show a purple 'core'. Rubies from the northern areas of Vietnam resemble Burmese stones quite closely and this is not surprising as the geology is similar. However 'Burma' is a better designation than 'Vietnam' in a catalogue description.

The difference in colour between Burma and Thai ruby and how Thai stones may be heated to more closely resemble Burma ones is dealt with in the descriptive section.

Rubies from Sri Lanka may sometimes have locality information in their sale catalogue description but this happens more with the finest Sri Lankan blue sapphires. Nonetheless exceptional rubies may be recognized by very fine rutile needles intersecting less closely than those found in Burma stones: the presence of brown flakes of biotite, a mica group mineral, usually indicates Sri Lanka. Zircon crystals appearing as grains surrounded by tension haloes are also clues to Sri Lankan origin.

Blue Burmese sapphires may contain rutile needles arranged in directions of crystallographic significance as in Burma ruby but it is also possible to find rutile as brown grains arranged in straight rows. Fluid inclusions resembling feathers are common and crystals of apatite are often found.

The chief characteristic noticed in the now rare and celebrated Kashmir sapphires is the haziness caused by profuse small fissures and exsolved material: sometimes structures like brush-strokes can be seen, scattered at random. Occasionally crystals of zircon or tourmaline have been reported. Kashmir sapphires are usually noted in sale catalogues.

Blue Sri Lanka sapphire usually gets a catalogue mention provided the stone is large and fine enough. Long threads of rutile and liquid drops are prominent, the two forming the familiar 'silk'. Close examination of the rutile needles shows that they form arrow-like twin forms. Brown and red mica group minerals (muscovite and phlogopite) are characteristic of Sri Lanka sapphire. Black graphite and hematite also combine with the other inclusions to build up a picture which clearly, together with negative crystals, indicates the origin of the sapphire.

The desired location for emerald is 'Colombia' but it has not yet become the custom for sale catalogues to distinguish between Chivor and Muzo specimens, which are described in detail in Chapter 4 on emerald. The gemmologist or cataloguer will look for three-phase inclusions above all as they are common to emeralds from both the two main mining districts. While the observer will note in passing the presence of pyrite or calcite crystals which could place the stone more precisely at present they serve merely to confirm the identification.

While details of the inclusions upon which locality information is based do not appear in sale catalogues they may need to be on letters accompanying certificates from a gem testing laboratory so that familiarity with the general and specific inclusion scene is vital.

Glossary

In general, terms explained in the glossary have not been explained in the main text. *Gemmology* (Read, 1999) should be consulted for further and more detailed information.

absorption spectrum: in the visible region, a continuous spectrum of red, orange, yellow, green, blue, indigo and violet upon which may be superimposed a usually irregular pattern of dark lines or bands, the different patterns indicating the presence in a specimen of a particular element or elements. The effect is viewed through a direct-vision spectroscope which examines light having traversed or been reflected from the specimen. Absorptions outside the visible region are also available for examination by spectroscopic methods. See also **emission spectrum**.

adularescence: the moonstone glow which has also been called **schiller**.

allochromatic: of a mineral coloured by an impurity element.

amorphous: not possessing a regular internal atomic structure – opal and glass are the only known examples in our field.

anisotropic: said of a substance through which an incident ray on entering at an angle to the surface travels as two rays each of a different velocity. The adjective applies only to crystalline substances.

asterism: a four- or six-rayed star effect seen in some cabochon-cut specimens. Occasionally even more rays have been reported.

birefringence or **double refraction:** the measured difference, expressed in dimensionless figures, between the two refractive indices of an anisotropic substance.

Brewster's angle (meter): testing instrument measuring the angle of incidence at which light reflected from a polished flat surface of a transparent material reaches maximum polarization parallel to that surface. This angle is related to the refractive index of the reflecting medium by $n = \tan i$.

carat: 1 ct = 0.2 g.

cat's-eye effect: see **chatoyancy**.

chatoyancy: the cat's-eye effect when in a cabochon-cut stone – a line of light along the apex of the stone crosses a differently coloured background.

Chelsea filter: a filter passing only red and green light and used for the detection of chromium in green gemstones, many of which will show red through it. Cobalt-bearing blue stones will also pass red light.

cleavage: property possessed only by crystalline substances of breaking more or less cleanly along certain crystallographic directions significant in that substance. May be perfect (leaving smooth surfaces behind) and easy or difficult, according to its ease of initiation.

colouring elements: the elements titanium, vanadium, chromium, manganese, iron, cobalt, nickel and copper give rise to colour in some gemstones when present either as trace impurities or as part of their normal chemical composition, though not all coloured stones are coloured in this way. These elements are known as the outer transition elements.

composite: a gemstone made up of more than a single component. See **doublet**, **triplet.**

contact liquid: the liquid used to ensure optical contact between stone and glass on the refractometer (q.v.). At the time of writing the liquid has an RI of 1.79: though older texts give 1.81 the composition of the liquid has been changed in recent years on health and safety grounds.

crossed filters: test for the presence of chromium or cobalt during which the specimen is examined through a red filter while illuminated in blue light. A red glow from the specimen indicates the presence of chromium or cobalt.

crown: the upper part of a faceted stone, above the girdle.

crypto-crystalline: made up of individual crystals too small to be detected with an optical microscope.

crystalline: possessing a regular internal atomic structure.

crystallites: very small crystals, sometimes not easily identified.

diaphaneity: the degrees of diaphaneity are transparent, translucent and opaque.

dichroism: coloured anisotropic crystals may show different colours or different strengths of the same colour when viewed in different directions.

Either two or three colours may be seen according to the optics of the crystal concerned. The term *pleochroism* is used when more than two colours are present.

dichroscope: an instrument employing a highly birefringent crystal of Iceland spar (optically clear calcite) to view dichroism or pleochroism.

diffraction: the breaking-up of white light into its component spectrum colours on traversing a regular three-dimensional array.

dispersion: the breaking-up of white light into its component spectrum colours. Presented as a dimensionless figure denoting the difference between refractive indexes measured under different colours of light, as near as possible to red and violet.

doping, dopant: an element foreign to its normal composition added to a substance for the purpose of achieving a different colour or fluorescent response.

double refraction: see **birefringence**.

doublet: a two-part composite stone. Seen especially in opal where a thin layer of fine-quality material is fashioned with its host rock (matrix) to strengthen it.

emission spectrum: bright coloured lines superimposed on the visible spectrum and indicating the presence of different elements, notably chromium (and also arising from strip lighting in the laboratory, if present). See also **absorption spectrum**.

fluorescence: emission of energy usually in the form of visible light when a material is stimulated by higher energies such as ultra-violet and x-radiation. See also **phosphorescence**.

flux: a compound or compounds used to dissolve the starting material of a desired crystalline substance before melting and cooling begin. The temperature of the growth process is lowered with beneficial effects on the apparatus used.

fracture: breakage not being the result of cleavage.

hardness table: mineralogists use the table of relative scratch hardness devised by Mohs (Mohs' scale) in which diamond at 10 is followed successively by corundum, topaz, quartz, feldspar, apatite, fluorite, calcite, gypsum and talc.

hydrothermal growth: growth of crystals, particularly quartz, from a melt in a sealed pressure vessel (autoclave) in the presence of water.

idiochromatic: of a mineral coloured by an element present in its normal composition.

imitation: a substance resembling or made to resemble another substance but differing from it in composition and thus in properties.

immersion liquid: transparent liquid with a known refractive index in which an unknown's outline and facet edges will show bold or faint according to the nature of the difference between the RI of the specimen and the liquid in which it is immersed.

inclusions: solid, liquid, gaseous or multi-phase bodies within a gemstone.

interference colours: rainbow-like colours seen when white light crosses a thin film such as a crack or cleavage. Colours are paler than those of dispersion or diffraction.

interference figure: distinctive pattern in anisotropic specimens viewed along the optic axis when placed between crossed polars. The pattern may be used to distinguish uniaxial specimens from biaxial ones and quartz from other gem minerals.

iridescence: the presence of interference colours, often indicating a crack or cleavage.

isotropic: said of a substance within which incident light entering at an angle to the surface traverses it at reduced velocity *and as a single ray.*

lanthanides: a group of elements, some of which have been used as dopants to cause colour in gemstones. They include cerium which gives yellow, praseodymium (green), neodymium (pink or lilac, doped examples often showing a colour change from lilac to pale blue), erbium (pink), holmium (yellow or pink), dysprosium (yellow). Some of the lanthanides give striking fine-line absorption spectra in the visible.

lapidary: engages in the fashioning of coloured stones rather than diamond which is the preserve of the diamond polisher.

lustre: the quality of a surface as seen by reflected light. Metals show the brightest lustre and diamond's adamantine lustre is characteristic of the mineral. Most gemstones have a vitreous (glassy) lustre.

Mohs' scale: see **hardness table.**

nanometre: 1 nm = 10^{-9} m.

negative crystal: common inclusions in some species, negative crystals are hollow but take the form of their host. They may or may not have contents.

negative reading: when a specimen has a refractive index higher than that of the glass or contact liquid of the refractometer and so cannot be measured thereon.

opalescence: a milky effect as seen in opal but nothing to do with play of colour.

optic axis: a direction of single refraction in a doubly refractive crystal. Inadvertent examination of a gemstone along this direction and along no other may trick the gemmologist into overlooking signs of birefringence or pleochroism.

pavilion: the lower part of a faceted stone, below the girdle.

phosphorescence: persisting fluorescence after removal of stimulating radiations.

piezoelectric: of a substance that will develop an electrical charge on compression or which will undergo a change in mass when an electric current is passed through it.

play of colour: a diffraction effect as seen in opal. All the spectrum colours may be seen.

pleochroism: see **dichroism**.

polariscope: instrument for the production of polarized light and consisting of two polarizing filters, polarizer and analyser which, when placed in opposite positions and a specimen introduced between them and rotated, will indicate in most cases whether the stone is doubly (*anisotropic*) or singly refractive (*isotropic*), together with other useful indications.

polarized light: light vibrating in one direction only at right angles to the direction of travel. *Unpolarized light* vibrates in all directions at right angles to the direction of travel.

pyroelectric: of a substance which will develop an electric charge when heated.

rare earth elements: see **lanthanides**.

reflectivity meter: assesses the surface reflectivity of a substance and indicates whether it is diamond/not diamond: occasionally indicates other diamond-like species.

refraction: occurs when an incident ray of light enters a transparent or translucent substance at an angle to the surface (i.e. not perpendicular to it): the ray is diverted from the straight path it would otherwise have taken and traverses the medium at reduced velocity.

refractive index: the measurable ratio, expressed in dimensionless figures, of the velocity of the refracted ray to its velocity when unrefracted.

refractometer: measures refractive index and presents it directly in numerical form to the observer.

sheen: light reflected from sub-surface structures within a transparent or translucent stone give a particular and recognizable quality to the light by which the stone is viewed, as in moonstone.

silk: name given to an effect caused by the intersection at 60° of sets of acicular crystals of rutile, most commonly in ruby.

simulant: a substance resembling another substance and either a synthetic or imitation version of it.

specific gravity: the ratio of the weight of a substance in air to its weight when completely immersed in pure water at 4°C, expressed in dimensionless figures. The figures are the same as those denoting relative density.

synthetic: said of a man-made substance with a natural counterpart.

thermal conductivity: the ability of a substance to conduct heat away from it.

thermal conductivity meter (tester)(probe): instrument measuring or otherwise presenting the ability of a tested substance to conduct heat. Normally designed to indicate whether an unknown is diamond or not diamond.

thermal reaction tester (hotpoint): provides a localized source of heat to a portion of the surface, the reaction being noted usually under magnification.

trigons: triangular markings on a crystal face and occasionally on the unpolished girdle of faceted diamonds.

triplet: a three-part composite stone. Most commonly seen when an opal doublet (q.v.) is capped by transparent material.

twinning: two or more individual crystals may grow together either adventitiously (interpenetrant crystals) or according to the operation of twinning laws. Notable in diamond, spinel and fluorite crystals.

The literature of gemstone identification

Further reading

Those interested in pursuing the technique of gemstone identification must immediately join the appropriate gemmological body for their area so that you will receive their journal (join more than one and you will get two – they are all respectable (the journals), at present). Membership of such bodies also brings you into contact with others working in the field. The best way to find out what is going on is by word of mouth – of course you find out a lot of other things as well!

With the Internet all this is made easy or at least easier. Details of how it operates are outside the scope of this book but all the gemmological bodies will eventually be represented and many papers abstracted. Determinative studies like gemmology inevitably handle a great amount of numerical data and these will be made both retrievable and updated by the companion web page to this book.

Journals

The journals to which it would be wise to subscribe if a good deal of gemstone identification is likely to come your way are the *Journal of Gemmology*, published quarterly by the Gemmological Association and Gem Testing Laboratory of Great Britain (now known for publicity purposes as Gem-A and used subsequently here). The journal (ISSN 1355-4565) is free to members of the Association whose address is 27 Greville St, London EC1N 8TN, UK, telephone (+44) (0) 20 7404 3334, fax (+44) (0) 20 7404 8843, email gagtl@btinternet.com, website: www.gagtl.ac.uk/gagtl. The journal carries original papers and the most comprehensive section of gemmological abstracts and reviews in the world.

The corresponding organization for gemmology in the United States is the Gemological Institute of America, 5345 Armada Drive, Carlsbad CA 92008, telephone (subscriptions) (1) 800 421-7250 ext 7142, fax (1) 760 603-4595. The subscriptions email site is dortiz@gia.edu and the website is www.gia.edu/gandg.

Note that US 800 numbers are not always obtainable outside North America. The official journal of GIA is the quarterly *Gems & Gemology* (ISSN 0016-626X).

The *Australian Gemmologist* (ISSN 0004-9174) is the quarterly official journal of the Gemmological Association of Australia, PO Box 477, Albany Creek, Queensland 4035, Australia, telephone 61 7 32646854 (international), website http://www.austgem.gil.com.au, email austgem@gil.com.au.

The German gemmological association is Deutsche Gemmologische Gesellschaft, PO Box 12 22 60 D-55714, Idar-Oberstein, Germany, telephone (49) 6781 43011, fax (49) 6781 41616, email info@dgemg.com, internet www.dgemg.com. The quarterly journal, *Gemmologie*, carries the ISSN 0948-7395.

Revue de gemmologie AFG (Association française de gemmologie) with the ISSN 0398-9011 is published quarterly from 7 rue Cadet, 75009, Paris, France. The e-mail address is gemmes@animasoft.fr.

Monographs

Textbooks are quite numerous but it is best to examine several before you decide what you want. For example, you might need a simple gemmology book to take you through the rudiments of gem testing: for this the second edition of Peter Read's *Gemmology* will give you an excellent start (it is also the main commercially published textbook for the Fellowship Diploma course of Gem-A). *Gemmology*, second edition, Oxford, Butterworth-Heinemann, 1999: ISBN 0 7506 4411 7.

Gemstone enhancement

The highly topical gemstone enhancement is covered by Kurt Nassau, *Gemstone Enhancement*, second edition (Butterworth-Heinemann, 1994: ISBN 0 7506 1797 7).

The Heat Treatment of Ruby and Sapphire by Ted Themelis can be enquired for at http://www.themelis.com. The book is essential reading for workers in this field.

Kurt Nassau's *The Physics and Chemistry of Colour* (Wiley, 1983: ISBN 0 471 86776 4) is the best introduction to this relevant subject and explains the basis of gemstone coloration among others as well as the phenomena of fluorescence and phosphorescence.

Synthetic and imitation gemstones

Identifying Man-made Gemstones (NAG Press, 1983: ISBN 7198 0111 7) and *Synthetic, Imitation and Treated Gemstones* (Butterworth-Heinemann, 1997: ISBN 0 7506 3173 2) both by Michael O'Donoghue, cover the same area as the present work but include only those products stated in the title.

Gemstones in general

Another text by the present writer, *Gemstones*, London, Chapman & Hall, 1988, ISBN 0 412 27390 X, covers the major and most of the less-common gemstones with synthetic and imitation products from the geological/mineralogical aspect rather than as a testing guide.

Gemstones and earth science literature in general

Professional mineralogists, realizing that the journal literature is vital and that there are far more titles than can easily be routinely scanned, rely on such abstracting services as *Mineralogical Abstracts* (published by the Mineralogical Society of Great Britain) which carries a gemstone section and a section on experimental mineralogy in which details of synthetic materials are sometimes included. Subscriptions to this are expensive at more than £200 annually but less for those elected Fellows of the Society which is based at 42 Queen's Gate, London SW7 5HR.

The same can be said for geologists: although their discipline covers a much wider field and the learned societies publish more journals only parts of which the gemmologist will find relevant (though they are useful for prospecting studies), the subject being wider is better funded and supports some comprehensive libraries. Though access to them is not always easy the Internet has helped a lot and some of them have early (i.e. non-current) literature which may throw useful light on old locality names and details. Wood *et al.*, *Sources of Information in the Earth Sciences* (Wiley, 1989: ISBN 0 408 01406 7) to which the present writer (MO'D) contributed a survey of early earth science bibliographies, is one of the best guides to this field.

Studies of different gem species

The best way to find out which these are is to read the reviews in the *Journal of Gemmology* and *Gems & Gemology* (see above). Serious reviews have appeared in both journals for as long as any reader of this text will want to search. Shortcomings are always noted. Since the 1970s the standard of single-subject studies has greatly improved.

Older textbooks

Though it would be unkind to single out specific titles for particular mention there are, still in print, several titles with serious faults, those including glossaries being especially prone to uncritical listings of never-used gemstone variety names. My advice is not to buy any glossary/dictionary work on gemstones but to keep to the journal literature. Citations are still given to unworthy monographs and readers should always try to find original papers.

Getting at the literature

Readers should by now have got the message that some of the monographs recommended by teaching bodies are inadequate vehicles in which to travel into the science of gemmology today and that access to the journal literature is enjoined. The question is then – how to get access to it? There is more of it than readers may imagine and few libraries hold sets. Those that do are usually open only to accredited ticket-holders or members of the larger and older universities. To some extent access is helped by very efficient inter-library loans and by Internet searching.

Probably the best way to get hold of obscure data is to consult specialists at gemmological teaching centres or at national museums. Staff will always be able to tell you the name of a likely individual to whom you should then write. The motto should be 'someone always knows what I want'.

The learned societies (the Geological Society of London and the Mineralogical Society are those most relevant) may often offer access to obscure papers or monographs. The Geological Society has its own library in

Burlington House, London W1V 0JU: use is free to Fellows of the Society though other visitors are charged. The Mineralogical Society, at 42 Queen's Gate, London SW7 5HR, has no library but uses the Department of Mineralogy Library at the nearby Natural History Museum.

Antiquarian books

Gemmology has had its own bibliography since the publication of the two volumes of *Gemology, An Annotated Bibliography* in 1993 (Scarecrow Press, Metuchen, New Jersey: ISBN 0 8108 2652 6). This is much more than a book list as the items are fully described with biographical notes on some of the writers. The author is the late John Sinkankas.

Index

The nature of the subject and the need for repetition makes it impossible for the index to note every occurrence of a particular species or variety of gemstone. Readers should look first for the species in which they are interested and then turn from the relevant part of the text to the sections dealing with materials with which there may be confusion.

References are not given for properties in isolation or to testing methods as they are mentioned only in passing. Details of tests are given in Read, *Gemmology*, second edition, 1999.